Springer Optimization and Its Applications

VOLUME 88

Aims and Scope
Optimization has been expanding in all directions at an astonishing rate during the past several decades. New algorithmic and theoretical techniques have been developed, the diffusion into other disciplines has proceeded at a rapid pace, and our knowledge of all aspects of the field has grown even more profound. At the same time, one of the most striking trends in optimization is the constantly increasing emphasis on the interdisciplinary nature of the field. Optimization has been a basic tool in all areas of applied mathematics, engineering, medicine, economics, and other sciences.

The series Springer Optimization and Its Applications publishes undergraduate and graduate textbooks, monographs and state-of-the-art expository work that focus on algorithms for solving optimization problems and also study applications involving such problems. Some of the topics covered include nonlinear optimization (convex and nonconvex), network flow problems, stochastic optimization, optimal control, discrete optimization, multiobjective programming, description of software packages, approximation techniques and heuristic approaches.

For further volumes:
http://www.springer.com/series/7393

Natali Hritonenko • Yuri Yatsenko

Mathematical Modeling in Economics, Ecology and the Environment

Natali Hritonenko
Department of Mathematics
Prairie View A&M University
Prairie View, TX, USA

Yuri Yatsenko
School of Business
Houston Baptist University
Houston, TX, USA

ISSN 1931-6828 ISSN 1931-6836 (electronic)
ISBN 978-1-4614-9310-5 ISBN 978-1-4614-9311-2 (eBook)
DOI 10.1007/978-1-4614-9311-2
Springer New York Heidelberg Dordrecht London

Library of Congress Control Number: 2013953869

Mathematics Subject Classification (2010): 44–01, 49–01, 49N90, 91B76, 93A, 97M40

Printed on acid-free paper

Springer is part of Springer Science+Business Media (www.springer.com)

Preface

Mathematical modeling is the art of describing real-world phenomena in the terms of mathematical concepts. The increasing importance of mathematical modeling in solving economic–environmental problems is a relevant trend in modern research. Relations between human society and the environment are interdisciplinary and include technological, scientific, economic, biological, demographic, social, and political aspects. This textbook presents various mathematical models used in economics, ecology, and environmental sciences and discusses connections among them. Topics include economic growth and technological development, population dynamics and human impact on the environment, resource extraction and scarcity, air and water contamination, rational management of economy and the environment, climate change, and global dynamics. The authors focus on deterministic models and their investigation techniques, including discrete and continuous models, differential and integral equations, optimization and optimal control, and steady-state and bifurcation analyses.

This expository textbook offers an attractive collection of a wide range of models ranging from the classic Cobb–Douglas production function, Solow models of economic growth, Lotka–Volterra and McKendrick–McCamy population models, Hotelling and Dasgupta–Heal models of exhaustible resource, and Forrester and Meadows models of world dynamics to modern models of technological change and environmental protection that have so far appeared only in scientific journals. The authors demonstrate that the same models can be used to describe different economic and environmental processes and similar investigation methods are applicable to analyze various models.

The main goals of this textbook are:

- To expose modern practice of applied mathematical modeling in economics, population biology, and environmental sciences
- To describe relations among various economic, population, and environmental models
- To demonstrate how integrated mathematical models are built from simple components

- To explain investigation techniques for considered models and to provide an interpretation of the obtained results

This textbook is intended for graduate and upper-division undergraduate students, faculty, academics, and industry practitioners in economics and environmental sciences as well as for a wide mathematical audience. It also presents a self-contained introduction for researchers coming into the field for the first time.

Textbook Features

Since the publication of the first edition of this book by *Kluwer Academic* in 1999, the authors have been regularly contacted by universities from across the world about using it as a textbook. The first edition was republished in China in 2006 by *Science Press* as Volume 23 of their "Series of Mathematical Masterpieces Abroad" and translated entirely into Chinese by the *Renmin University of China Press* in 2011.

The second edition is entirely revised and updated compared to the first edition. Obsolete material has been replaced with new and more relevant models, references have been essentially updated, and exercises have been added to the end of every chapter. Solutions to all end-of-chapter exercises and other supplement materials are not included due to space constraints but will be available to instructors who adopt this textbook into their courses (contact the authors at nahritonenko@pvamu.edu or the publisher). The present edition has been classroom tested. The authors have successfully used its draft in teaching undergraduate and graduate courses in mathematical modeling at several universities in the United States, Europe, and Asia.

Most of the material is modular to allow for various course configurations, emphasizing certain economic, biological, or environmental applications. The authors strive to give an instructor substantial flexibility in designing a syllabus and using their preferred mathematical tools. The majority of chapters are relatively independent and can be covered in full or partially and in an arbitrary order. Some exceptions are the following:

- Chapter 2 and Sect. 3.1 are recommended for a better understanding of Chaps. 4, 5, and 10–12.
- Chapter 5 continues Chap. 4 and is mathematically more advanced.
- Chapter 7 can be covered after Chap. 6 and is more advanced.
- Chapter 9 can be covered after Chap. 8 and is more advanced.

To better understand a specific modeling problem, students need an integrated understanding of mathematical modeling. To address this need, the textbook explores a variety of diverse mathematical models from different applied areas and provides elements of their analysis, rather than merely focusing on a complete analysis of few problems. It includes theorems with occasional proofs where they are reasonable and effective. Special attention is given to the step-by-step construction of models, choice of control variables, analysis of arising mathematical problems and their interaction, qualitative behavior of model trajectories, and applied interpretation.

The set of considered economic–environmental models is representative enough to demonstrate how new problems and processes under study determine the choice of mathematical tools. The models of Chaps. 2, 3, 6, 10, and 11 use mostly ordinary differential equations, whereas Chaps. 7–9 consider partial differential equations and Chaps. 4 and 5 use integral equations.

This textbook explains how complex models are constructed from common simple modules that describe elementary economic and environmental processes. The economic models of Chaps. 2 and 3 are used as blocks in the models of Chaps. 4, 5, 10, 11, and 12. The models of resource extraction of Chap. 10 are used as blocks in the models of Chaps. 11 and 12. The models with environmental control of Chap. 11 and the world models of Chap. 12 consider population models of Chap. 4 and environmental pollution models of Chaps. 8 and 9 in an aggregated form.

Recommended Courses

The present edition is designed to serve as a textbook for one- and/or two-semester courses in mathematical modeling. The entire content of this textbook covers a two-semester graduate course in *Applied Mathematical Modeling* or *Mathematical Models and Methods*. In one-semester courses, some chapters and sections can be omitted without affecting the logical development of the material. Depending on the chapters chosen, this textbook can fit both undergraduate and graduate courses. Specifically, it can be used for several undergraduate courses in Departments of Mathematics, Environmental Sciences, Environmental Research, Ecosystem Science and Management, Management Sciences, Science and Technology, and so on. The table below lists some of the courses for which this textbook is recommended.

Sample course	Level	Content
Applied Mathematical Modeling	Undergraduate	Chaps. 1–4, 6, 10, 12
Mathematical Methods in Economics and Environment	Undergraduate	Chaps. 1–4, 8, 10, 11
Mathematical Modeling in Economics	Undergraduate	Chaps. 1–5, 10, 12
Models of Biological Systems	Undergraduate	Chaps. 1, 6, 7, 10, 12
Environmental Models	Undergraduate	Chaps. 1, 2, 6, 8–12
Applied Optimization Models	Undergraduate	Chaps. 1–7, 10, 11
Mathematical Models (for mathematics majors)	Undergraduate	Chaps. 1, 2, 4, 5, 6–10
Mathematical Modeling (for non-mathematics majors)	Undergraduate	Chaps. 1–3, 6, 10, 12
Applied Mathematical Modeling	Graduate	Chaps. 1–4, 6, 8, 10–12
Mathematical Models and Methods (two semesters)	Graduate	Chaps. 1–12

The presentation level requires mathematical knowledge of basic university mathematics courses, including *Calculus* and, ideally, *Differential Equations*. The authors avoid using advanced mathematical concepts or provide them as Appendices, e.g., in Chaps. 2 and 5.

Review of Textbook Content

Chapter 1 explores the steps of applied mathematical modeling and provides a brief overview of its concepts, notations, and tools. The remaining chapters are divided into three parts.

Part I "Mathematical Models in Economics" (Chaps. 2–5) is devoted to the mathematical modeling of economic systems. This area of modeling is well established with its own terminology, classification, and investigation methods. The considered models are used later in Part III as components in more sophisticated models of integrated systems. Chapters 2 and 3 analyze aggregate nonlinear economic–mathematical models based on production functions. Chapters 4 and 5 concentrate on the models of economic and technological development under improving technology, described by integral or partial differential equations. Part I focuses on the qualitative analysis and optimization in considered models. An appendix to Chap. 2 contains a review of extremum conditions (maximum principle) for the optimal control problems studied in this and subsequent chapters.

Part II "Models in Ecology and Environment" (Chaps. 6–9) explores various mathematical models used in population and environmental problems. It covers two large topics: models of biological communities and their rational exploitation (Chaps. 6 and 7) and models of pollution propagation in the atmosphere and water reservoirs (Chaps. 8 and 9). Some basic models of Chaps. 2–10 are only briefly discussed because they can be found in more specialized textbooks. However, more complex models constructed from these components are explained in detail.

Part III "Models of Economic–Environmental Systems" is devoted to integrated models of economic and environmental dynamics. Chapter 10 describes various models of nonrenewable resource extraction, including the well-known Hotelling's rule and Dasgupta–Heal model of economic growth with an exhaustible resource. Chapter 11 focuses on economics of climate change and explores aggregate optimization models of economic–environmental interactions such as pollution accumulation and abatement and adaptation to environmental damage. Chapter 12 offers a brief glance at the history and mathematical structure of famous models of global change, from the Club of Rome models to the modern integrated assessment models, their specifics, achievements, and limitations.

Prairie View, TX, USA Natali Hritonenko
Houston, TX, USA Yuri Yatsenko

Acknowledgements

The authors would like to express their special thanks to their dear colleagues and collaborators Profs. Dauren Adilbekov, Noël Bonneuil, Seilkhan Boranbaev, Raouf Boucekkine, Thierry Bréchet, Benito Chen, Renan Goetz, Elina Grigorieva, David Greenhalgh, Aliakbar Haghighi, Boyan Jovanovic, Nobuyuki Kato, Andre de Korvin, Sergey Lyashko, Janos Turi, Angels Xabadia, and George Zaccour for long discussions and friendly advice.

The authors are also grateful to Profs. Reza Ahangar, Shair Ahmad, Linda Allen, Sebastian Anita, Ted Barton, Alain Bensoussan, George Bitros, Steve Bleiler, John Boland, Alberto Bucci, Michael Caputo, Felix Chernousko, Constantin Conduneanu, Allen Donald, Saber Elaydi, Gustav Feichtinger, Jerzy Filar, Rafail Gabasov, Anahit Galstyan, Mary Ann Horn, Richard B. Howarth, Mimmo Iannelli, Peter Kort, Yuri Ledyaev, Urszula Ledzewicz, Maria Leite, Suzanne Lenhart, Omar Licandro, Klaus Puhlman, Roy Radner, Gerald Rambally, José Ramón Ruiz-Tamarit, S. Trivikrama Rao, Catherine Roberts, Suresh Sethi, Katherine Schubert, Olli Tahvonen, Cuong Le Van, Vladimir Veliov, Jay Walton, and David Zilberman for their kind assistance, useful remarks, and conversations that they may have very well forgotten.

This textbook benefited from the support of NSF, NATO, International Soros Foundation, Isaac Newton Institute for Mathematical Sciences (University of Cambridge, UK), Mathematical Science Research Institute (University of Berkeley, CA), Institute for Mathematics and its Applications (University of Minnesota, MN), American Institute of Mathematics (Palo Alto, CA), and Prairie View A&M University.

Finally, our deep appreciation goes to our students Marcia Brown, Frankson Collins, Idrissa Diarra, Esmaeli Djavidi, Michelle Jackson, Khavansky Johnson, Santos Pedraza, Sheri Stewart, Kassoum Traore, and Olga Yatsenko, who carefully studied the original manuscript, asking numerous questions and making suggestions that improved our work.

We received many helpful comments on the first edition and highly appreciate any new correspondence from the readers of the current edition. We retain responsibility for any remaining errors.

July 2013 Natali Hritonenko and Yuri Yatsenko

Contents

Chapter 1
Introduction: Principles and Tools of Mathematical Modeling

This chapter provides a brief overview of the goals, general principles, and specific tools of mathematical modeling, specifically for economic and environmental systems. Section 1.1 analyzes the role and structure of the modeling process in scientific research and decision-making practice. Section 1.2 explores and compares different types of mathematical models (deterministic and stochastic, continuous and discrete, linear and nonlinear). Section 1.3 outlines selected concepts and results of calculus and the theory of differential and integral equations used in this textbook.

1.1 Role and Stages of Mathematical Modeling

Mathematical modeling is a vital component of scientific research and policy making. Its effectiveness has been proven for centuries. The modeling provides an explanation and prediction of the behavior of complex economic and environmental systems and helps to obtain new theoretical knowledge about the nature and society. The concept of the economic–environmental system assumes the influence of both the economy and environment on each other and the possibility of human control in the system [7]. The importance of modeling of such systems increases proportionally to the scale of human impact on the environment.

Mathematical modeling and computer simulation have a special place among scientific methods. The advantages of modeling as compared to experimentation are as follows:

- Universal availability and applicability of modeling tools.
- Low costs and short timeline of the modeling process.
- Multiple simulations on a wide range of model parameters ("what-if" analysis).
- Possibility of making various model modifications and improvements.
- Evading negative outcomes of real experiments, and others.

N. Hritonenko and Y. Yatsenko, *Mathematical Modeling in Economics, Ecology and the Environment*, Springer Optimization and Its Applications 88,
DOI 10.1007/978-1-4614-9311-2_1,

Modeling should begin at the early stage of a study, just after initial observation or experimentation. It can take decades to notice visible changes in environmental systems, by which time the changes may have already become irreversible. Mathematical modeling can predict negative changes in such systems and recommend measures to prevent them. The analysis of early modeling results can also suggest what kind of additional information is required and what model modifications can be made to achieve a better correspondence with the real-life picture.

A mathematical model is not a copy of the real world: it is always a simplification of the reality, which assists in revealing principal features of real phenomena. In theoretical research or decision-making practice, people use models because they do not possess an absolute knowledge of reality. The models initially emerge in the human brain. Scientific research improves and justifies such *mental models*, which become *conceptual models* in corresponding areas of science. Mathematical and computer modeling methods are based on the conceptual models and, therefore, cannot be more informative than these models. Formal mathematical models are secondary with respect to the conceptual models; however, they allow for finding new insights that are impossible to obtain by other scientific methods.

1.1.1 Stages of Mathematical Modeling

The diagram in Fig. 1.1 demonstrates links and interactions between major stages of mathematical modeling, though any such classification is incomplete and biased. The provided diagram reflects many years of experience by the authors and their colleagues in applied mathematical modeling and is helpful for the purpose of this textbook. We will follow this diagram in our classification, construction, and investigation of specific economic, ecological, and environmental models. The topics of this textbook are related to the stages of *Conceptual Model*, *Mathematical Model*, and *Analysis of the Problem* of Fig. 1.1.

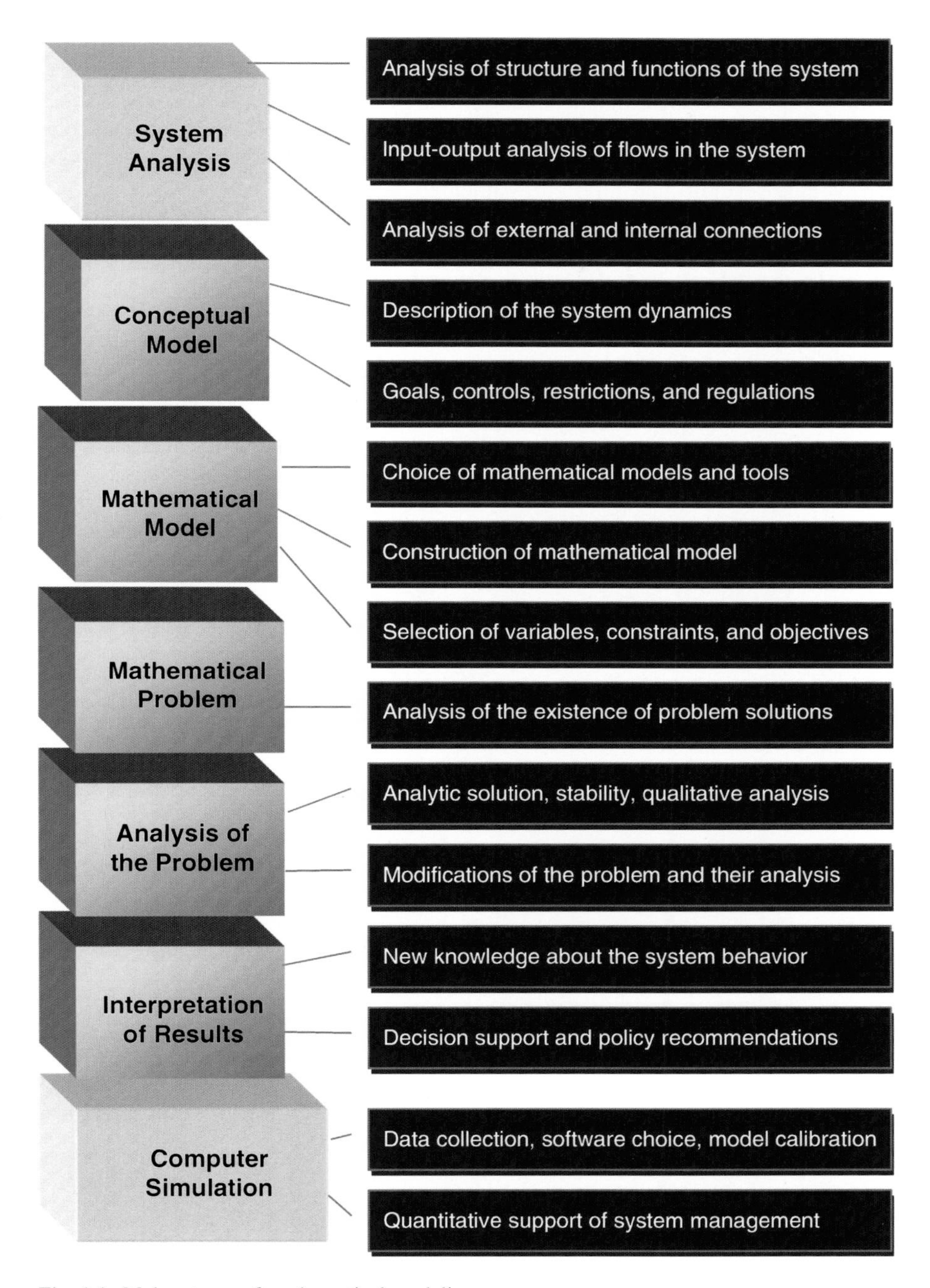

Fig. 1.1 Major stages of mathematical modeling

Figure 1.2 highlights the relationship between more detailed steps of the *Conceptual Model* and *Mathematical Problem*. We distinguish the *Mathematical Problem* as a more specific and advanced modeling stage compared to the *Mathematical Model*. This distinction is clearly visible in subsequent chapters devoted to concrete economic and environmental models. In particular, a mathematical model can include several quite different mathematical problems. After a mathematical model has been developed, the next important steps are to choose modeling goals, analyze constraints and objectives of the process under study, indentify given and unknown characteristics of the process, and others.

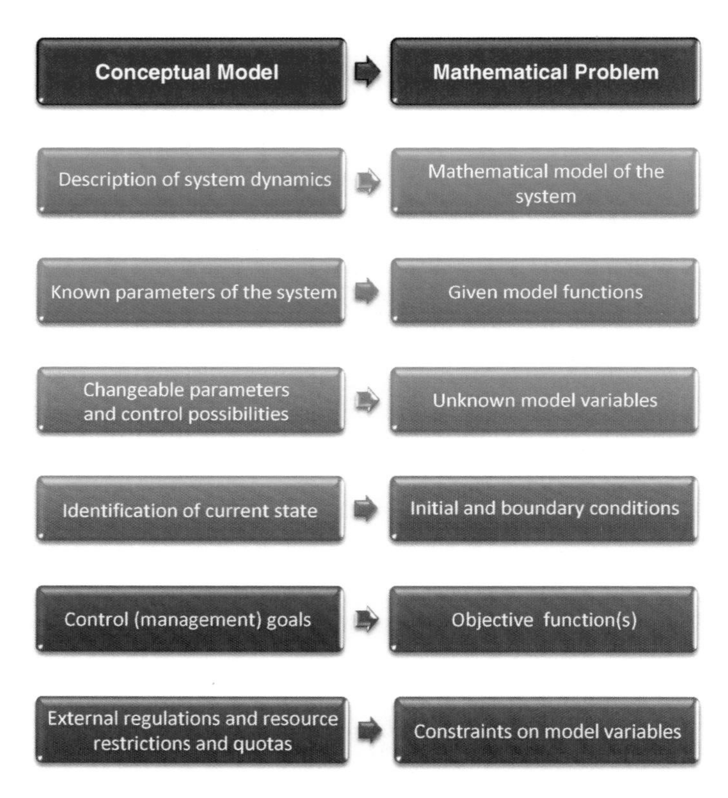

Fig. 1.2 Relations between *Conceptual Model* and *Mathematical Problem*

Relations among different categories of mathematical problems explored in this textbook are illustrated in Fig. 1.3. Solving the identification, prediction, and control problems requires using different mathematical techniques and tools. Some of them are discussed in Sects. 1.2 and 1.3.

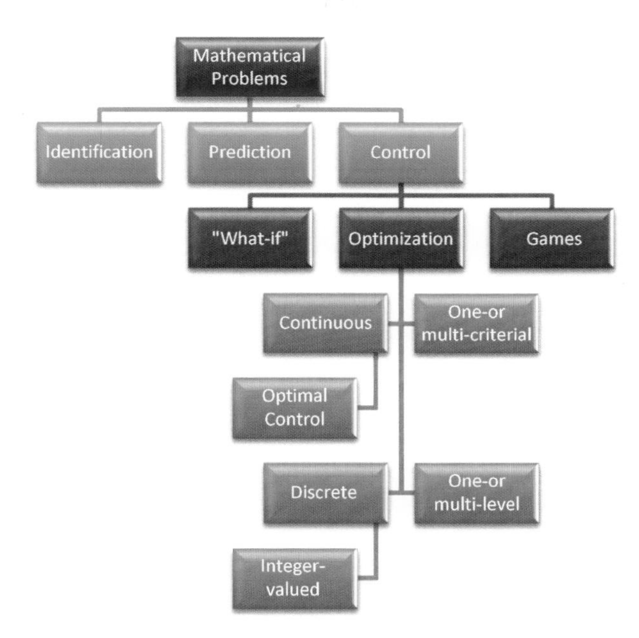

Fig. 1.3 Different categories of mathematical problems

1.1.2 Mathematical Modeling and Computer Simulation

Computer modeling complements and extends traditional analytic forms of mathematical modeling. Modern computers are able to process vast amounts of data, including various choices of system evolution in a quick and efficient manner. Therefore, computer simulation has become a common additional or even primary modeling technique, especially when analytic solution is challenging or impossible to obtain. Computers are widely used in interactive modeling of complex environmental problems, such as weather prediction and global climate change.

Modeling gives a quantitative description of a real system and its connections with the external environment in the presence of unpredicted or inaccessible factors. Both traditional and computer models meet major challenges related to principal impossibility to obtain complete ecological information for modeling. At the same time, the increasing capabilities and reasonable prices of modern computers lead to the appearance of new modeling concepts entirely based on computer information processing, such as the *agent-based modeling*. Such models are populated by millions of computer-simulated agents that act as predicted for living organisms. In economics, the agent-based models try to simulate elementary transactions that occur in an actual economy. This area of research is emerging, but it has not yet delivered convincing breakthroughs.

The possibilities of computer modeling and simulation should not be overestimated because computer models are also based on original conceptual models of specific disciplines. In any scenario, traditional mathematical modeling keeps its place and relevance in the predictable future, as both a learning and decision-support tool.

1.2 Choice of Models

Two trends currently exist in applied mathematical modeling:

- The construction of simple models and applying them to available data without a detailed look into the system under study. This approach is rather popular in applied areas of modeling and leads to acceptable results in many applied problems.
- The construction of mathematical models that reflect the internal structure of the system, taking into account its delicate features. It leads to rather complicated mathematical problems that can be challenging to investigate. Such models are not always convenient in practice; nevertheless, their elaboration reflects the internal logic of scientific research: both pure and applied mathematics benefit from new models.

The economic and environmental modeling uses various mathematical tools, from linear algebraic equations to nonlinear stochastic multi-criteria optimization and game theory [6]. One of the major quality criteria of applied mathematical models is their successful approbation on real-life problems, which, however, does not decrease the importance of their theoretical analysis and comparison with other alternative models. Other quality criteria of models include their internal consistency and capability to consider different control factors and aspects of the process under study. Scientific research and decision-making practice often require analyzing substantially new features of real processes. To address such needs, new mathematical models are developed or known models are essentially modified. Such cases are illustrated throughout the textbook.

Below we briefly explore various types of mathematical models used in this textbook. These notes are not exhaustive but reflect our modeling goals.

1.2.1 Deterministic and Stochastic Models

Deterministic models operate with certain quantitative characteristics of systems and processes without assuming their probabilistic nature. Deterministic models are helpful in many realistic situations that involve relatively few sources of uncertainty inside the system. In modeling practice, deterministic models can deal with the averaged probabilistic characteristics of processes under study (an average "concentration of pollutant" instead of the real concentration, the expected value of "equipment lifetime" instead of the real equipment lifetime, and so on) and are based on the approximation of a real process.

Economic and environmental systems belong to complex systems with high dimensionality and uncertainty of the relationships inherent in them. Nevertheless, the subsequent chapters demonstrate that deterministic models are commonly used for their description. In many cases, increasing complexity of mathematical description using stochastic factors does not lead to substantial insight into the nature of a problem.

Stochastic models describe connections among stochastic (probabilistic) characteristics of systems and processes under study. They are useful for the analysis of repetitive processes and usually require a large amount of data to start modeling. Implementation of economic and environmental processes is unique and accompanied by a shortage of data (especially for large-scale systems). A comprehensive analysis of all available information should be the first step of the system analysis. A majority of the models in this textbook are deterministic. Stochastic models are used in Sect. 10.2 due to a substantially stochastic nature of the natural resource discovery.

1.2.2 Continuous and Discrete Models

Depending on the type of available data and process description, the two major categories of mathematical models are *continuous* and *discrete models*. The *continuous models* operate with continuous variables, while the *discrete models* operate with discrete variables. More specifically, a discrete model involves a finite number n, $n \geq 1$, of the unknown (endogenous, sought-for) scalar variables y_1, y_2, ..., y_n. A general form of a discrete model is

$$F_j(y_1, y_2, \ldots, y_n) = 0, \quad j = 1, \ldots, m, \tag{1.1}$$

where $F_j(.)$ are some functions of n scalar variables. In this textbook, we assume that each variable y_i is a real number: $y_i \in \mathbf{R}^1$. Models with the integer-valued variables y_i are less common and harder to analyze.

A continuous model uses a continuous (scalar or vector) independent variable x defined on some domain $D \subset \mathbf{R}^n$, $n \geq 1$, and operates with scalar- or vector-valued functions $y(x)$. Continuous dynamic models include time as one of the independent variables. A general form of continuous models is

$$\Phi_j(y) = 0, \quad j = 1, \ldots, m, \tag{1.2}$$

where $\Phi_j(y)$ is a *functional* that sets a real value for each function y from a certain functional space Ω. Common examples of the functional space Ω are:

- the space $\boldsymbol{C}[a,b]$ of all continuous functions defined on the interval $[a,b]$
- the space $\boldsymbol{L}^{\infty}[a,b]$ of all measurable functions bounded almost everywhere on $[a,b]$.

A discrete analogue can usually be constructed for a continuous model, and vice versa. Discrete analogues are known for the most of continuous models of economic and ecological systems considered in this textbook. Computer simulation commonly uses discrete models or discrete analogues of continuous models in numeric algorithms. The choice between continuous and discrete models, and among their particular types, depends on the specifics of the real-life process under study. Models that combine discrete and continuous variables are known as *hybrid models*.

1.2.2.1 Differential and Integral Models

Depending on the type of the functional $\Phi(.)$ in (1.2), the major categories of continuous models are the *differential* and *integral equations*, explored in Sect. 1.3.4 below. The integral models are more general, but the differential models are simpler and more common in analytical and numerical study. The *selection rule* is this: if a process can be efficiently described by differential equations with sufficient accuracy, then there is no need to construct an integral model.

1.2.2.2 Optimization

If the number of unknown variables in the model (1.1) or (1.2) is larger than the number of equations, then the model usually allows for more than one solution. In such cases, an *optimization* problem can be introduced to find the best possible solution. Optimization problems minimize or maximize a certain *objective function* under the equality constraints (1.1) or (1.2) and other possible restrictions (see Sect. 1.3.5 below). The optimization problems for continuous models are known as the *optimal control problems*.

1.2.3 Linear and Nonlinear Models

The choice between linear and nonlinear models depends on the nature of the process under study and/or on the desired level of the process approximation. Many real-life processes are nonlinear but are commonly described by approximate linear models because the latter are simpler and have better theory and investigative techniques. Other processes are substantially nonlinear and their linearization leads to oversimplified description and incorrect modeling outcomes.

Linear discrete model is a *system of linear algebraic equations*:

$$\sum_{i=1}^{n} a_{ij}y_j = b_i, \quad i = 1, \ldots, m, \quad \text{or} \quad A\mathbf{y} = \mathbf{b}, \tag{1.3}$$

where

$\mathbf{y} = (y_1, y_2, \ldots, y_n) \in \mathbf{R}^n$, $\mathbf{b} = (b_1, b_2, \ldots, b_m) \in \mathbf{R}^m$, and $A = \{a_{ij}\}$ is an $m \times n$ *matrix*.

Model (1.3) represents a convenient and well-investigated mathematical object. If $m = n$ and the determinant $\det A \neq 0$, then the system (1.3) has a unique solution $\mathbf{y}$ (under the given A and $\mathbf{b}$).

Linear continuous model is the model (1.2) with linear functionals Φ_j, $j = 1, \ldots, m$. The *linear functional* Φ keeps the linear operations of *addition* and *scalar multiplication* for any elements y and z from a functional space Ω:

$$\Phi(y + z) = \Phi(y) + \Phi(z), \cdots \Phi(\alpha y) = \alpha\Phi(y) \quad \text{for} \quad \alpha \in \mathbf{R}^1. \tag{1.4}$$

Theories of the linear differential and integral equations are well developed and provide a good background for modeling many real systems and processes.

Nonlinear continuous model is the model (1.2) when at least one functional $\Phi_j(.)$ is nonlinear. There is no complete general theory for such equations, although fundamental breakthroughs are obtained for many specific nonlinear problems. The most studied categories of such models are nonlinear differential and integral equations. The theory of such equations is intensively investigated and possesses numerous essential results. Some of these results are reviewed in Sect. 1.3..

Nonlinear discrete models of the form (1.1) with nonlinear functions F_j also do not possess a general theory, and investigation of a specific system of nonlinear equations often runs into great theoretical or numeric difficulties. The solution may be nonunique or not existing in the nonlinear models, both discrete and continuous. The famous example is the polynomial equation $a_n x^n + a_{n-1} x^{n-1} + \ldots + a_1 x + a_0 = 0$ of one scalar variable x, which allows for a complete analytic solution at $n = 2$, 3, and 4, but not for n larger than 4. However, there are special classes of nonlinear discrete models, for example *difference equations* [4], which have well-developed theory and applications.

1.3 Review of Selected Mathematical Tools

This section briefly reviews some mathematical notations, techniques, and results used in the textbook.

1.3.1 Derivatives and Integrals

Let $f(x)$ be a real-valued function of one scalar variable $x \in \mathbf{R}^1$. The textbook uses the common notations

$$f'(x) \quad \text{and} \quad \frac{\mathrm{d}f(x)}{\mathrm{d}x}$$

for the derivative of $f(x)$. The "prime" notation $f'(x)$ was introduced by Joseph-Louis Lagrange (1736–1813), an Italian mathematician and astronomer, while $\mathrm{d}y/\mathrm{d}x$ was suggested by Gottfried Wilhelm Leibniz (1646–1716), a German diplomat and mathematician. Other known notations for the derivative are $\dot{y}$ proposed by Isaac Newton (1642–1727), an English scholar, and $D_x y = Dy$ suggested by Leonhard Euler (1707–1783), a Swiss mathematician and physicist.

1.3.1.1 Basic Rules of Differentiation and Integration

Let $u(x)$ and $v(x)$ be differentiable functions and $c \in \mathbf{R}^1$. Then

- The sum–difference rule: $(cu \pm v)' = cu' \pm v'$
- The product rule: $(u \cdot v)' = u' \cdot v + u \cdot v'$
- The quotient rule: $\left(\frac{u}{v}\right)' = \frac{u' \cdot v - u \cdot v'}{v^2}$
- The sum–difference rule of integration: $\int (cu \pm v)\mathrm{d}x = c\int u\mathrm{d}x \pm \int v\mathrm{d}x$
- The integration by parts: $\int u\mathrm{d}v = uv - \int v\mathrm{d}u$

1.3.1.2 The Fundamental Theorem of Calculus

Part I: If the function f is continuous on a closed interval $[a,b]$ and the function F is defined for all x in $[a,b]$ as $F(x) = \int_a^x f(t)\mathrm{d}t, \quad a \le x \le b$, then $F(x)$ is continuous on $[a,b]$ and differentiable on (a,b) and $F'(x) = f(x)$ for all x in (a,b).

Part II: If f and F are real-valued functions on $[a,b]$ such as $F'(x) = f(x)$, then

$$\int_a^b f(x)\mathrm{d}x = F(b) - F(a). \tag{1.5}$$

The first proof of this theorem was published by James Gregory (1638–1675), a Scottish mathematician and astronomer, and its more complete version was offered in by Isaac Barrow (1630–1677), an English scholar and a teacher of Isaac Newton. Part II is often referred to as the *Newton–Leibniz axiom*. A more general form of Part I of the Fundamental Theorem of Calculus is

1.3.1.3 Leibniz's Formula for Derivatives

If $f(x,t)$ and its partial derivative $\partial f(x,t)/\partial x$ are continuous in x and t, and functions $a(x)$ and $b(x)$ are differentiable, then

$$\frac{\mathrm{d}}{\mathrm{d}x}\int_{a(x)}^{b(x)} f(x,t)\mathrm{d}t = \\ = \int_{a(x)}^{b(x)} \frac{\partial f(x,t)}{\partial x}\mathrm{d}t + f(x,b(x))\frac{\mathrm{d}b(x)}{\mathrm{d}x} - f(x,a(x))\frac{\mathrm{d}a(x)}{\mathrm{d}x}. \tag{1.6}$$

1.3.1.4 Taylor Series

The Taylor series of a real-valued function $f(x)$ infinitely differentiable in the neighborhood of $x = a$ is the power series

$$f(x) = \sum_{n!}^{\infty} \frac{f^{(n)}(a)}{n!}(x-a)^n \\ = f(a) + \frac{f'(a)}{1!}(x-a) + \frac{f''(a)}{1!}(x-a)^2 + \frac{f''(a)}{1!}(x-a)^3 + \dots. \tag{1.7}$$

In particular,

$$e^x = \sum_{n=0}^{\infty} \frac{x^n}{n!} = 1 + x + \frac{x^2}{2!} + \frac{x^3}{3!} + \dots. \tag{1.8}$$

1.3.1.5 Implicit Function Theorem

For clarity, we consider the case of the scalar dependent variable $y \in \mathbf{R}^1$ and scalar independent variable $x \in \mathbf{R}^1$. The *implicit function* is a function defined by the equality

$$g(x, y) = 0. \tag{1.9}$$

Ideally, an implicit function can be converted into the *explicit function* $y = f(x)$ by solving (1.9) for y in terms of x. In practice, it is often difficult or not possible. The following theorem provides a convenient tool for analyzing the implicit functions.

The implicit function theorem in a scalar case: Let the function $g(x,y)$ have continuous partial derivatives on an open set $S \subset \mathbf{R}^2$ containing the point (x_0, y_0) such that $g(x_0, y_0)$. If $\partial g/\partial y \neq 0$ at (x_0, y_0), then there exists an interval $\Delta =$

$(x_0 - \delta, x_0 + \delta)$ such that for any $x \in \Delta$ there is a unique value $y = y(x)$ such that $g(x, y) = 0$. The derivative of this explicit function $y = y(x)$ is

$$\frac{\mathrm{d}y}{\mathrm{d}x} = -\frac{\partial g}{\partial x} / \frac{\partial g}{\partial y}. \tag{1.10}$$

1.3.2 Vector Algebra and Calculus

Let us consider the Cartesian coordinate system $\mathbf{x} = (x_1, x_2, x_3)$ in the three-dimensional space $\mathbf{R}^3$. The vectors $\mathbf{i} = (1,0,0)$, $\mathbf{j} = (0,1,0)$, and $\mathbf{k} = (0,0,1)$ are called the *fundamental vectors* or the *basis*.

The *dot product* (*scalar product, inner product*) of two three-dimensional vectors $\mathbf{x}$ and $\mathbf{y}$ is a scalar

$$\mathbf{x} \cdot \mathbf{y} = (\mathbf{x}, \mathbf{y}) = x_1 y_1 + x_2 y_2 + x_3 y_3. \tag{1.11}$$

The dot product is used to find the angles between the two vectors, determine an orthogonal basis, find a normal to a plane, find work done by a force, and for others purposes (see Chap. 9).

The *cross product* (*vector product, outer product*) of two three-dimensional vectors $\mathbf{x}$ and $\mathbf{y}$ is the vector

$$\mathbf{x} \times \mathbf{y} = \begin{vmatrix} \mathbf{i} & \mathbf{j} & \mathbf{k} \\ x_1 & x_2 & x_3 \\ y_1 & y_2 & y_3 \end{vmatrix}. \tag{1.12}$$

Applications of the cross product are to find the moment of a force, the velocity of a rotating body, the volume of solids, and others.

The *gradient* of a scalar differentiable function $f(x_1,x_2,x_3) \in \mathbf{R}^1$ is the vector

$$\nabla f = \operatorname{grad} f = \frac{\partial f}{\partial x_1}\mathbf{i} + \frac{\partial f}{\partial x_2}\mathbf{j} + \frac{\partial f}{\partial x_3}\mathbf{k}. \tag{1.13}$$

It defines the direction and magnitude of the maximum rate of increase of the function f at the point $\mathbf{x} = (x_1,x_2,x_3)$. The gradient is a normal vector to the surface $f(x_1,x_2,x_3)$ at point $\mathbf{x}$.

The differential operator ∇ (nabla) is $\nabla = \frac{\partial}{\partial x_1}\mathbf{i} + \frac{\partial}{\partial x_2}\mathbf{j} + \frac{\partial}{\partial x_3}\mathbf{k}$.

The Laplace operator Δ (delta) is $\Delta = \nabla^2 = \frac{\partial^2}{\partial x_1^2} + \frac{\partial^2}{\partial x_2^2} + \frac{\partial^2}{\partial x_3^2}$.

The *Laplacian* of a scalar function $S(x_1,x_2,x_3)$ is the scalar

$$\Delta S = \operatorname{div} \operatorname{grad} S = \nabla \cdot (\nabla S) = \nabla^2 S = \frac{\partial^2 S}{\partial {x_1}^2} + \frac{\partial^2 S}{\partial {x_2}^2} + \frac{\partial^2 S}{\partial {x_3}^2}. \tag{1.14}$$

Let $x_1 = x_1(t)$, $x_2 = x_2(t)$, $x_3 = x_3(t)$. Then, the *total derivative* of a scalar function $S(x_1, x_2, x_3, t)$ with respect to t is

$$\frac{dS}{dt} = \frac{\partial S}{\partial t} + \frac{\partial S}{\partial x_1}\frac{dx_1}{dt} + \frac{\partial S}{\partial x_2}\frac{dx_2}{dt} + \frac{\partial S}{\partial x_3}\frac{dx_3}{dt}. \tag{1.15}$$

The *partial derivative* of a *vector-function* $\mathbf{V}(\mathbf{x}) = V_1\mathbf{i} + V_2\mathbf{j} + V_3\mathbf{k} \in \mathbf{R}^3$ with respect to x_i is the vector

$$\frac{\partial \mathbf{V}}{\partial x_i} = \frac{\partial V_1}{\partial x_i}\mathbf{i} + \frac{\partial V_2}{\partial x_i}\mathbf{j} + \frac{\partial V_3}{\partial x_i}\mathbf{k}.$$

The *divergence* of a vector function $\mathbf{V}(x_1, x_2, x_3)$ is the scalar

$$\text{div } \mathbf{V} = \nabla \cdot \mathbf{V} = \frac{\partial V_1}{\partial x_1} + \frac{\partial V_2}{\partial x_2} + \frac{\partial V_3}{\partial x_3}. \tag{1.16}$$

1.3.3 Differential Equations

The *differential equation* is an equation that contains derivatives of an unknown function. The unknown function, called the *dependent variable*, depends on one or several continuous *independent variables*. A differential equation that includes the derivatives in one independent variable x is called an *ordinary differential equation* (ODE) [1]. Its general form is

$$F\left(x, y, \frac{dy}{dx}, \ldots, \frac{dy^m}{d^m x}\right) = 0. \tag{1.17}$$

A *partial differential equation* (PDE) is a differential equation that involves the partial derivatives with respect to several independent variables $x_1, x_2, \ldots, x_n$:

$$F\left(x_1, \ldots, x_n, \frac{\partial y}{\partial x_1}, \frac{\partial y}{\partial x_2}, \ldots, \frac{\partial y}{\partial x_n}, \frac{\partial^2 y}{\partial {x_1}^2}, \frac{\partial^2 y}{\partial x_1 \partial x_2}, \ldots\right) = 0. \tag{1.18}$$

The *order* of a differential equation is the order of the highest-order derivatives in the equation. Thus, (1.17) is an ordinary differential equation of the m-th order.

A differential equation is *linear* when the function F in (1.17) or (1.18) is linear with respect to the dependent variable and its derivatives. Otherwise, the equation is *nonlinear*.

A *solution* of a differential equation can exist in the *explicit form* $y = f(x)$ or the *implicit form* as $F(x,y) = 0$. To verify a solution, it is enough to substitute it to the original differential equation.

1.3.3.1 Initial Value Problem

The *initial value problem* (or the *Cauchy problem*) for the ordinary differential equation (1.17) of the m-th order is as follows:

- Find a solution to the differential equation (1.17) on the interval $[x_0, X)$, $x_0 < X \leq \infty$, that satisfies at $x = x_0$ the following m initial conditions

$$\begin{aligned} & y(x_0) = y_0, \\ & \frac{\mathrm{d}y}{\mathrm{d}x}(x_0) = y_1, \\ & \ldots \\ & \frac{\mathrm{d}^{m-1}y}{\mathrm{d}x^{m-1}}(x_0) = y_{m-1}, \end{aligned} \tag{1.19}$$

where $x_0 \in I$ and $y_0, \ldots, y_{m-1}$ are given constants.

In general, m conditions of the form (1.19) should be known to determine a unique solution of an ODE of the m-th order. If such conditions are known at more than one point x_0, then we have a *boundary problem* for the corresponding ODE. The theory of initial value problems is well developed and contains many results about the existence and uniqueness of their solutions. For instance,

Existence and Uniqueness of Solution to First-Order Differential Equation: If f and $\partial f/\mathrm{d}y$ are continuous in some rectangle $\{(x,y) : a < x < b, c < y < d\}$ that contains the point (x_0, y_0), then the initial value problem for the nonlinear first-order ODE

$$\mathrm{d}y/\mathrm{d}x = f(x, y), \quad y(x_0) = y_0, \tag{1.20}$$

has a unique solution $y(x)$ in the interval $x_0 - \delta < x < x_0 + \delta$ for a certain $\delta > 0$.

There are several classic and modern methods for solving differential equations. Below we review some methods that are used in this textbook.

1.3.3.2 Linear First-Order Ordinary Differential Equation

$$S(x)y' + P(x)y = Q(x) \tag{1.21}$$

has the general solution of the form

$$y = \frac{1}{\mu(x)}\left(\int \mu(x)\frac{Q(x)}{S(x)}\mathrm{d}x + C\right), \quad \mu(x) = \mathrm{e}^{\int \frac{P(x)}{S(x)}\mathrm{d}x}, \tag{1.22}$$

where C is any constant. The constant C is determined from the initial condition $y(x_0) = y_0$ (when this condition is known).

Frequently used special cases of the general formula (1.22) are

$$P(x) \equiv 0 \quad \Rightarrow \quad y = \int \frac{Q(x)}{S(x)}\mathrm{d}x + C, \tag{1.23}$$

$$P(x) \equiv S'(x) \quad \Rightarrow \quad y = \frac{1}{S(x)}\left(\int Q(x)\mathrm{d}x + C\right), \tag{1.24}$$

$$Q(x) \equiv 0 \quad \Rightarrow \quad y = C\mathrm{e}^{-\int \frac{P(x)}{S(x)}\mathrm{d}x}. \tag{1.25}$$

1.3.3.3 Separable Ordinary Differential Equations

If the right-hand side of the first-order differential equation (1.20) can be represented as a product of a function in x and a function in y:

$$y' = f(x, y) = g(x)h(y), \tag{1.26}$$

then (1.26) can be written in the form $\frac{\mathrm{d}y}{\mathrm{d}x} = g(x)h(y)$ or

$$\frac{\mathrm{d}y}{h(y)} = \frac{\mathrm{d}x}{g(x)}. \tag{1.27}$$

The integration of both sides of (1.27) leads to the solution of (1.26) in an implicit form.

1.3.3.4 Partial Differential Equations

Linear partial differential equations (PDE) of the first and second orders are widely used in applications. The linear PDE of the first order in two variables x and t

$$\frac{\partial u}{\partial t} + \frac{\partial u}{\partial x} = a + bu \tag{1.28}$$

with respect to the unknown function $u(t,x)$ is used in Chaps. 4 and 6.

A general form of the *linear PDE of the second order* in two space variables (x, y) and time t is

$$\frac{\partial^2 u}{\partial x^2} + \frac{\partial^2 u}{\partial y^2} = a + bu + c\frac{\partial u}{\partial t} + d\frac{\partial^2 u}{\partial t^2}. \tag{1.29}$$

Equation (1.29) includes well-known special PDEs used in the textbook. They are listed in Table 1.1.

Table 1.1 Major linear partial differential equations of the second order

Case of (1.29)	Name of equation	Equation
$a = 0, b = 0, c = 0, d = 0$	Laplace's equation	$\frac{\partial^2 u}{\partial x^2} + \frac{\partial^2 u}{\partial y^2} = 0$
$a \neq 0, b = 0, c = 0, d = 0$	Poisson's equation	$\frac{\partial^2 u}{\partial x^2} + \frac{\partial^2 u}{\partial y^2} = a$
$a = 0, b = -k^2u \neq 0, c = 0, d = 0$	Helmholtz equation	$\frac{\partial^2 u}{\partial x^2} + \frac{\partial^2 u}{\partial y^2} = -k^2 u$
$a = 0, b = 0, c \neq 0, d = 0$	Diffusion equation	$\frac{\partial^2 u}{\partial x^2} + \frac{\partial^2 u}{\partial y^2} = c\frac{\partial u}{\partial t}$
$a = 0, b = 0, c = 0, d = 1/v^2 \neq 0$	Wave equation	$\frac{\partial^2 u}{\partial x^2} + \frac{\partial^2 u}{\partial y^2} = \frac{1}{v^2}\frac{\partial^2 u}{\partial t^2}$

1.3.4 Integral Equations

Integral equations contain integrals with unknown functions in their integrands. This textbook uses the following types of the integral equations:

- *Volterra integral equation of the first kind* with respect to the unknown function $x(t)$ of the one-dimensional independent variable $t\in[a, b]$:

$$\int_a^t K(t,\tau)x(\tau)\mathrm{d}\tau = f(t), \quad t\in[a,b], \tag{1.30}$$

where the function $f(t)$ and the *kernel* $K(t,\tau)$ are given functions.

- *Volterra integral equation of the second kind*

$$x(t) = \int_a^t K(t,\tau)x(\tau)\mathrm{d}\tau + f(t), \quad t\in[a,b], \tag{1.31}$$

is more common as compared to (1.30) of the first kind.

Equations (1.30) and (1.31) are named after Vito Volterra (1860–1940), a famous Italian mathematician and physicist, who introduced them and developed their theory and applications. The Volterra integral equations are widely used in population biology, physics, engineering, economics, and demography [3].

These equations are well suited for the description of dynamic processes. Indeed, if the variable t is time, then the current state of a *dynamic system* (process) depends on the past states and cannot depend on the future. Hence, $K(t,\tau) \equiv 0$ at $\tau > t$ for dynamic systems, which is reflected in the integrals in (1.30) and (1.31).

Another major type of the linear integral equations is the *Fredholm integral equations*, which are described by the same expressions (1.31) and (1.31) where the upper integration limit t is replaced with b. This small change causes significant differences in the qualitative dynamics of their solutions.

Despite the similarity of the Fredholm and Volterra integral equations, their properties are quite different. The Volterra integral equations (1.30) and (1.31) generalize the initial value problems for differential equations considered in Sect. 1.3.3, whereas the Fredholm integral equations correspond to boundary problems (not considered in this textbook).

Linear integral equations have well-developed theories. There is a variety of theorems about the existence and uniqueness of solutions for these equations [2]. The theorems differ in smoothness requirements and forms of the equation. In particular, the Volterra integral equation of the second kind has a unique solution under natural assumptions. A classic existence result is as follows:

1.3.4.1 Existence and Uniqueness Theorem for Volterra Integral Equation

If $K(t,\tau)$ is measurable on $[a,b]\otimes[a,b]$ and $f(t)$ is continuous on $[a,b]$, then a unique continuous solution $x(t)$ of the Volterra integral equation (1.31) of the second kind exists on $[a,b]$ and can be determined as

$$x(t) = \int_a^t R(t,\tau)f(\tau)d\tau + f(t), \tag{1.32}$$

where the *resolvent kernel* $R(t,\tau)$, $\tau \in [a,b]$, $t \in [a,b]$, is a solution of the following linear Volterra integral equation:

$$R(t,\tau) = \int_\tau^t R(t,u)K(u,\tau)\mathrm{d}u + K(t,\tau). \tag{1.33}$$

In particular, if $K(t,\tau) = K = \text{const}$, then

$$R(t,\tau) = K\mathrm{e}^{K(t-\tau)}. \tag{1.34}$$

After discretization by the variable t, the linear integral equations (1.30) and (1.31) are reduced to systems of linear algebraic equations (1.4). The analogy between continuous integral models and their discrete analogues (1.4) is useful for better understanding and interpretation of the linear integral equations.

However, theory of linear continuous models is much more complex as compared to linear discrete models. In particular, a significant difference exists between the integral equations of the first and the second kind.

Nonlinear Volterra integral equation of the second kind

$$x(t) = \int_a^t F(t,\tau,x(\tau))\mathrm{d}\tau + f(t), \quad t\in[a,b], \tag{1.35}$$

is the generalization of the linear integral equation (1.31) with the integrand $K(t,\tau)x$ replaced with a nonlinear function $F(t,\tau, x)$ of x.

Hammerstein–Volterra integral equation is the special case of the nonlinear equation (1.35) with the nonlinearity $F(t,\tau, x) = K(t,\tau)G(x)$:

$$x(t) = \int_a^t K(t,\tau)G(x(\tau))\mathrm{d}\tau + f(t), \quad t\in[a,b]. \tag{1.36}$$

A key condition for the existence and *uniqueness* of the solution x to the nonlinear integral equation (1.35) is the Lipschitz condition for the function $F(t,s,x)$ with respect to x: $|F(t,s,x) - F(t,s,y)| \leq L(t,s)|x - y|$. If it holds, then (1.35) possesses a unique solution x, at least, for continuous f and L.

1.3.4.2 Volterra Integral Equations with Variable Delay

The classic integral equations (1.30), (1.31), (1.35) take into account the distributed delay (*after-effect, hereditary effects*) on the interval $[a, t]$. The distributed delay means that a continuous sequence of the past states of a dynamical system affects the future evolution of the system. The *integral equations with variable delay* occur if the distributed delay exists at the initial time $t = a$ as well. Specifically, the linear *integral equation with variable delay*

$$x(t) = \int_{a(t)}^t K(t,\tau)x(\tau)\mathrm{d}\tau + f(t), \quad t\in[t_0,T], \tag{1.37}$$

with the initial condition

$$x(\tau) = x_0(\tau), \quad \tau\in[\tau_0, \ t_0], \tag{1.38}$$

means that the solution x depends on its known behavior x_0 over a certain prehistory interval $[\tau_0, t_0]$. The lower integration limit $a(t)$ in (1.37) is a given function such as $\tau_0 \leq a(t) < t$.

Equation (1.37) can be solved by reducing it to the standard Volterra integral equation

$$x(t) = \int_{t_0}^{t} K(t,\tau)x(\tau)\mathrm{d}\tau + f_1(t), \quad t \in [t_0, t_1], \tag{1.39}$$

on some interval $[t_0, t_1]$ such that $a(t) \leq t_0$, $t \in [t_0, t_1]$. Then,

$$f_1(t) = \int_{a(t)}^{t_0} K(t,\tau)x_0(\tau)\mathrm{d}\tau + f(t) \tag{1.40}$$

is a given function on $[t_0, t_1]$. Next, this solution process is repeated on a new interval $[t_1, t_2]$ with the updated initial condition $x(\tau) = x_0(\tau)$, $\tau \in [\tau_0, t_1]$, and so on.

The integral equations with variable delay are used for the modeling of economic development in Chaps. 4 and 5. Moreover, in some economic applications, the lower integration limit $a(t)$ can be an unknown control. Then, the model (1.39) leads to nonlinear *integral equations with controllable delay* (see Chap. 5).

1.3.5 Optimization and Optimal Control

The objective of an *optimization problem* is to maximize or minimize a function f called the *objective function*. The objective function $f(x_1, x_2, \ldots, x_n)$ depends on the unknown variables $x_1, x_2, \ldots, x_n$, which can be subject to *constraints* or *restrictions* that follow from applications of the problem. If the constraints include equality-constraints, then some of the unknown variables can be chosen as the independent (*control*) *variables*, and the rest become the *dependent (state) variables*.

1.3.5.1 Unconstrained Optimization

Let us consider the maximization (or minimization) of $f(x_1, x_2, \ldots, x_n)$ with no constraints. If the objective function $f(x_1, \ldots, x_n)$ is differentiable and has an extremum at a point $(\hat{x}_1, \ldots, \hat{x}_n)$, then the partial derivatives of f are zero at this point:

$$\partial f(\hat{x}_1)/\partial x_1 = 0, \quad \ldots, \quad \partial f(\hat{x}_n)/\partial x_n = 0. \tag{1.41}$$

In the one-dimensional case, this condition is reduced to $f' = 0$ and is known as the *Fermat Theorem*. The condition (1.41) is necessary for an extremum of f. It is not sufficient, as the simple example $f(x) = x^3$ demonstrates: the derivative $f' = 3x^2$ satisfies (1.41) at $x = 0$, but the point $x = 0$ brings neither minimum nor maximum to the function. Equations (1.41) are often nonlinear and their analysis is challenging; however, in many cases it is simpler than the analysis of the complete optimization problem.

1.3.5.2 Optimization Problems with Constraints

Solving an optimization problem with constraints is more difficult compared to the unconstrained optimization. One of powerful techniques for finding local maxima and minima of a function subject to constraints is the *method of Lagrange multipliers* [5]. To illustrate it for a problem with *equality-constraints*, let us consider the following simple discrete optimization problem: find x_1, x_2 that

$$\text{maximize } f(x_1,\ x_2) \tag{1.42}$$

subject to the equality-constraint

$$f(x_1,\ x_2) = c, \tag{1.43}$$

where both functions f and g have continuous first partial derivatives.

The *Lagrange function* (or *Lagrangian*) of the optimization problem (1.42)–(1.43) is determined as

$$L(x_1,\ x_2,\ \lambda) = f(x_1,\ x_2) + \lambda(g(x_1,\ x_2) - c), \tag{1.44}$$

where the unknown variable λ is called the *Lagrange multiplier* (dual variable, or adjoint variable). The second term in (1.44) is zero along a solution to the problem. Thus in order to solve (1.42)–(1.43), we can find the maximum of (1.44). The maximization of the Lagrangian (1.44) includes one more unknown variable but does not involve the equality-constraint (1.43). By construction of (1.44), if (x_{10}, x_{20}) brings a maximum to the original problem (1.42)–(1.43), then there exists λ_0 such that ($x_{10}, x_{20}, \lambda_0$) is a stationary point ($\partial L/\partial\lambda = 0$) of the Lagrange function (1.38). Note that $\partial L/\partial\lambda = 0$ implies (1.37).

The method of Lagrange multipliers yields necessary conditions for optimality. Sufficient conditions for optimality are also possible but are more difficult to obtain. The Lagrangian can be reformulated in the terms of Hamiltonian for many specific optimization problems. In particular, the method of Lagrange multipliers can be used to derive the maximum principle for the optimal control of differential equations provided in Sect. 2.4.

1.3.5.3 Continuous Optimization

Optimization problems in the continuous models of Sect. 1.2.2 are known as *continuous-time optimization problems* or the ***optimal control problems***. The *control variables* in such problems are scalar- or vector-valued functions of a continuous independent variable and the *objective function* is a functional that depends on the control variables.

Historically, *calculus of variations* is the first classic technique for the continuous-time optimization developed over 200 years mainly for geometric and physical applications. A *variational problem* minimizes a certain functional on a set of *smooth functions* in an open domain. Further extension of the variational techniques to the *non-smooth* unknown functions and closed domains leads to the modern *optimal control theory* and its main tools, the *principle of maximum* of L. Pontryagin and the *dynamic programming* method of R. Bellman.

Exercises

1. Rewrite the general discrete model (1.1) in a vector form.
   HINT: introduce vectors $\mathbf{F} = (F_1,\ldots,F_m) \in \mathbf{R}^m$ and $\mathbf{0} = (0,\ldots,0) \in \mathbf{R}^m$.
2. Prove the formula (1.10) for the derivative of an implicit function using formula (1.15).
   HINT: The total derivative of $g(x,y(x))$ in x should be equal to zero.
3. Derive the formula (1.24) from the general formula (1.22) at $P(x) \equiv S'(x)$
   HINT: If $P(x) \equiv S'(x)$, then (1.21) can be rewritten as $S(x)y' + S'(x)y = g(x) \Leftrightarrow (S(x)y)' = Q(x) \Leftrightarrow S(x)y = \int Q(x)\mathrm{d}x$, from which follows (1.24).
4. Derive the third formula (1.25) from general formula (1.22) at $Q(x) \equiv 0$
   HINT: If $Q(x) \equiv 0$, then (1.21) becomes $\frac{y'}{y} = \frac{P(x)}{S(x)}$ and the integration of both sides leads to (1.25).
5. At $K(t,\tau) = K(\tau)$, find the solution of the linear Volterra integral equation (1.30) of the first kind.
   HINT: Differentiate (1.30).
6. At $K(t,\tau) = K(t)$, find the solution of the linear Volterra integral equation (1.30) of the first kind and explain how and why it differs from the Exercise 5.
7. At $K(t,\tau) = K(\tau)$, convert the linear Volterra integral equation (1.31) of the second kind to an initial problem for a linear ordinary differential equation.
   HINT: Differentiating (1.31) leads to an ODE. The initial condition $x(a) = f(a)$ is obtained from (1.31) at $t = a$.
8. Show that the linear Volterra integral equation (1.33) for the resolvent kernel R has the solution (1.34) at $K(t,\tau) = K = \text{const}$.
   HINT: Differentiate (1.33) in t and solve the initial problem for the obtained linear ODE with the initial condition $R(\tau,\tau) = K(\tau,\tau)$ using the formula (1.25).
9. Find an exact solution of the linear Volterra integral equation (1.31) at $K(t,\tau) = \text{const}$ and $f(t) = \text{const}$ from the resolvent formula (1.32) and (1.34).
10. Find a solution of the linear Volterra integral equation (1.31) at $K(t,\tau) = \text{const}$ and $f(t) = \text{const}$ without using the resolvent kernel.
    HINT: Differentiate the integral equation (1.31) and solve the obtained initial problem for a linear ODE from Exercise 7 using the formula (1.25).

References[1]

1. 📖 Boyce, W.E., DiPrima, R.C.: Elementary differential equations, 9th edn. Wiley, New York (2009)
2. 📖 Burton, T.A.: Volterra integral and differential equations. Academic, Orlando, FL (1983)
3. 📖 Corduneanu, C.: Integral equations and applications. Cambridge University Press, Cambridge (1991)
4. Elaydi, S.: An introduction to difference equations, 3rd edn. Springer, New York (2010)
5. 📖 Luenberger, D.G.: Introduction to dynamic systems: theory, models, and applications. Wiley, New York (1979)
6. Kapur, J.: Mathematical modeling. Wiley, New York (1988)
7. 📖 Saaty, T.L., Joyce, M.: Thinking with models: mathematical models in the physical, biological, and social sciences. Pergamon, Oxford (1981)

[1] The book symbol 📖 means that the reference is a textbook recommended for further student reading.

Part I
Mathematical Models in Economics

Chapter 2
Aggregate Models of Economic Dynamics

This chapter explores aggregate optimization models of the neoclassic economic growth theory, which are based on the concept of production functions. The models are described by ordinary differential equations and involve static and dynamic optimization. Section 2.1 analyzes production functions with several inputs, their fundamental characteristics, and major types (Cobb–Douglas, CES, Leontief, and linear). Special attention is given to two-factor production functions and their use in the neoclassic models of economic growth. Sections 2.2 and 2.3 describe and analyze the well-known Solow–Swan and Solow–Ramsey models. Section 2.4 contains maximum principles used to analyze dynamic optimization problems in this and other chapters.

2.1 Production Functions and Their Types

A *production function* describes a relationship

$$y = f(x_1, \ldots, x_n) \tag{2.1}$$

between the aggregate product output y and the *productive inputs* $x_1, \ldots, x_n$ that can include labor, capital, knowledge (human capital), energy consumption, raw materials, natural resources (land, water, minerals), and others. The output y and inputs x_i are assumed to be identical. For example, the labor is the quantity of workers indistinguishable in a productive sense.

Henceforth, we will often use the following definition. The function $r(t) = f'(t)/f(t)$ is the *relative rate* of the function $f(t)$ and is often referred to as the *growth rate* of $f(t)$. If $r \equiv$ const, then $f(t) = C\exp(rt)$.

Economists often use the notation $\dot{f}$ for the derivative of a function f in time. We will keep the standard notation f'.

N. Hritonenko and Y. Yatsenko, *Mathematical Modeling in Economics, Ecology and the Environment*, Springer Optimization and Its Applications 88,
DOI 10.1007/978-1-4614-9311-2_2, © Springer Science+Business Media New York 2013

2.1.1 Properties of Production Functions

Commonly accepted properties of production functions are:

1. **Essentiality of inputs**: If at least one $x_i = 0$, then $y = 0$, i.e., production is not possible without any of the inputs.
2. **Positive returns**: $\partial f/\partial x_i > 0$, $i = 1, \ldots, n$, i.e., the output increases if an input increases.
3. **Diminishing returns**: The Hessian matrix

$$H = \begin{bmatrix} \partial^2 f/\partial {x_1}^2 & \ldots & \partial^2 f/\partial x_1 \partial x_n \\ \ldots & \ldots & \ldots \\ \partial^2 f/\partial x_n \partial x_1 & \ldots & \partial^2 f/\partial {x_n}^2 \end{bmatrix} \tag{2.2}$$

   is negatively definite. It means that if only one input x_i increases and the other inputs x_j, $j \neq i$, remain constant, then the efficiency of using the input x_i decreases.
4. **Proportional returns to scale**: $f(\mathbf{x})$ is a *homogeneous function* of degree $\gamma > 0$, i.e.,

$$f(l\mathbf{x}) = l^{\gamma} f(\mathbf{x}), \quad l \in R^1, \quad l > 0, \quad \mathbf{x} = (x_1, \ldots, x_n). \tag{2.3}$$

The production function $f(\mathbf{x})$ exhibits *increasing returns to scale* at $\gamma > 1$, *decreasing returns to scale* at $\gamma < 1$, and *constant returns to scale* at $\gamma = 1$. The increasing returns mean that a 1 % increase in the levels of all inputs leads to a greater than the 1 % increase of the output y.

In the case of *constant returns to scale*, the function $f(\mathbf{x})$ is *linearly homogeneous*: $f(l\mathbf{x}) = lf(\mathbf{x})$, and the output increases linearly with respect to a proportional increase of all inputs: a 1 % increase of all inputs produces exactly the 1 % increase of the output. Then, the condition (2.2) is reduced to

$$\partial^2 f/\partial {x_i}^2 < 0, \quad i = 1, \ldots, n. \tag{2.4}$$

2.1.2 Characteristics of Production Functions

The major characteristics of production functions are

- The *average product* $f(x_1, \ldots, x_n)/x_i$ of the i-th input is the output per one unit of the input x_i spent, $i = 1, \ldots, n$.
- The *marginal product* $\partial f/\partial x_i$ of the i-th input describes the additional output obtained due to the increase of the i-th input quantity by one unit.
- The *isoquant* is the set of all possible combinations of inputs $\mathbf{x} = (x_1, \ldots, x_n)$ that yield the same level of the output $y = f(\mathbf{x})$. Along an isoquant, the differential of the function $f(\mathbf{x})$ is zero: $\sum_{i=1}^{n} (\partial f/\partial x_i) dx_i = 0$.

- The *marginal rate of substitution* between the inputs i and j

$$h_{ij} = (\partial f / \partial x_i)/(\partial f / \partial x_j) \tag{2.5}$$

shows how many units of the j-th input are required to substitute one unit of the i-th input in order to produce the same level of the output y.

- The *partial elasticity of output* with respect to the input i

$$\varepsilon_i(\mathbf{x}) = (\partial f(\mathbf{x})/\partial x_i)/(f(\mathbf{x})/x_i) = \partial \ln f(\mathbf{x})/\partial \ln x_i \tag{2.6}$$

is the ratio between the marginal product and the average product of the i-th input. It describes the increase of the output y when the i-th input increases by 1 %.

- The *total output elasticity* $\varepsilon(\mathbf{x}) = \sum_{i=1}^{n} \varepsilon_i(\mathbf{x})$ describes the output increase under a proportional production scale extension. For a homogeneous production function (2.3), $\varepsilon(\mathbf{x}) = \gamma$.
- The *elasticity of substitution* is a quantitative measure of a possibility of changes in the input combination to produce the same output. It is equal to the relative change in the ratio of the i-th and j-th inputs divided by the relative change in their marginal rate of substitution h_{ij}:

$$\sigma_{ij} = \frac{\mathrm{d}(x_i/x_j)}{(x_i/x_j)} \times \frac{h_{ij}}{\mathrm{d}h_{ij}} = \frac{\mathrm{d}\ln(x_i/x_j)}{\mathrm{d}\ln h_{ij}}. \tag{2.7}$$

This characteristic shows the percentage change of the ratio x_i/x_j of these inputs along an isoquant in order to change their marginal substitution rate by one percent. The larger the σ_{ij}, the greater the *substitutability* between the two inputs. The inputs i and j are *perfect substitutes* at $\sigma_{ij} = \infty$ and they are not substitutable at all at $\sigma_{ij} = 0$. The elasticity of substitution is used for classification of various production functions.

2.1.3 Major Types of Production Functions

2.1.3.1 Linear Production Function

$$y = a_1 x_1 + \ldots + a_n x_n, \quad a_i > 0, \quad i = 1, \ldots, n, \tag{2.8}$$

has the following characteristics:

$$\partial f / \partial x_i = a_i, \quad h_{ij} = a_j / a_i = \text{const}, \quad \varepsilon = 1, \quad \sigma_{ij} = \infty, \quad i,j = 1, \ldots, n,$$

i.e., constant returns to scale, constant marginal rates of substitution, all inputs are perfectly substitutable. Despite its mathematical simplicity, the linear production function with several inputs is rarely used in the economic theory because it violates the fundamental economic property of the input essentiality (Sect. 2.1.1). However, the linear production function with one input (capital) known as the *AK production function* has been intensively investigated, see (2.30) below.

2.1.3.2 Cobb–Douglas Production Function

$$y = A{x_1}^{\alpha_1} \times \ldots \times {x_n}^{\alpha_n} \tag{2.9}$$

has the following characteristics:

$$\partial f/\partial x_i = \alpha_i(y/x_i), \quad h_{ij} = \alpha_j x_i/\alpha_i x_j, \quad \varepsilon_i = \alpha_i, \quad \varepsilon = \alpha_1 + \ldots + \alpha_n,$$
$$\sigma_{ij} = 1, \quad i,j = 1, \ldots, n,$$

i.e., the elasticity of substitution for any pair (i, j) of inputs is equal to one. The *returns to scale* are *increasing* in the case $\alpha_1 + \ldots + \alpha_n > 1$, *decreasing* in the case $\alpha_1 + \ldots + \alpha_n < 1$, and *constant* at $\alpha_1 + \ldots + \alpha_n = 1$.

By taking the logarithm of both sides of (2.9), we obtain a linear expression

$$\ln y = \ln A + \sum_{i=1}^{n} \alpha_i \ln x_i,$$

that after differentiation becomes

$$y'/y = \alpha_1\left(x_1'/x_1\right) + \ldots + \alpha_n\left(x_n'/x_n\right), \tag{2.10}$$

i.e., the growth rate of the output in the Cobb–Douglas production function is equal to the weighted sum of the growth rates of the inputs.

2.1.3.3 Production Function with Fixed Proportions

$$y = A\min(x_1, \ldots, x_n), \tag{2.11}$$

has the following characteristics:

$$h_{ij} = \begin{cases} 0, & x_j > x_i \\ \infty, & x_j < x_i \end{cases}, \quad \varepsilon = 1, \quad \sigma_{ij} = 0, \quad i,j = 1, \ldots n,$$

i.e., the elasticity of substitution for all inputs is zero (the inputs are not substitutable). This production function is also known as the *piecewise-linear production function* and the *Leontief production function*.

2.1.3.4 Production Function with Constant Elasticity of Substitution (CES)

$$y = A\,[\beta_1 x_1{}^{\rho} + \ldots + \beta_n x_n{}^{\rho}]^{\gamma/\rho}, \tag{2.12}$$

where $\rho < 1$, $\rho \neq 0$, $\beta_i > 0$, $i = 1, \ldots, n$, $\beta_1 + \ldots + \beta_n = 1$, has the following characteristics:

$$h_{ij} = (\beta_i/\beta_j)\,(x_j/x_i)^{1-\rho}, \quad \varepsilon = \gamma, \quad \sigma_{ij} = 1/(1-\rho), \quad i,j = 1, \ldots, n,$$

i.e., the elasticity of substitution is a positive constant. The CES production function is also known as the *Solow production function*.

The CES production function is the most general among the production functions considered above: it leads to the linear production function as $\rho \to 1$, to the Cobb–Douglas production function as $\rho \to 0$, and to the production function with fixed proportions as $\rho \to -\infty$.

The Cobb–Douglas and CES production functions are frequently used in low-sector aggregate economic models (see Sects. 2.2–2.4, 3.3, 3.4, 10.2), whereas the production function with fixed proportions is used in the multi-sector input–output models and determines fixed sets of productive technologies in specific industries.

2.1.4 Two-Factor Production Functions

Production functions with two inputs, called *two-factor production functions*, are the most common in economics and are usually written as

$$Q = F(K,L), \tag{2.13}$$

where

Q is the output,

K is the amount of capital used,

L is the amount of labor used.

The capital K reflects the total cost of the equipment, machines, buildings, etc., used in production process. Such production functions are characterized by single values of the marginal rate of substitution h and the elasticity of substitution σ between capital and labor.

A two-factor production function is called the *neoclassical production function*, if it satisfies the following properties:

1. *Essentiality of inputs*:

$$F(K,0) = F(0,L) = 0. \tag{2.14}$$

2. *Positive and diminishing returns*:

$$\partial F/\partial K > 0, \quad \partial F/\partial L > 0, \quad \partial^2 F/\partial K^2 < 0, \quad \partial^2 F/\partial L^2 < 0. \tag{2.15}$$

3. *Constant returns to scale*: $F(K, L)$ is a linearly homogeneous function,

$$F(lK, lL) = lF(K,L) \quad \text{for } l > 0. \tag{2.16}$$

4. *The Inada conditions*: the marginal products of capital and labor satisfy

$$\lim_{K\to 0}\frac{\partial F}{\partial K} = \infty, \quad \lim_{L\to 0}\frac{\partial F}{\partial L} = \infty, \quad \lim_{K\to\infty}\frac{\partial F}{\partial K} = 0, \quad \lim_{L\to\infty}\frac{\partial F}{\partial L} = 0. \tag{2.17}$$

The Inada conditions mean that the production increases very fast if the production input (capital or labor) is low and increases slowly, whereas the production increase is very slow if the production input has been already abundant and more is added. Property 1 holds if the other three properties hold.

Per capita variables. At condition (2.16), the production function (2.13) can be rewritten as $Q = LF(K/L, 1)$ or in the so-called *intensive form* (or *per capita* form) as

$$q = f(k), \quad f(k) = F(k, 1) \tag{2.18}$$

where

$q = Q/L$ is the *output per worker* or the *productivity*,

$k = K/L$ is the *capital per worker* or the *capital–labor ratio*.

The intensive form (2.18) of production functions is more convenient for analysis and illustration because it reduces the number of variables. Then, the marginal products of capital and labor are

$$\partial F/\partial K = f'(k), \quad \partial F/\partial L = f(k) - kf'(k), \tag{2.19}$$

the marginal rate of substitution between labor and capital is

$$h = \frac{\partial F/\partial L}{\partial F/\partial K} = \frac{f(k) - kf'(k)}{f'(k)}, \tag{2.20}$$

and *Properties 1–4 of the neoclassical production function* become

$$f(0) = 0, \quad f'(k) > 0, \quad f''(k) < 0, \quad \lim_{k \to 0} f'(k) = \infty, \quad \lim_{k \to \infty} f'(k) = 0. \tag{2.21}$$

The two-factor versions of the major production functions are provided below.

2.1.4.1 Two-Factor Cobb–Douglas Production Function

$$Q = AK^{\alpha}L^{\beta}, \quad \alpha > 0, \quad \beta > 0, \tag{2.22}$$

where the *total factor productivity* A reflects the level of technology. In the general case when $\alpha + \beta \neq 1$, the Cobb–Douglas production is not neoclassical because it does not satisfy Property 3 of constant returns. The Cobb–Douglas production at $\alpha + \beta = 1$ is neoclassical and can be presented in the standard and intensive forms as

$$Q = AK^{\alpha}L^{1-\alpha} \text{ or } q = Ak^{\alpha}, \quad 0 < \alpha < 1. \tag{2.23}$$

Then, the marginal products of capital and labor of (2.22) are

$$\partial Q/\partial K = \alpha A k^{\alpha-1}, \quad \partial Q/\partial L = (1 - \alpha)Ak^{\alpha}, \tag{2.24}$$

the marginal rate of substitution is $h = k\,(1-\alpha)/\alpha$, the output elasticity of capital is $\varepsilon_K = \alpha$, the total output elasticity is $\varepsilon = 1$, and the elasticity of substitution is $\sigma = 1$. The graph of the Cobb–Douglas production function (2.23) in the intensive form is shown Fig. 2.1 with a black curve and is typical for the neoclassical production functions. The output $f(k)$ increases indefinitely when the capital per capita $k \to \infty$, which reflects the Inada condition (2.17). Some economists consider such an increase to be unrealistic.

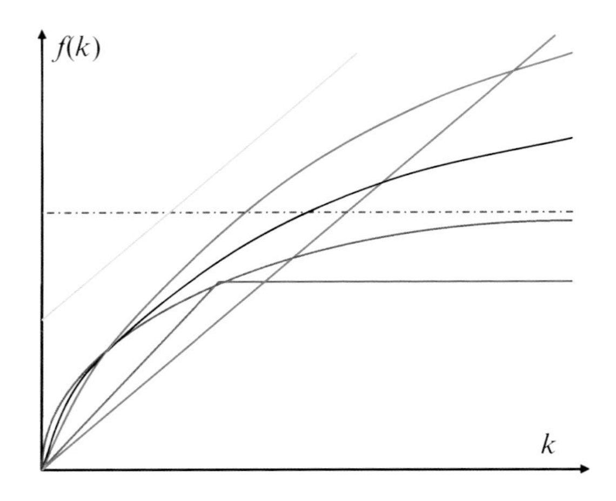

Fig. 2.1 The major types of production functions in per capita form: two-factor Cobb–Douglas (*black curve*), two-factor CES with $\sigma < 1$ (*brown curve*), two-factor CES with $\sigma > 1$ (*red curve*), two-factor Leontief (*blue curve*), two-factor linear (*yellow curve*), and one-factor linear or AK production function (*green curve*)

2.1.4.2 Two-Factor CES Production Function

$$Q = A[\alpha(bK)^{\rho} + (1-\alpha)((1-b)L)^{\rho}]^{1/\rho}, \quad \rho < 1 \tag{2.25}$$

$$\text{or} \quad q = A[\alpha(bk)^{\rho} + (1-\alpha)(1-b)^{\rho}]^{1/\rho}. \tag{2.26}$$

Here, the marginal product of capital (2.19) is

$$\partial Q/\partial K = A\alpha b^{\rho}[\alpha b^{\rho} + (1-\alpha)(1-b)^{\rho}k^{-\rho}]^{(1-\rho)/\rho}, \tag{2.27}$$

$$h = (1-\alpha)(1-b)^{\rho}k^{1-\rho}/(\alpha b^{\rho}), \quad \sigma = 1/(1-\rho)$$

The CES production function is not neoclassical because the Inada conditions are violated. It is visible in Fig. 2.1. At a low degree of substitution $\sigma < 1$ ($\rho < 0$), its graph has a horizontal asymptote (see the brown curve in Fig. 2.1). When $\rho \to 0$, the CES production function approaches the Cobb–Douglas production function. At a high degree of substitution $\sigma > 1$ ($0 < \rho < 1$), this function increases faster than the Cobb–Douglas one (see the red curve in Fig. 2.1). At $\sigma = \infty$ ($\rho = 1$), the CES function becomes linear: $Q = A\alpha bK + A(1-\alpha)(1-b)L$. When $\rho \to -\infty$ ($\sigma \to 0$), this production function approaches the Leontief production function $Q = \min[bK, (1-b)L]$ discussed next. There is essential economic evidence that the CES production function better fits many economic processes than the Cobb–Douglas production function. For this reason, the CES production function currently dominates in applied economic research.

We shall notice that some textbooks introduce the CES production function in a slightly different form as $Q = A[\alpha K^{\rho} + (1-\alpha)L^{\rho}]^{1/\rho}$ and/or with the parameter ρ replaced by $-\rho$ (then the new $\rho > -1$).

2.1.4.3 Two-Factor Leontief Production Function (with Fixed Proportions)

$$Q = \min\{aK, bL\} \quad \text{or} \quad q = \min\{ak, b\}. \tag{2.28}$$

Here, the marginal rate of substitution is $h = \infty$ at $k > b/a$ and $h = 0$ at $k < b/a$, and there is no substitution between capital and labor: $\sigma = 0$. This function was introduced by the famous American economist W. Leontief in 1940, well before other types of production functions. It can be obtained from the two-factor function CES (2.25) at $\rho \to -\infty$. This function is built on the suggestion that there is a unique reasonable value $k_0 = b/a$ of the capital labor ratio $k = K/L$ such that all workers and machines are fully employed. An additional capital is useless at $k > b/a$, whereas some part of labor is not used at $k < b/a$. The Leontief function is shown Fig. 2.1 with a blue curve.

2.1.4.4 Two-Factor Linear Production Function

$$Q = AK + BL \quad \text{or} \quad q = Ak + B. \tag{2.29}$$

The marginal and average products of capital and labor of this production function are constant and equal to A, $h = B/A = \text{const}$, and capital and labor are perfect substitutes: $\sigma = \infty$. A primary weakness of the two-factor linear production function (2.29) is that the input essentiality property (2.14) fails, i.e., the production is possible without capital or labor (see the yellow line in Fig. 2.1). This shortcoming disappears in the *one-factor linear production function*

$$Q = AK \quad \text{or} \quad q = Ak, \tag{2.30}$$

also known as the *AK production function* and commonly used in mathematical economics. It is shown Fig. 2.1 with a green line. Both linear production functions (2.29) and (2.30) do not satisfy the property (2.15) of diminishing returns and, thus, do not belong to neoclassical functions. Some modern economists have considered the property (2.15) as obsolete and not applicable to the capital in a broad sense that includes the *human capital* (see Sect. 3.4).

2.2 Solow–Swan Model of Economic Dynamics

The one-sector model explored below is one of the most celebrated models in the economic growth theory [9]. It has become a foundation for further successful studies.

2.2.1 Model Description

Let us consider an economy described by the following dynamic characteristics in the continuous time t:

$Q(t)$—the total *output* produced at time t,
$C(t)$—the amount of *consumption*,
$I(t)$—the amount of gross *investment*,
$L(t)$—the amount of *labor*,
$K(t)$—the amount of *capital*.

The Solow–Swan model is described by the following equations:

$$Q(t) = F(K(t), L(t)), \tag{2.31}$$

i.e., the output Q is determined by a neoclassical production function $F(K, L)$,

$$Q(t) = C(t) + I(t), \tag{2.32}$$

i.e., the output Q is distributed between the consumption C and the investment I,

$$K'(t) = I(t) - \mu K(t), \quad \mu = \text{const} > 0, \tag{2.33}$$

i.e., the capital K depreciates at a constant rate $\mu > 0$ (a constant fraction of the capital leaves a production process at each point of time),

$$L'(t) = \eta L(t), \quad \eta = \text{const} \geq 0 \tag{2.34}$$

i.e., the labor $L(t) = L_0\exp(\eta t)$ grows at a constant exogenous rate η.

The structure of the Solow–Swan model is shown in Fig. 2.2. The part of the investment in the total product is known as the *saving rate*:

$$s(t) = I(t)/Q(t)$$

The saving rate is assumed to be constant in the classic Solow–Swan model:

$$I(t) = sQ(t), \quad 0 < s < 1, \quad s = \text{const}. \tag{2.35}$$

This assumption simplifies the investigation of the model and leads to a number of essential economic results. More advanced economic models (see next sections) consider the saving rate $s(t)$ as an endogenous control function.

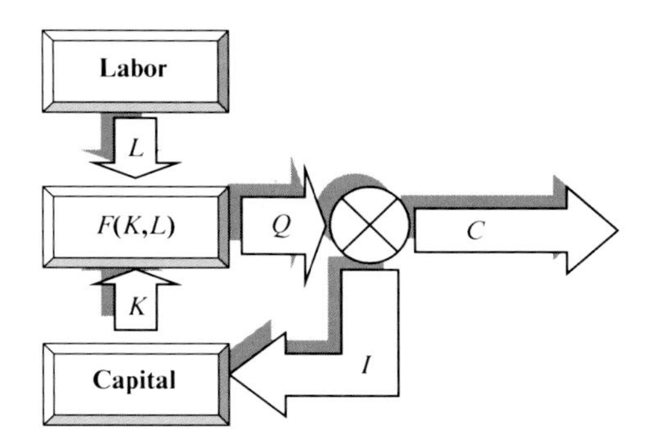

Fig. 2.2 The flow diagram of the Solow–Swan model

2.2.2 Analysis of Model

2.2.2.1 Fundamental Equation of Model (2.31)–(2.35)

Because the production function $F(K, L)$ is neoclassical and, therefore, linearly homogeneous, then $F(K, L) = Lf(k)$ and the equation (2.33) leads to

$$K'(t)/L(t) = f(k(t)) - \mu k(t),$$

where the capital–labor ratio $k = K/L$ is defined as in (2.18). On the other side,

$$k'(t) = K'(t)/L(t) - \eta k(t)$$

by (2.34). Combining the last two equalities, we obtain the *fundamental equation of the Solow–Swan model*

$$k'(t) = sf(k) - (\mu + \eta)k(t). \tag{2.36}$$

Thus, the dynamics of the model (2.31)–(2.35) is reduced to one *autonomous* (not dependent on t explicitly) differential equation (2.36) with respect to k.

2.2.2.2 Steady-State Analysis

The goal of a steady-state analysis is to find possible *steady states*, which can be

- A *stationary trajectory* (unknown variables are constant in time) or
- A *balanced growth path* (all variables grow at the same constant rate).

The steady-state analysis plays an important role in economics and is mathematically simpler than a complete dynamic analysis.

Let us find and analyze possible balanced growth paths in the model (2.31)–(2.35). It is easy to see that the original variables $Q(t)$, $C(t)$, $I(t)$, and $K(t)$

of the model grow with the same rate only if the capital–labor ratio $k(t)$ is constant. Indeed, substituting k = const into (2.33), (2.32), and (2.35), we obtain

$$\begin{aligned} &K(t) = kL(t), \quad I(t) = (\mu + \eta)K(t), \quad Q(t) = (\mu + \eta)K(t)/s, \\ &C(t) = Q(t) - I(t), \end{aligned} \tag{2.37}$$

i.e., all these functions increase with the same rate η as the labor $L(t) = L_0\exp(\eta t)$. Therefore, to find steady states, we should assume $k(t)$ = const. Then $k'(t) = 0$ and the equation (2.36) produces the equation

$$sf(k) = (\mu + \eta)k \tag{2.38}$$

for possible steady states $k \equiv$ const. Because $f(0) = 0, f'(k) > 0, \lim_{k\to 0} f'(k) = \infty$, and $\lim_{k\to\infty} f'(k) = 0$, the equation (2.38) has a unique solution $\hat{k} = \hat{k}(s)$ = const > 0 for any given value $s > 0$. The steady-state capital–labor ratio $\hat{k}(s)$ increases when the saving rate s increases.

2.2.2.3 Static Optimization

For a given saving rate s, the steady-state consumption per capita $c = C/L$ is determined by the formula

$$c(s) = f(\hat{k}(s)) - (\mu + h)\hat{k}(s), \tag{2.39}$$

where the corresponding steady-state capital–labor ratio $\hat{k}(s)$ is determined by (2.38). Because $f(0) = 0$, $f'(k) > 0$ and $f''(k) < 0$, the composite function (2.39) increases for smaller values of s and decreases for larger s.

Then, we can determine the saving rate s^* = const and the corresponding steady-state $k^* = \hat{k}(s^*)$ that maximizes the consumption per capita (2.39):

$$\max_{0<s\le 1} c(s) = f(\hat{k}(s)) - (\mu + \eta)\hat{k}(s).$$

The maximization of (2.39) is an optimization problem with one scalar variable s. The necessary extremum condition for (2.39) at an interior $0 < s < 1$ is $c'(s) = 0$ or

$$\mathrm{d}\left[f(\hat{k}(s)) - (\mu + \eta)(\hat{k}(s))/\mathrm{d}s = f'(\hat{k})\right) - (\mu + \eta)\right]\mathrm{d}\hat{k}/\mathrm{d}s = 0.$$

Hence, the optimal capital–labor ratio k^* should satisfy

$$f'(k^*) = \mu + \eta. \tag{2.40}$$

The relation (2.40) is known as the *golden rule of capital accumulation*. It implies that the marginal product of capital should be equal to the sum of the depreciation and labor growth rates. After determining the optimal k^* from (2.40), the corresponding *golden-rule saving rate* is found from (2.38) as

$$s^* = (\mu + \eta)k^*/f(k^*) = k^* f'(k^*)/f(k^*), \tag{2.41}$$

i.e., the optimal saving rate s^* is equal to the output elasticity of the capital ε_K (2.6) for the corresponding k^*. The formulas (2.40) and (2.41) for the optimal s^* and k^* are known as the *golden rule of economic growth*.

In the case of the Cobb–Douglas production function (2.22) $F(K, L) = AK^{\alpha}L^{1-\alpha}$, $0 < \alpha < 1$, the function $f(k) = Ak^{\alpha}$ and the golden rule is

$$s^* = \alpha, \quad k^* = [As^*/(\mu + \eta)]^{1/(1-\alpha)}. \tag{2.42}$$

At the optimal steady state (s^*, k^*) and the given labor $L(t) = \overline{L}\mathrm{e}^{\eta t}$, the original variables $Q(t)$, $C(t)$, $I(t)$, and $K(t)$ of the model (2.31)–(2.35) grow with the given rate η as

$$K(t) = \overline{K}\mathrm{e}^{\eta t}, \quad I(t) = \overline{I}\mathrm{e}^{\eta t}, \quad Q(t) = \overline{Q}\mathrm{e}^{\eta t}, \quad C(t) = \overline{C}\mathrm{e}^{\eta t}, \tag{2.43}$$

$$\begin{aligned} &\overline{K} = \overline{L}k^*, \overline{I} = (\mu + \eta)\overline{L}k^*, \quad \overline{Q} = \frac{(\mu + \eta)\overline{L}k^*}{s}, \\ &\overline{C} = \frac{(1 - s)(\mu + \eta)\overline{L}k^*}{s}. \end{aligned} \tag{2.44}$$

The constants $\overline{K}, \overline{I}, \overline{Q}, \overline{C}$ in exponential functions of the form (2.43) are often called in economics the *level variables*. In the case of constant labor $L(t) = \overline{L}$ (i.e., $\eta = 0$), the steady state is given by (2.44) and known as a *stationary point*.

Because the aggregate output Q, consumption C, investment I and capital K increase with the same rate η as the exogenous labor L, the Solow–Swan model is classified in the economic theory as the *exogenous growth model*.

2.3 Optimization Versions of Solow–Swan Model

Modern models of economic dynamics often include the dynamic optimization. Mathematically, such problems belong to the optimal control area. This section introduces *the dynamic optimization* into the Solow–Swan model and considers its two optimization versions on finite and infinite planning periods [3].

2.3.1 Optimization over Finite Horizon (Solow–Shell Model)

The Solow–Shell model is the Solow–Swan model (2.31)–(2.34) considered on a finite planning horizon $[0, T]$ in the case when the saving rate $s = I/Q$ depends on the time t and is endogenous [7]. To determine this rate, we consider the following *one-sector optimization* problem:

- Maximize the present value

$$\int_0^T e^{-rt} c(t) dt$$

of the *consumption per capita* $c = C/L$ over a given finite *horizon* $[0, T]$, subject to (2.31)–(2.35) and certain initial and terminal conditions.

In this problem, the given *discount rate* $r > 0$ reflects the planner's subjective rate of the decreasing utility of the output produced in more distant future. We still use the same aggregate variables of the Solow–Swan model: the output Q, consumption C, capital K, labor L, and investment I. For simplicity, let the labor $L(t)$ be constant, that is, $\eta = 0$ in (2.34). Switching the model (2.31)–(2.34) to the per capita variables $k = K/L$, $q = Q/L$, $c = C/L$, $i = I/L$, and excluding q, c, and i, the optimization problem under study becomes:

- Find the function $s(t)$, $0 \leq s(t) \leq 1$, and the corresponding $k(t)$, $k(t) \geq 0$, $t \in [0, T]$, which maximize

$$\max_{s,k} \int_0^T e^{-rt} (1 - s(t)) f(k(t)) dt \tag{2.45}$$

under the equality-constraint:

$$k'(t) = s(t) f(k(t)) - \mu k(t), \tag{2.46}$$

and the initial and terminal conditions:

$$k(0) = k_0, \quad k(T) \geq k_T. \tag{2.47}$$

The value of $k(T)$ cannot be arbitrary because the economy will continue after the end of the planning period. The terminal condition $k(T) \geq k_{\mathrm{T}}$ keeps a minimal acceptable level of capital at the end of the finite horizon.

The problem (2.45)–(2.47) is an optimal control problem, in which the function $s(t)$, $t \in [0, T]$, is unknown (rather than the scalar s = const as in the static optimization of Sect. 2.2). In the optimal control terminology, the *independent* unknown function $s(.)$ is referred to as the *control variable* and the corresponding *dependent* unknown $k(.)$ is the *state variable*.

2.3.1.1 Steady-State Analysis

The results of the steady-state analysis of the Solow–Swan model (2.31)–(2.35) remain true for the Solow–Shell model because the fundamental equation (2.36) is the same. In particular, it means that the dynamic optimization problem (2.45)–(2.47) possesses the constant solution $s(t) = s^*$ and $k(t) = k^*$, if the initial condition is $k_0 = k^*$ and the terminal value is $k_T = k^*$ in (2.47). In the general case when $k_0 \neq k^*$ and/ or $k_T \neq k^*$, the solution of the model will have more complicated structure, which is the subject of the dynamic analysis that follows.

2.3.1.2 Dynamic Analysis

A dynamic analysis of such problems is more complex and requires sophisticated mathematical tools. The results provided below are obtained employing the *maximum principle* from Sect. 2.4.

Necessary Condition for an Extremum: If the function $s(t)$, $t\in[0, T]$, is a solution of the optimization problem (2.45)–(2.47), then:

(a) There exists a continuous function $\hat{\lambda}(t)$, $t\in[0, T]$, called the *dual* or *adjoint variable*, that satisfies the *dual equation*

$$\hat{\lambda}'(t) = (\mu + r)\hat{\lambda}(t) - \left[1 - s(t) + \hat{\lambda}(t)s(t)\right]f'(k(t)), \tag{2.48}$$

with the terminal *transversality condition*

$$\left[k(T) - k_T\right]e^{-rT}\hat{\lambda}(T) = 0, \tag{2.49}$$

where the corresponding state variable $k(t)$, $t\in[0, T]$, is found from (2.46).

(b) $s(t)$ maximizes $\left[1 - s(t) + \hat{\lambda}(t)s(t)\right]$ at each point $t\in[0, T]$.

The proof of this result follows from Corollary 2.1 of Sect. 2.4. Namely, the current-value Hamiltonian (2.69) for the optimal control problem (2.45)–(2.47) is constructed as

$$\hat{H}\left(s, k, \hat{\lambda}\right) = f(k)(1 - s) + \hat{\lambda}\left[sf(k) - \mu k\right], \tag{2.50}$$

and, then, the dual equation (2.48) is obtained from (2.70) as $\hat{\lambda}' = r\hat{\lambda} - \partial H/\partial k$, the state equation (2.46) fits $k' = \partial H/\partial\hat{\lambda}$, and $s(t)$ maximizes $H\left(s, k, \hat{\lambda}\right)$.

Extremum Condition for an Interior Solution. The maximum principle is constructed specifically to handle the case of boundary solutions: $s(t) = 0$ or $s(t) = 1$ in the domain $0 \leq s(t) \leq 1$ at some instants t. The possibility of boundary (or corner) solutions essentially complicates the optimal control dynamics. If a solution is known to be interior in the domain, then the optimality conditions become simpler. Namely, by Corollary 2.2 from Sect. 2.4, if $0 < s(t) < 1$, then the optimal $s(t)$ satisfies $\partial H/\partial s = 0$.

Let us utilize this optimality condition for the optimization problem (2.45)–(2.47). Taking the derivative of (2.50) in s, we obtain $\partial \hat{H} / \partial s = f(k)\left(\hat{\lambda} - 1\right)$. If a priori $0 < s(t) < 1$ for $t \in [0, T]$, then $\partial H/\partial s = 0$ and, therefore, $\hat{\lambda}(t) = 1$. Substituting $\hat{\lambda}$ to (2.48), we obtain

$$0 = \mu + r - f'(k(t)), \tag{2.51}$$

which is the same *golden rule of capital accumulation* (2.40) as obtained during static optimization in the Solow–Swan model of Sect. 2.2.

Structure of Solution: Using the extremum condition (2.48) and (2.49) and rewriting (2.50) as $\hat{H}\left(s, k, \hat{\lambda}\right) = s\left(\hat{\lambda} - 1\right)f(k) - \hat{\lambda}\mu k + f(k)$, we can show that $s(t) = 0$ maximizes $\hat{H}\left(s, k, \hat{\lambda}\right)$ at $\hat{\lambda}(t) < 1$ and $s(t) = 0$ maximizes $\hat{H}\left(s, k, \hat{\lambda}\right)$ at $\hat{\lambda}(t) > 1$. If $\hat{\lambda}(t) = 1$, then $\hat{H}\left(s, k, \hat{\lambda}\right)$ does not depend on s and the optimal k^* is found from (2.48), which is the same as the golden rule of capital accumulation (2.40). Thus, the solution $s(t)$, $t \in [0, T]$, of the optimization problem (2.45)–(2.47) is

$$s(t) = \begin{cases} 0 & \text{when} \quad \hat{\lambda}(t) < 1 \\ s^* & \text{when} \quad \hat{\lambda}(t) = 1\,, \\ 1 & \text{when} \quad \hat{\lambda}(t) > 1 \end{cases} \tag{2.52}$$

where $0 < s^* < 1$ is the optimal (golden-rule) saving rate (2.41) in the Solow–Swan model. When $s(t) = s^*$, the corresponding trajectory is $k(t) = k^*$, where the unique k^* is found from (2.40).

2.3.1.3 Long-Term and Transition Dynamics

Because of the specifics of economic optimization problems, their dynamic analysis is usually split into two steps: the investigation of a *long-term* dynamics and the investigation of the *transition* dynamics. In many problems, the long-term dynamics is independent of initial conditions of the problem and coincides with the steady state solution of the model. Then, the transition dynamics describes how the optimal trajectory approaches the steady state.

The solution $s(t)$, $k(t)$, $t \in [0, T]$, of the optimization problem (2.45)–(2.47) in the case $k_0 < k^* < k_T$ is illustrated in Fig. 2.3.

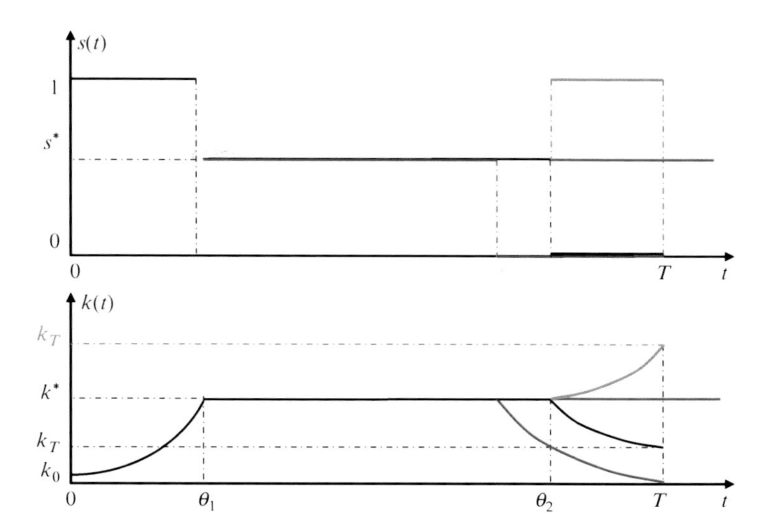

Fig. 2.3 Optimal trajectories in the Solow–Shell model at $k^* > k_0$ in the cases $k^* > k_T$ (*black curves*), $k^* < k_T$ (*black–orange curves*), and $k_T = 0$ (i.e., no terminal condition, *black–red curves*). The lines $s \equiv s^*$ and $k \equiv k^*$ depict the golden rule trajectory. The *blue lines* represent the optimal regime in the infinite-horizon Solow–Ramsey model of Sect. 2.3.2

The transition (short-term) dynamics of the problem (2.45)–(2.47) is common for well-formulated economic problems. The optimal trajectory $s(t)$, $k(t)$ approaches the best steady state solution (s^*, k^*) on the initial interval $[0, \theta_1]$ and the transition dynamics ends at the instant θ_1 such that $k(\theta_1) = k^*$.

The optimal trajectory $s(t)$, $k(t)$ leaves the steady state solution (s^*, k^*) at some instant $\theta_2 < T$ near the right end of the planning horizon $[0, T]$. This behavior illustrates the so-called *end-of-horizon effect* and is also common in economic problems. Even if the terminal condition is absent, such effects still take place and even become more substantial. In particular, if $k_T = 0$, then there is no investments at the end $[\theta_2, T]$ of planning horizon.

Mathematically, this end-of-horizon effect appears because the optimal trajectory $k(t)$ must satisfy the transversality condition (2.49). This condition becomes less restrictive at $T = \infty$. Non-importance of the transversality condition for the infinite-horizon problem (2.31)–(2.35) was pointed out by *K*. Shell in [8]. It will be shown in the next section that the end-of-horizon effect is absent in the infinite-horizon problem.

The trajectory $s(t) \equiv s^*$, $k(t) \equiv k^*$ over $[\theta_1, \theta_2]$ represents the *long-term* dynamics of the optimization problem. The optimal saving rate $s(t)$ coincides with the constant golden-rule saving rate s^* in the Solow–Swan model on a certain interior part $[\theta_1, \theta_2]$ of the planning period $[0,T]$. The length of $[\theta_1, \theta_2]$ becomes larger when T increases. It means that *a turnpike property* holds for the optimization problem (2.45)–(2.47), where the *turnpike* trajectory is $s_T \equiv s^*$.

2.3.1.4 Turnpike Properties

The described structure of solutions to the Solow and Solow–Shell models provides a typical example of *turnpike properties*. A turnpike property states that the optimal trajectory on long planning horizons approaches in certain sense a *turnpike trajectory*, which is independent of the length of planning horizon and the initial state of an economic model. The turnpike trajectory usually has a simpler structure than the optimal trajectory; for example, the turnpike trajectory is simply constant $s_T(t) = s^*$ in the model (2.31)–(2.35). There are several categories of turnpike properties known as turnpike theorems in the *weak, normal, strong, and strongest forms*. A turnpike theorem in the normal form will appear during the optimization analysis of vintage capital models in Chap. 6.

The turnpike analysis is an important tool of the theory of linear multi-sector economic models, such as the Neumann–Gale model. Turnpike properties also appear in the nonlinear economic-mathematical models. The solution structure (2.52) demonstrates *the turnpike theorem in the strongest form*: the optimal trajectory $s(t)$ coincides with a unique turnpike trajectory $s_T(t) = s^*$, except for certain initial and final intervals of the fixed length.

Mathematically speaking, the turnpike properties reflect the stability and robustness of optimal regimes. The turnpike properties are not a universal feature of economic models. They appear only when a certain balance exists among various controls in models. It is often easier to find turnpikes and analyze their properties than to solve an optimization problem directly. In general, the turnpike theorems reflect some fundamental tendencies and laws of economic dynamics.

2.3.2 *Infinite-Horizon Optimization (Solow–Ramsey Model)*

The Solow–Shell model (2.31)–(2.35) with the optimization over the infinite planning horizon $[0,\infty)$ is known as the *Solow–Ramsey model with linear utility*. Namely, we consider the following optimization problem:

- A benevolent central planner determines the optimal saving rate $s(t)$, $t \in [0,\infty)$, to maximize the present value of the consumption per capita over the infinite planning horizon $[0,\infty)$:

$$\max_{s} \int_{0}^{\infty} \mathrm{e}^{-rt} c(t) \mathrm{d}t \tag{2.53}$$

subject to the state equation (2.46) with the initial condition $k(0) = k_0$. No terminal conditions are imposed at ∞.

2.3.2.1 Steady-State Analysis

The results of the steady-state analysis of the Solow–Ramsey model (2.53) remain the same as those in the Solow–Swan model (2.45)–(2.47) with the exogenous constant saving rate.

2.3.2.2 Dynamic Analysis

The dynamic analysis of the Solow–Ramsey model includes new mathematical challenges such as the convergence of the improper integral (2.53) along the optimal trajectory c. The condition for this convergence in our model (2.31)–(2.34), (2.53) is simply

$$r > \eta, \tag{2.54}$$

where η is the given growth rate of labor in (2.35). However, finding such conditions becomes more complicated in more advanced models (see for example Chap. 3).

Under (2.54), the extremum conditions remain the same, (2.48)–(2.52), as in the Solow–Shell model. Using (2.48) and (2.52), we can show that the solution $s(t)$, $t\in[0,\infty)$, of the problem (2.53) in the model (2.31)–(2.34) is

$$s(t) = \begin{cases} 1 & \text{at} \quad 0 \le t < \theta_1 \\ s^* & \text{at} \quad \theta_1 \le t < \infty \end{cases}, \tag{2.55}$$

$$k(t) = \begin{cases} k_{tr}(t) & \text{at} \quad 0 \le t < \theta_1 \\ k^* & \text{at} \quad \theta_1 \le t < \infty \end{cases}, \tag{2.56}$$

with the golden-rule saving rate s^* and capital per capita k^* in the Solow–Swan model. Also, it can be shown that the transversality condition (2.49) is reduced to the inequality (2.54). So the transversality condition is less important in the infinite-horizon problem in the sense that it does not directly affect the solution dynamics.

On the qualitative side, the behavior of the optimal trajectories appears to be simpler than in the finite-horizon Solow–Shell model (2.45)–(2.47). The solution $(s(t), k(t))$, $t\in[0, \infty)$, of the optimization problem (2.53) in the case $k_0 < k^*$ is illustrated in Fig. 2.3 by blue curves.

The transition dynamics of the problem (2.53) over the interval $[0, \theta_1]$ is the same as for the Solow–Shell model. The optimal trajectory $(s(t), k(t))$ approaches the steady state (s^*, k^*) and the transition dynamics ends at the instant θ_1 such that $k(\theta_1) = k^*$.

The long-term dynamics is $s(t) \equiv s^*$, $k(t) \equiv k^*$ over $[\theta_1, \infty)$, i.e., the optimal saving rate $s(t)$ coincides with the constant golden-rule saving rate s^* in the Solow–Swan model starting with the time θ_1. As shown in Fig. 2.3, the optimal trajectory $s(t)$, $k(t)$ does not leave the steady state (s^*, k^*) because the *end-of-horizon effects* are absent in infinite-horizon problems.

The considered optimization versions of the Solow–Swan, Solow–Shell, and Solow–Ramsey models, are classified by the economic theory as the *models of exogenous growth* because they cannot generate an endogenous growth when the labor $L(t)$ is constant. However, even small modifications of these models can lead to the *endogenous growth*. For instance, if we replace the neoclassic production function in the model equation (2.31) with the CES or *AK* production function, then the corresponding models are able to generate an endogenous growth. Models with endogenous growth are discussed in Sect. 3.4.

2.3.3 Central Planner, General Equilibrium, and Nonlinear Utility

The optimization problem (2.53), (2.31)–(2.34) describes the Solow–Ramsey model with linear utility in the *central planner setup*. The alternative economic environment is the *general equilibrium setup*. It assumes a decentralized economy with competitive firms and households with optimizing behavior, which interact on competitive markets. The households own capital assets, provide labor, receive wages, and choose their consumption over saving ratio to maximize their overall utility. The firms hire capital and labor and use them to produce goods to sell in order to maximize their profit. The perfect market *equilibrium* equalizes the supply and demand and determines the relative wages and prices of the capital and produced goods.

This *general equilibrium setup* has many modifications and simplifications. In particular, more players can be added, such as government, resource extraction firms, R&D firms, and others. On the other side, the separation of firms and households is not mandatory because households can perform the functions of firms. In the central planner setup, an economy is managed by a benevolent central planner that maximizes the utility of households. In many economic models, fundamental equations obtained in the central planner problem will be the same as in the general equilibrium framework. This textbook focuses on the productive side of the economy, so general equilibrium models are omitted. We refer the interested reader to Chaps. 1 and 2 of [1].

2.3.3.1 Utility Functions

Optimization problems in the central planner and general equilibrium frameworks often maximize a so-called *individual or social utility* that nonlinearly depends on consumption, rather than the direct amount of consumption. For example, the

Solow–Ramsey model with nonlinear utility maximizes the present value of the consumer's utility over $[0,\infty)$:

$$\max_{s} \int_{0}^{\infty} \mathrm{e}^{-rt} u(c(t))\mathrm{d}t \tag{2.57}$$

instead of (2.53). The nonlinear function $u(c)$ in (2.57) is called the *utility function* and describes the value of the future consumption product c for consumers. The function $u(c)$ is smooth, increasing and concave downward: $u'(c) > 0$, $u''(c) < 0$. Its concavity reflects the property of *diminishing marginal utility*: the product is more valuable for households when its amount is small. The utility function is said to satisfy the *Inada conditions* if $\lim_{c\to 0} u'(c) = \infty$ and $\lim_{c\to\infty} u'(c) = 0$.

Two most common utility functions are:

- The *isoelastic* (or *power*) utility function $u(c) = c^{1-\gamma}/(1-\gamma)$, where $0 < \gamma < 1$.
- The *logarithmic* utility function $u(c) = \ln c$.

2.4 Appendix: Maximum Principle

A maximum principle is the most popular type of extremum conditions for the optimal control of differential and integral equations. Below we provide the standard maximum principle for the optimal control of a scalar nonlinear ordinary differential equation, which is used for analyzing dynamic optimization models in Sect. 2.3 and other chapters.

Let us consider the following *optimal control* problem:

- Find the control function $u(t)\in\mathbf{R}^1$ and the corresponding $x(t)\in\mathbf{R}^1$, $t\in[0, T)$, which maximize

$$\max_{u,x} \int_{0}^{T} f(x(t), u(t), t)\mathrm{d}t, \tag{2.58}$$

subjected to the *state equation*

$$x'(t) = g(x(t), u(t), t), \tag{2.59}$$

the inequality-constraint

$$u_{\min}(t) \le u(t) \le u_{\max}(t), \tag{2.60}$$

and the initial and terminal conditions

$$x(0) = x_0, \quad x(T) \geq x_T. \tag{2.61}$$

The functions $f(x, u, t)$ and $g(x, u, t)$ are differentiable in x and u and continuous in t. The presence of the inequality-constraint (2.60) is a distinguished feature of the optimal control problems as opposed to the calculus of variations. The closed interval $U(t) = [u_{\min}(t), u_{\max}(t)] \subset \mathbf{R}^1$ is called the *control region* of the problem (2.58)–(2.61).

Definition. The *Hamiltonian* of the optimal control problem (2.58)–(2.61) is the function

$$H(x, u, \lambda, t) = f(x, u, t) + \lambda g(x, u, t), \tag{2.62}$$

where

λ is the *dual* (*costate, adjoint*) *variable*.

The dual variable reflects the change in the objective function due to changes in the constraints.

Then, the state equation (2.59) can be rewritten as

$$x'(t) = \frac{\partial H(x(t), u(t), \lambda, t)}{\partial \lambda}. \tag{2.63}$$

Statement 2.1 (***Pontryagin maximum principle***)**:** If the function $u^*(t)$, $t \in [0, T]$, is a solution of the optimal control problem (2.58)–(2.61), then:

(a) The dual variable $\lambda(t)$, $t \in [0, T]$, exists and satisfies the *dual equation*

$$\lambda'(t) = -\frac{\partial H(x(t), u(t), \lambda(t), t)}{\partial x} \tag{2.64}$$

with the *transversality conditions* at the right end $t = T$:

$$\lambda(T) \geq 0, [x(T) - x_T]\lambda(T) = 0; \tag{2.65}$$

(b) The corresponding state variable $x(t)$, $t \in [0, T]$, is found from (2.59) for the given $u^*(t)$.

(c) For each $t \in [0, T]$, $u^*(t)$ maximizes $H(x, u, \lambda, t)$:

$$H(x(t), u(t), \lambda(t), t) = \max_{v \in U(t)} H(x(t), v(t), \lambda(t), t). \tag{2.66}$$

The proof of this statement is out of the scope of this textbook. It is available in textbooks on the optimal control and mathematical economics, e.g., [2, 4, 5, 6]. The maximum principle delivers only necessary condition for an extremum.

Necessary and sufficient condition for an extremum: If the $H(x,u,\lambda,t)$ is concave in u and x for each $t\in[0, T]$, then the conditions (2.64)–(2.66) are also sufficient for the function u to be a solution of the optimal control problem (2.58)–(2.61).

Extremum conditions of the form (2.64)–(2.66) are known as the *maximum principle* because they reduce an optimal control problem to maximization of the function H of one or several variables. There are numerous modifications of the Pontryagin maximum principle for various extensions of the problem (2.58)–(2.61). Below we discuss some of them that are applicable to specific models in the textbook.

2.4.1 *Scalar Controls*

The maximum principle is powerful when the control variable is a vector function. In our problem (2.58)–(2.61) with one scalar control u, the maximum condition (2.66) can be easily resolved and leads to the following structure of the optimal control:

$$u(t) = \begin{cases} u_{\min}(t) & \text{if} \quad \dfrac{\partial H(x(t), u(t), \lambda(t), t)}{\partial u} < 0 \\ u_{\min}(t) < \tilde{u}(t) < u_{\max}(t) & \text{if} \quad \dfrac{\partial H(x(t), u(t), \lambda(t), t)}{\partial u} = 0 \\ u_{\max}(t) & \text{if} \quad \dfrac{\partial H(x(t), u(t), \lambda(t), t)}{\partial u} > 0 \end{cases}. \tag{2.67}$$

The formula (2.67) demonstrates that the optimal control $u(t)$ can be piecewise continuous at natural conditions. In more complicated problems, optimal controls possess many (even, indefinitely many) jumps and, therefore, are supposed to be measurable functions.

2.4.2 *Discounted Optimization*

In many economic and environmental models, optimization problems appear in a special form (2.58)–(2.61), where the state equation (2.59) is autonomous, the control region $[u_{\min}, u_{\max}]$ does not depend on t, and the time t explicitly appears only in the function f in (2.58) as the multiplier e^{-rt}:

$$g(x, u, t) = \hat{g}(x, u), \quad f(x, u, t) = e^{-rt}\hat{f}(x, u). \tag{2.68}$$

Then, the maximum principle can be simplified by introducing the so-called *current-value dual variable* $\hat{\lambda}(t) = \lambda(t)e^{rt}$ and the *current-value Hamiltonian* of the problem (2.58)–(2.61), (2.68)

$$\hat{H}\left(x, u, \hat{\lambda}\right) = \hat{f}(x, u, t) + \hat{\lambda}\, g(x, u). \tag{2.69}$$

Then, the dual variable λ in (2.62) is called the *present-value* dual variable.

Corollary 2.1 (*Current-value Maximum Principle*): If the optimal control problem (2.58)–(2.61) is of the form (2.68) and u is its solution, then:

(a) The dual variable $\hat{\lambda}(t)$, $t \in [0, T]$, exists and satisfies the *dual equation*

$$\hat{\lambda}' - r\hat{\lambda} = -\frac{\partial \hat{H}\left(x, u, \hat{\lambda}\right)}{\partial x}, \tag{2.70}$$

with the *transversality conditions* at the right end $t = T$

$$\hat{\lambda}(T) \geq 0, \quad [x(T) - x_T]e^{-rT}\hat{\lambda}(T) = 0; \tag{2.71}$$

(b) The corresponding $x(t)$, $t \in [0, T]$, is found from (2.59).

(c) For each $t \in [0, T]$, $u(t)$ maximizes $\hat{H}\left(x, u, \hat{\lambda}\right)$:

$$\hat{H}\left(x(t), u(t), \hat{\lambda}(t)\right) = \max_{v \in U} \hat{H}\left(x(t), v(t), \hat{\lambda}(t)\right). \tag{2.72}$$

In contrast to (2.64), the dual differential equation (2.70) is autonomous and much easier to solve. The maximization problem (2.72) is the same for all t.

2.4.3 *Interior Controls*

In the case of *interior controls*, the maximum principle leads to the following simpler optimality condition for *interior controls*.

Corollary 2.2. Let Statement 1.1 hold. If a priori $u_{\min}(t) < u(t) < u_{\max}(t)$ for $t \in [0, T]$, then the optimal $u(t)$ satisfies

$$\partial H(x(t), u(t), \lambda(t), t)/\partial u = 0, \quad t \in [0, T]. \tag{2.73}$$

In the case of the discounted problem (2.58)–(2.61), (2.68), the condition (2.73) is

$$\partial \hat{H}\left(x(t), u(t), \lambda(t)\right)/\partial u = 0, \quad t \in [0, T]. \tag{2.74}$$

The condition (2.73) is simpler to deal with than (2.66) or (2.72). Corollary 2.2 is often used in the steady-state analysis of economic optimization problems, where steady-state trajectories are naturally interior in the domain of admissible controls. Economists often call the formulas (2.64), (2.65), (2.73) the *first-order optimality conditions*.

2.4.4 Transversality Conditions

The transversality conditions represent a necessary part of the optimality conditions, which is often overlooked. Their relevance depends on the specifics of the problem under study. We provide several transversality conditions for different versions of the optimal control problem (2.58)–(2.61).

Problems with a fixed right end: If the terminal condition in (2.61) is strengthened to the equality $x(T) = x_{\mathrm{T}}$, then the corresponding maximum principle does not involve the transversality condition (2.65) on $\lambda(T)$.

Problems with free right end: If the optimal control problem (2.58)–(2.61) does not have a terminal condition on $x(T)$ at all ("the right end of x is free"), then the transversality condition (2.65) becomes

$$\lambda(T) = 0. \tag{2.75}$$

Free terminal-time problems: If the right boundary point T in the optimal control problem (2.58)–(2.61) is not specified, then the corresponding transversality condition is

$$H(x(T), u(T), \lambda(T), T) = 0. \tag{2.76}$$

The infinite planning horizon $[0,\infty)$: If $T = \infty$ in the problem (2.58)–(2.61), then the corresponding transversality condition (2.65) becomes

$$\lim_{T\to\infty} \lambda(T) \geq 0, \; \lim_{T\to\infty} [x(T) - x_T]\lambda(T) = 0. \tag{2.77}$$

2.4.5 Maximum Principle and Dynamic Programming

An alternative approach for solving optimization problems is the *dynamic programming* method developed by R. Bellman. Although the maximum principle and dynamic programming are related, they have their essential differences, strengths and shortages. In a certain sense, the maximum principle is more practical for simple deterministic problems explored in this textbook. In particular, the maximum principle does not imply Bellman's equation while the dynamic programming conditions imply the maximum principle. In mathematical economics, the maximum principle is often considered to be a more powerful method for analytic solution [5].

Exercises

1. Fill in the table below:

Two-factor production function (PF)					
	Linear PF	Leontief PF	Cobb–Douglas PF	Cobb–Douglas PF	CES PF
Properties			$\alpha > 0, \beta > 0$	$\alpha + \beta = 1$	
Essentiality of inputs					
Positive returns					
Diminishing returns					
Homogeneity					
Returns to scale					
Marginal rate of substitution					
Total output elasticity					
Elasticity of substitution					
Inada conditions					
Neoclassical PF (yes, no)					

2. Prove that the total output elasticity of a homogeneous production function is equal to the degree of homogeneity: $\varepsilon(\mathbf{x}) = \gamma$.
3. Justify that the CES production function is more general as compared to the linear, Leontief, and Cobb–Douglas production functions.
4. Is the three-factor production function $F(K, L, N) = AK^{\alpha}L^{\beta}N^{\gamma}$, $\alpha > 0, \beta > 0$, $\gamma > 0$, homogeneous? If yes, then what is its degree of homogeneity? If no, then what condition should be implemented for the homogeneity?
5. The two-factor CES production function can be presented as $F(K,L) = A[\alpha K^{\rho} + (1 - \alpha)L^{\rho}]^{1/\rho}$. Is this CES function a neo-classical production function?
6. Prove that the first property *Essentiality of inputs* of the neoclassical production functions holds if the other three properties (*Positive returns*, *Diminishing returns*, and *Proportional returns to scale*) are valid.
7. Derive the formula (2.19): $\partial F/\partial K = f'(k)$, $\partial F/\partial L = f(k) - kf'(k)$, where $f(k) = F(k, 1)$, and $k = K/L$.

8. Find $\partial Q/\partial K$ and $\partial Q/\partial L$ for the CES production function (2.23) and prove that its marginal rate of substitution is $h = (1-\alpha)(1-b)^{\rho}\ k^{1-\rho}/(\alpha\ b^{\rho})$ and the elasticity of substitution is $\sigma = 1/(1-\rho)$.
9. Show that the variables $Q(t)$, $C(t)$, $I(t)$, and $K(t)$ of the Solow–Swan model (2.31)–(2.35) at the optimal steady state (s^*, k^*) are given by formulas (2.43) and (2.44).
10. Provide the steady-state analysis of the Solow–Shell model (2.45)–(2.47) and show that its golden rule of capital accumulations (2.51) and optimal steady state (s^*, k^*) are the same as in the Solow–Swan model (2.31)–(2.35).

References[1]

1. 🕮 Barro, R.J., Sala-i-Martin, X.: Economic Growth. The MIT Press, Cambridge, MA (2003)
2. 🕮 Caputo, M.R.: Foundations of Dynamic Economic Analysis: Optimal Control Theory and Applications. Cambridge University Press, New York (2005)
3. Cass, D.: Optimum growth in an aggregative model of capital accumulation. Rev. Econ. Stud. **32**, 233–240 (1965)
4. 🕮 Chiang, A.C., Wainwright, K.: Fundamental Methods of Mathematical Economics, 4th edn. McGraw-Hill, New York (2005)
5. 🕮 Intriligator, M.D.: Mathematical Optimization and Economic Theory. Society for Industrial and Applied Mathematics, Philadelphia (2002)
6. 🕮 Leonard, D., Van Long, N.: Optimal Control Theory and Static Optimization in Economics. Cambridge University Press, New York (1992)
7. Shell, K.: Optimal programs of capital accumulation for an economy in which there is exogenous technical change. In: Shell, K. (ed.) Essays on the Theory of Optimal Economic Growth, Chapter I, pp. 1–30. MIT Press, Cambridge (1967)
8. Shell, K.: Applications of Pontryagin's maximum principle to economics. In: Kuhn, H.W., Szegö, G.P. (eds.) Mathematical Systems Theory and Economics, pp. 241–292. Springer, Berlin (1969)
9. Solow, R.: A contribution to the theory of economic growth. Quart. J. Econ. **70**, 65–94 (1956)

[1] The book symbol 🕮 means that the reference is a textbook recommended for further student reading.

Chapter 3
Modeling of Technological Change

Economic data for the last two centuries has demonstrated the presence of a self-sustaining mechanism of cumulative productivity growth, known as *technical progress* or *technological change*. In modern times, the technical progress affects not only the efficiency of the economy, but also the natural environment and entire lifestyle of human society. Section 3.1 provides a comprehensive review of major directions and trends in the modeling of technical progress, including: autonomous, induced, exogenous and endogenous, embodied and disembodied technological change, and technological change as a separate sector of the economy. Section 3.2 analyzes the classic Solow–Swan, Shell, and Ramsey models of economic dynamics with exogenous technological change. Section 3.3 explores modern one- and two-sector models with endogenous technological change, physical and human capital, and knowledge accumulation. Substitution, diffusion, and evolution models of technological innovations are briefly discussed in Sect. 3.4.

3.1 Major Concepts of Technological Change

In mathematical economics, *technological change (technical change, technical progress)* refers to a combination of all effects that lead to an increasing production output without increasing the amounts of used productive inputs (capital, labor, resources). Such a concept of technological change includes the acquisition of new superior technologies as well as a progress in production management methods.

Major types of technological change include the following:

- *Exogenous technological change* is introduced into an economic system from outside.
- *Endogenous technological change* is a consequence of focused economic activities, such as research and development (R&D) efforts of profit-maximizing firms and governmental policies.

N. Hritonenko and Y. Yatsenko, *Mathematical Modeling in Economics, Ecology and the Environment*, Springer Optimization and Its Applications 88, DOI 10.1007/978-1-4614-9311-2_3,

- *Embodied (investment-specific) technological change* is introduced into the economic system with more efficient capital or better qualified labor.
- *Autonomous (disembodied) technological change* impacts the entire production process evenly.
- *Output-augmenting technological change* increases the labor productivity.
- *Resource-saving technological change* increases the efficiency of converting resources into useful work.
- *Induced technological change* is a result of previous economic development and is caused by other economic processes or regulations.
- *Technological change as a separate sector of* economy, whose product is the technological change.

Different categories from this classification use various modeling tools and lead to different conclusions because of different understanding of sources, causes, and effects of the technological change [4, 6, 9, 10].

3.1.1 Exogenous Autonomous Technological Change

Neoclassic economic growth models of the 1960s and 1970s assumed the autonomous and exogenous nature of technological change. Economists noticed the existence of an unexplained factor (other than the accumulation of capital), which was accounted for almost 90 % of growth in the output per capita (GDP) over the past half-century. They referred to this exogenous driver as "technical progress".

The *exogenous technological change* presumes that continuous technical improvements occur independently of economic activities and are exogenous with respect to an economy under study. This approach provides no economic explanation of why the change occurs. The feedback mechanisms between technological change and economic growth are not revealed, i.e., the technological change is considered as a costless and uncontrolled process.

The *autonomous technological change* assumes that the production inputs are homogeneous and affected by the technological change in the same way. The autonomous technological change does not analyze factors that influence the technological change dynamics and, thus, is exogenous. The efficiency of used inputs is supposed to increase independently of capital investments and dynamics of labor force. The autonomous technological change hypothesis is common in economic theory and practice because of its simplicity. Some well-known models with autonomous technological change are analyzed in Sects. 3.2 and 3.3.

3.1.2 Embodied and Disembodied Technological Change

The *embodied* (also known as *investment-specific*) technological change focuses on relations between the dynamics of technological change and capital investments. It takes into account the heterogeneity of capital assets (vintages) under improving technology and assumes that the technological change is introduced into an economic system with more efficient capital or better qualified labor.

In economic reality, both autonomous and embodied changes are presented simultaneously. The autonomous technological change is also referred to as the *disembodied technological change* to emphasize the fact that it affects all capital vintages and workers in the same way. It describes a progress in management techniques and methods, e.g., installing new enterprise-wide software. More than half (52 %) of the growth of the US economy during the post-war time was due to the embodied technological change, so the rest can be attributed to the disembodied change.

The models of economic growth under *embodied technological change* are known as the *vintage capital models*. Vintage capital models provide a united description of separate processes of investing in new efficient capital and scrapping (disinvestment) of the capital vintages with low efficiency. In many vintage models, the improving efficiency of capital vintages is given as a function of time. So the embodied technological change can be *exogenous*, where the source of technological change is still unclear. The vintage capital models are explored in Chaps. 4 and 5.

3.1.3 Endogenous Technological Change

Models of *endogenous* technological change were introduced to explain the driving forces behind technological change.

The majority of technological improvements results from *research and development* (*R&D*) activities carried out and financed by government and/or private firms. The concept of endogenous technological change attempts to explain economic reasons and sources of technological change. Corresponding economic models describe technological innovations as determined by economic actors and suggest economic reasons for firms to innovate, specific mechanisms and directions of inventive activity, drivers of incremental improvements that occur during technology diffusion, and so on. These mechanisms are endogenous with respect to economic activities and, thus, are determined inside the model. Some classic models of endogenous technological change are explored in Sect. 3.4.

Induced Technological Change

An early concept of the endogenous technological change is known as the *induced technological change* that links technological change to previous economic development. It was the result of incorporating technological change into the neoclassical growth framework. The description of induced technological change was based on various hypotheses about relations between the technological change intensity and other aggregated economic characteristics, see Sect. 3.4.1. However, the early hypotheses of induced technological change could not explain the need of purpose-directed investments into science and technology.

In modern economics, the induced technological change commonly refers to additional technological improvements caused by other economic processes or governmental regulations, for instance by more restrictive environmental policies.

3.1.4 Technological Change as Separate Sector of Economy

Considering the technological change as a separate sector of economy is a prospective approach for the description of economic development that dominates in the modern economic theory. It can be traced back to K. Shell's [11] specification of an inventive sector devoted to producing knowledge. The technological change is essentially formed inside economic systems through R&D investments, experimental design and production, and other focused activities. Due to importance of technological change, a prospective modeling approach is to consider technological change as a separate sector of economy with its own inputs and outputs. The output of this sector can be enhancement of human capital, production of new knowledge, and similar.

The most acclaimed models in this direction have been developed in the endogenous growth framework by P. Romer, R. Lucas, P. Aghion, P. Howitt, G. Grossman, E. Helpman, C. Jones, and their successors. They assume that not only capital and resources but also the third factor, *knowledge*, is a relevant production input. Production functions with the R&D technology differ from the classic production functions with physical capital and resources and require different types of investment.

Modern endogenous growth models include two separate sectors: the final output sector and the R&D sector. In general equilibrium settings, the R&D sector consists of R&D firms that develop ideas and sell them to production firms. More detailed models add sectors of intermediate products and link them to innovation quality and the development of new knowledge. Some of such models are explored in Sect. 3.4.3.

3.2 Models with Autonomous Technological Change

The autonomous technological change reflects an increase of the total efficiency of an economic system brought from outside, but it does not reveal a way it occurs and affects the economy. Following the mainstream of economic growth theory, we consider deterministic models with perfect foresight, which means that the whole evolution of future technology is already known at the present time. This theoretical assumption leads to important conclusions about the rational response of economy to technological advances.

As in Chap. 2, we use the following dynamic economic characteristics:

$Q(t)$—the total output produced at time t, $C(t)$—the amount of *consumption*, $I(t)$—the amount of *investment*, $L(t)$—the amount of *labor*, $K(t)$—the amount of *capital*.

The autonomous (disembodied) technological change is described by the *dynamic production function* (see Sect. 2.2):

$$Q = F(K, L, t), \tag{3.1}$$

that explicitly depends on the time t. The autonomous technological change is equivalent to an increase of (3.1) in t: $\partial F / \partial t \geq 0$. It means that the production output increases in time for the same combination of capital and labor.

Neutrality of Technological Change

There are three special types of the so-called *neutral technological change* that are described by special cases of the formula (3.1):

- *Hicks neutral technological change*:

$$Q = A(t)F(K, L), \quad A'(t) \geq 0, \tag{3.2}$$

when the efficiencies of both capital K and labor L increase proportionally. The function $A(t)$ is commonly interpreted as the *state of technology*.

- *Harrod neutral* or *labor-augmenting technological change*:

$$Q = F(K, A(t)L), \quad A'(t) \geq 0, \tag{3.3}$$

when only the efficiency of labor L increases. Such description is equivalent to an exogenous increase of the labor L.

- *Solow neutral* or *capital-augmenting technological change*:

$$Q = F(A(t)K, L), \quad A'(t) \geq 0, \tag{3.4}$$

when the efficiency of capital K increases. It is equivalent to an exogenous increase of the capital K.

All three types (3.2)–(3.4) of the *neutral technological change* are equivalent in the case of Cobb–Douglas production function with autonomous technological change

$$F(K, L, t) = A(t)K^{\alpha}L^{1-\alpha}, \tag{3.5}$$

that is, the autonomous technological change described by (3.5) is Hicks, Harrod, and Solow neutral at the same time. The Cobb–Douglas production function with an exponential autonomous technological change

$$F(K, L, t) = Ae^{gt}K^{\alpha}L^{1-\alpha} \tag{3.6}$$

is referred to as the *Cobb–Douglas–Tinbergen* production function. Here, $g > 0$ is called the *rate of the exponential technological change*.

The deficiency of the autonomous technological change is the following. In a static production function $Q = F(K, L)$, the variable K describes the amount of homogeneous capital assets (equipment, machines) of the same type and L is the number of identical workers. If the function F varies in time, then the assets created at different time may be not identical anymore. To address this issue, heterogeneous capital and labor need to be considered, which leads to the concept of the embodied technological change (see Sect. 3.1.2). The models with embodied technological change will be explored in Chap. 4.

Models considered below in Sects. 3.2.1–3.2.3 are modifications of the Solow models from Chap. 2. Adding the autonomous technological change to those models brings new essential qualitative properties.

3.2.1 *Solow–Swan Model*

The Solow–Swan model with autonomous technological change [12] is obtained from the Solow–Swan model (2.31)–(2.35) by replacing the production function $F(K, L)$ in the equation (2.31) with the dynamic production function (3.1):

$$Q(t) = F(K(t), L(t), t), \tag{3.7}$$

$$Q(t) = C(t) + I(t), \tag{3.8}$$

$$K'(t) = I(t) - \mu K(t), \quad \mu = \text{const} > 0, \tag{3.9}$$

$$L'(t) = \eta L(t), \quad \eta = \text{const} \geq 0. \tag{3.10}$$

The equations (3.8)–(3.10) remain the same as in the original Solow–Swan model.

Steady-State Analysis

The model (3.7)–(3.10) possesses a steady state and the golden rule of economic growth holds (see Sect. 2.2) *if and only if* the production function (3.7) reflects the labor-augmenting (Harrod neutral) exponential technological change (3.3) with a constant rate $g > 0$, i.e.,

$$Q(t) = F(K(t), e^{gt}L(t)). \tag{3.11}$$

Under the condition (3.11), we can introduce the *effective labor* as:
$\widetilde{L}(t) = e^{gt}L(t)$—the actual labor multiplied by its efficiency,
$\widetilde{k}(t) = K(t)/(e^{gt}L(t)) = e^{-gt}k(t)$—the *capital per effective labor*,
where $k(t) = K(t)/L(t)$ is the *capital–labor ratio* as in (2.18).

Then, the fundamental equation (2.36) of the Solow–Swan model is modified to the following equation with respect to $\widetilde{k}(t)$:

$$\widetilde{k}'(t) = sf\left(\widetilde{k}\right) - (\mu + \eta + g)\widetilde{k}(t), \tag{3.12}$$

where $s = I/Q$ is the *saving rate*. The equation (3.12) differs from (2.36) by the parameter g only. Correspondingly, all transformations of Sect. 2.2 for the Solow–Swan model (2.31)–(2.35) remain valid for the model (3.7)–(3.10) with autonomous technological change. In particular, a unique steady state $\widetilde{k} = \text{const} > 0$ for any given $s = \text{const}$ is determined from

$$sf\left(\widetilde{k}\right) = (\mu + \eta + g)\widetilde{k}. \tag{3.13}$$

As in Chap. 2, the static optimization problem determines the saving rate s and corresponding $\widetilde{k}$ that maximize the consumption per capita (2.39). The optimal $\widetilde{k}$ is determined from the modified *golden rule of capital accumulation*

$$f'\left(\widetilde{k}^*\right) = \mu + \eta + g, \tag{3.14}$$

and the *modified golden-rule saving rate* is

$$s^* = (\mu + \eta + g)\widetilde{k}^*/f\left(\widetilde{k}^*\right). \tag{3.15}$$

The formulas (3.14) and (3.15) coincide with (2.40) and (2.41) at $g = 0$. In the case of the Cobb–Douglas $F(K, L) = AK^{\alpha}L^{1-\alpha}$, $0 < \alpha < 1$, we obtain $f\left(\widetilde{k}\right) = A\widetilde{k}^{\alpha}$ and

$$s^* = \alpha, \quad \widetilde{k}^* = [A\alpha/(\mu + \eta + g)]^{1/(1-\alpha)}. \tag{3.16}$$

The formulas (3.13) and (3.14) mean that the unknown variable $\widetilde{k} = K/(e^{gt}L)$ is constant (is a stationary trajectory) at the steady state of the model (3.7)–(3.10). Correspondingly, the unknown per capita variables $k = K/L$, $q = Q/L$, and $c = C/L$ increase in time t with the rate g of technological change. Respectively, the original variables $Q(t)$, $C(t)$, $I(t)$, and $K(t)$ of the model (3.7)–(3.10) with autonomous technological change increase with the given rate $\eta + g$:

$$K(t) = \overline{K}e^{(\eta+g)t}, \quad I(t) = \overline{I}e^{(\eta+g)t}, \quad Q(t) = \overline{Q}e^{(\eta+g)t}, \quad C(t) = \overline{C}e^{(\eta+g)t}, \tag{3.17}$$

$$\overline{K} = \overline{L}\widetilde{k}, \quad \overline{I} = (\mu + \eta + g)\overline{L}\widetilde{k}, \quad \overline{Q} = \overline{I}/s, \quad \overline{C} = \overline{I}(1 - s)/s. \tag{3.18}$$

3.2.2 Solow–Shell Model

Let us consider a modification of the dynamic *finite-horizon* optimization Solow–Shell model (2.45)–(2.47), in which the output is described by the Cobb–Douglas production function with the labor-augmenting technological change:

$$Q(t) = AK(t)^{\alpha}(\mathrm{e}^{gt}L(t))^{1-\alpha}, \quad 0 < \alpha < 1. \tag{3.19}$$

Similarly to Sect. 2.3.1, after converting to the per capita variables s and k, this model is represented by the following optimization problem.

Optimization Problem

Find the function $s(t)$, $0 \leq s(t) \leq 1$, and the corresponding $k(t)$, $k(t) \geq 0$, $t \in [0, T]$, which maximize

$$\max_{s,k} \int_0^T \mathrm{e}^{-rt}(1 - s(t))\mathrm{e}^{g(1-\alpha)t}Ak^{\alpha}(t)\mathrm{d}t \tag{3.20}$$

under the equality-constraint:

$$k'(t) = s(t)\mathrm{e}^{g(1-\alpha)t}Ak^{\alpha}(t) - \mu k(t), \tag{3.21}$$

and the initial and terminal conditions

$$k(0) = k_0, \quad k(T) \geq k_T. \tag{3.22}$$

The problem (3.20)–(3.22) coincides with the Solow–Shell model (2.45)–(2.47) at $g = 0$ (in the absence of technological change).

Steady-State and Dynamic Analysis

Similarly to Chap. 2, the steady state (3.16) of the Solow–Swan model (3.7)–(3.10) with autonomous technological change and Cobb–Douglas function represents a long-term regime in the Solow–Shell model (3.20)–(3.22) because the fundamental equation (3.12) appears to be the same in both models. In particular, the steady state capital per effective labor $\tilde{k}^* = (A\alpha/(\mu + g))^{1/(1-\alpha)}$ is constant and the corresponding capital per capita is $k^*(t) = \mathrm{e}^{gt}(A\alpha/(\mu + g))^{1/(1-\alpha)}$.

Using the maximum principle from Sect. 2.4, we can show that the solution $s(t)$, $t \in [0, T]$, of the optimization problem (3.20)–(3.22) is

$$s(t) = \begin{cases} \begin{cases} 1 & \text{at } k_0 < k^*(0) \\ 0 & \text{at } k_0 > k^*(0) \end{cases} & \text{at} \quad 0 \leq t \leq \theta_1 \\ \quad s^* & \text{at} \quad \theta_1 < t \leq \theta_2, \\ \begin{cases} 0 & \text{at } k_T < k^*(T) \\ 1 & \text{at } k_T > k^*(T) \end{cases} & \text{at} \quad \theta_2 < t \leq T \end{cases} \tag{3.23}$$

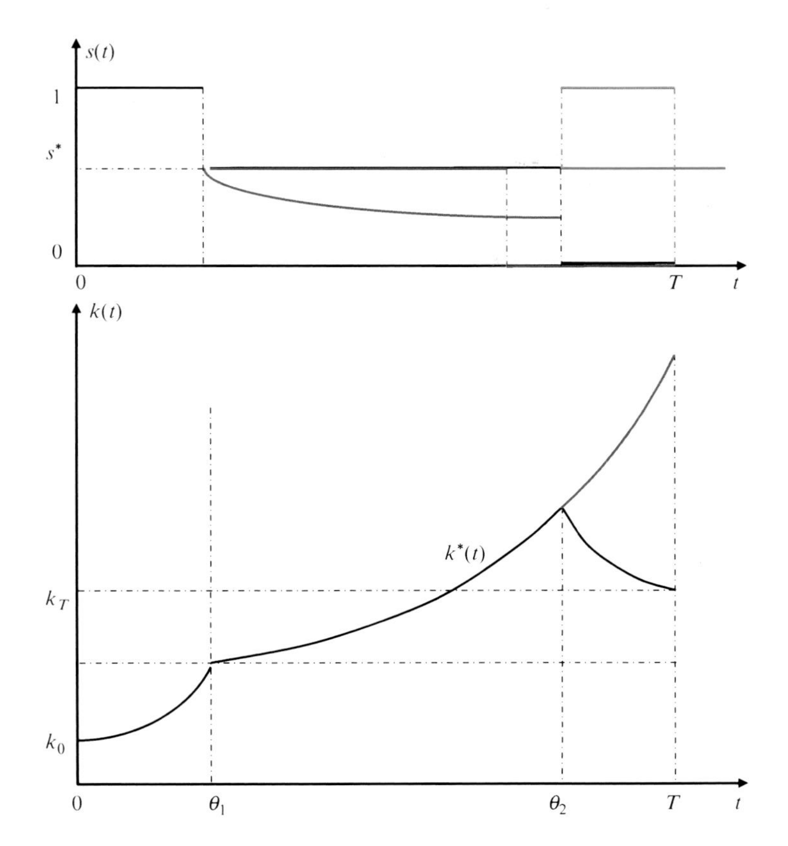

Fig. 3.1 Optimal trajectories in the Solow–Shell model with technological change in the case $k^*(0) > k_0$ and $k^*(T) < k_T$. The line $s(t) \equiv s^*$ and the increasing curve $k^*(t)$ depict the modified golden rule trajectory. The *blue lines* represent the optimal regime in the infinite-horizon Solow–Ramsey model with technological change of Sect. 3.2.2

where $0 < s^* = \alpha < 1$ is the optimal golden-rule saving rate (3.16) of the Solow–Swan model, and the optimal capital per effective labor $\widetilde{k}^* = (A\alpha/\mu)^{1/(1-\alpha)}$ is constant.

Thus, under the autonomous technological change, the optimal accumulation norm $s(t)$ is constant and equal to s^* during the *long-term* dynamics interval $[\theta_1, \theta_2]$ but the optimal capital labor ratio $k^*(t)$ increases exponentially with the rate g over $[\theta_1, \theta_2]$. Because the labor is constant ($\eta = 0$) as in the Solow–Shell model (2.45)–(2.47), the original variables $Q(t)$, $C(t)$, $I(t)$, and $K(t)$ also increase with the given rate g of technological change. The solution $s(t)$, $k(t)$, $t \in [0, T]$, of the optimization problem (3.20)–(3.22) in the case $k_0 < k^*(0)$, $k^*(T) > k_T$ is illustrated in Fig. 3.1.

The transition (short-term) dynamics of the problem (3.20)–(3.22) is $s(t) = 1$ at $k_0 < k^*(0)$ or $s(t) = 0$ at $k_0 > k^*(0)$ on an initial interval $[0, \theta_1]$. It aims the fastest possible switch of the optimal trajectory $s(t)$, $k(t)$ to the steady state trajectory (s^*, k^*). The transition dynamics ends at the instant θ_1 such that $k(\theta_1) = k^*(\theta_1)$. As in the finite-horizon model of Sect. 2.2 without technological change, the optimal trajectory leaves the steady state solution (s^*, k^*) at some instant $\theta_2 < T$ because of the *end-of-horizon effect*.

Importance of the labor-augmenting technological change. Karl Shell analyzed a more general version of the model (3.20)–(3.22) with the Hicks-neutral technological change (3.2) in [11]. His analysis shows that this model has a constant steady state only if the production function is of the Cobb–Douglas form. In this case, the Hicks-neutral technological change coincides with the labor-augmenting Harrod-neutral one. This result confirms a special role of the labor-augmenting technological change (3.11) in the Solow–Swan model, see also Sect. 3.2.1. Another proof that the autonomous technological change needs to be labor-augmenting is provided in [2, p. 53].

3.2.3 *Solow–Ramsey Model*

The infinite-horizon version

$$\max_s \int_0^\infty e^{-rt} c(t) dt, \tag{3.24}$$

$$k'(t) = s(t) e^{g(1-\alpha)t} A k^\alpha(t) - \mu k(t), \quad k(0) = k_0, \tag{3.25}$$

of the dynamic optimization problem (3.20)–(3.22) can be classified as the linear-utility Solow–Ramsey model with autonomous technological change (see Sect. 2.2).

Analysis

Similarly to Sect. 2.3.2, the transversality condition in this problem

$$r > \eta + g \tag{3.26}$$

guarantees the convergence of the improper integral in (3.24).

As in the model with no technological change, the behavior of the infinite-horizon optimal trajectories appears to be simpler than in the finite-horizon Solow–Shell model (3.20)–(3.22) because of the absence of the *end-of-horizon effects*. Namely, the solution of the optimization problem (3.24) and (3.25) is

$$s(t) = \begin{cases} \begin{cases} 1 & \text{at } k_0 < k^*(0) \\ 0 & \text{at } k_0 > k^*(0) \end{cases} & \text{at} \quad 0 \le t \le \theta_1 \\ s^* & \text{at} \quad \theta_1 < t < \infty \end{cases}, \tag{3.27}$$

$$k(t) = \begin{cases} k_{tr}(t) & \text{at} \quad 0 \leq t < \theta_1 \\ k^*(t) & \text{at} \quad \theta_1 \leq t < \infty \end{cases} \tag{3.28}$$

with the same constant golden-rule saving rate $s^* = \alpha$ and the capital per capita $k^*(t)$ as in the Solow–Shell model. So the optimal trajectory does not leave the long-term steady state (s^*, k^*) after reaching it at the time θ_1. The solution $s(t)$, $k(t)$, $t \in [0, \infty)$, of the optimization problem (3.24) and (3.25) in the case $k_0 < k^*(0)$ is illustrated in Fig. 3.1 by the red curve.

In all versions of the Solow model considered in this and previous chapters, the long-run growth rate appears to be exogenously determined by given growth rates of technological progress and labor. The source of technological progress is not described in these models. So all models considered above are *models of exogenous growth*. They cannot generate an economic growth if technological change is absent and labor is constant.

3.3 Models with Endogenous Technological Change

The role of commercial R&D increases in modern times of large multinational corporations and global markets for new products. The introduction of R&D department is one of the major innovations of the twentieth century. Industry leaders and policymakers stress the role of innovation for nation wealth and competitiveness. The way Japan and some Asian countries achieved rapid growth in the 1960s–1980s demonstrate that government policies and economic incentives can influence the technological change within a nation. As a result, economists view the technological change as the outcome of economic decisions rather than an unexplained phenomenon. The endogenous technological change occurs as a reaction to economic incentives and opportunities to develop new technologies, and models of *endogenous* technological change have been introduced to explain the sources and forces of technological change.

3.3.1 Induced Technological Change

The *induced technological change* is an early concept of the endogenous technological change. It links the technological change to previous economic development. First models of the induced technological change were based on various hypotheses about relations between the technological change intensity and such aggregated economic characteristics as capital stock, investment, knowledge accumulation, and others. One of popular hypotheses was that the technological change is induced by capital investment, i.e., there exists a nonlinear relationship between the total productivity and the total investment.

More popular and sophisticated *Arrow hypothesis* states that the improvement in technology depends on the experience within the production process [1]. Specifically, this hypothesis assumes that the technical knowledge (*learning by doing*) increases with the acquisition of new vintages of capital measured by investment. As a result, the efficiency of the new capital depends on the investment: the higher the investment, the greater the opportunity for learning, and the faster the rate of technical progress and the production output. The Arrow hypothesis was the first model with endogenous embodied technical change.

The implications of the induced technological change have been also explored in growth models with limited resources. The consequences of changes in the resource availability are better understood if the technology is endogenous rather than exogenous variable. When technological change is given, firms and individuals can react to changes in resource availability only by changing the allocation of economic activity. With the endogenous technological change, innovation may be intensified or redirected in response to economic changes. An increased shortage of resources may, through rising prices, stimulate firms to develop new technologies that save on the resource input.

The corresponding *induced-innovation hypothesis* about relationships between resource scarcity and innovation states that a higher price of production inputs leads to technological improvements to save the resource that becomes more expensive. It is still debated in modern economics. In the context of energy consumption, this hypothesis stipulates that rapidly rising energy prices make the development of energy-saving technologies more profitable. However, the induced technological change cannot explain the need of focused investment into science and technology. In modern economics, the term "induced technological change" usually refers to additional technological improvements caused by other economic processes and phenomena, governmental regulations, and more restrictive environmental policies.

3.3.2 One-Sector Model with Physical and Human Capital

The view of technological change has changed over the past few decades. New growth theories of 1980s and 1990s involve specific factors, such as physical capital, technology, research and development (R&D), human capital, or infrastructure, as sources of technological change. The long-run economic growth essentially depends on the source of technological change. For example, the technological change increases when there are more highly educated workers. The model of this section is an extension of the Solow–Swan model (2.31)–(2.35) and the Solow–Ramsey model (2.53) with the following major modifications:

- The dynamic characteristic $L(t)$ of the Solow–Swan model is interpreted as the *human capital* in a broad sense. We assume that the human capital can be increased by the additional investment $H(t)$ in education, professional training, public health, and other labor-enhancing activities. The actual labor is assumed to be *constant*.

- The investment $H(t)$ into human capital is a new control variable of the model. The aggregate characteristics $Q(t)$, $C(t)$, $I(t)$, and $K(t)$ retain the same meaning as in Sect. 3.2 and Chap. 2.
- The central planner chooses the optimal investments $I(t)$ and $H(t)$ into the physical and human capital, $t \in [0, \infty)$, to maximize the present value of a *nonlinear utility function* over the infinite planning horizon $[0, \infty)$:

$$\max_{I,H} \int_0^{\infty} e^{-rt} \frac{C^{1-\gamma}(t) - 1}{1 - \gamma} dt. \tag{3.29}$$

See more about the utility functions in Sect. 2.3.

The one-sector model with physical and human capital is described by the following identities:

$$Q(t) = F(K(t), L(t)), \tag{3.30}$$

$$Q(t) = C(t) + I(t) + H(t) \tag{3.31}$$

(the output Q is distributed among consumption and investments into physical and human capital),

$$K'(t) = I(t) - \mu K(t), \quad \mu = \text{const} > 0 \tag{3.32}$$

(the physical capital K increases with the investment I and depreciates at a constant rate $\mu > 0$),

$$L'(t) = H(t) - \eta L(t), \quad \eta = \text{const} > 0 \tag{3.33}$$

(the human capital L increases with the investment H and depreciates at a constant rate $\eta > 0$).

The structure of the model (3.30)–(3.33) is illustrated in Fig. 3.2.

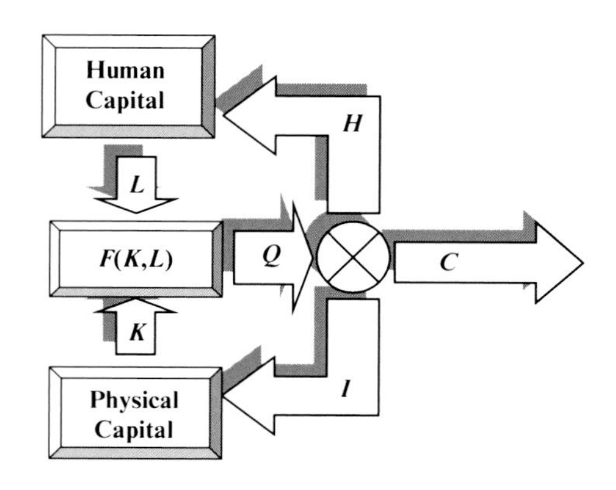

Fig. 3.2 The flow diagram of the one-sector model with physical and human capital

For clarity, we consider the model (3.29)–(3.33) under the Cobb–Douglas production function and the same depreciation rate of physical and human capital:

$$F(K,L) = AK^{\alpha}L^{1-\alpha}, \quad 0 < \alpha < 1, \quad \mu = \eta. \tag{3.34}$$

Steady-State Analysis

Let us examine a *balanced growth path* (BGP) in the model (3.29)–(3.34) when the variables $Q(t)$, $C(t)$, $I(t)$, $H(t)$, $K(t)$, and $L(t)$ grow with the same constant rate $g > 0$:

$$\begin{aligned} Q(t) &= \overline{Q}\,e^{gt}, \quad C(t) = \overline{C}\,e^{gt}, \quad I(t) = \overline{I}\,e^{gt}, \\ H(t) &= \overline{H}\,e^{gt}, \quad \overline{K}(t) = \overline{K}\,e^{gt}, \quad L(t) = \overline{L}\,e^{gt} \end{aligned} \tag{3.35}$$

Substituting (3.35) into (3.30)–(3.34), we can see that the BGP is possible. Moreover, using $\mu = \eta$, we express all level constants through $\overline{K}$ and $\overline{L}$ as

$$\begin{aligned} &\overline{I} = \overline{K}(g+\mu), \quad \overline{H} = \overline{L}(g+\mu), \\ &\overline{Q} = A\overline{K}^{\alpha}\overline{L}^{1-\alpha}, \quad \overline{C} = A\overline{K}^{\alpha}\overline{L}^{1-\alpha} - (\overline{K}+\overline{L})(g+\mu) \end{aligned} \tag{3.36}$$

So we need to determine three unknown variables $g, \overline{K}$, and $\overline{L}$ to specify the BGP.

Static Optimization

Let us find the optimal BGP that maximizes the functional (3.29) subject to (3.30)–(3.34). We assume a priori that the unknown rate g satisfies the transversality condition $g < r/(1-\gamma)$, under which the integral (3.29) along the BGP (3.35) is finite. Then, substituting (3.35) and (3.36) into (3.29) and integrating, we obtain a nonlinear optimization problem

$$\max_{g,\overline{K},\overline{L}} \Phi = \max_{g,\overline{K},\overline{L}} \frac{\left[A\overline{K}^{\alpha}\overline{L}^{1-\alpha} - (\overline{K}+\overline{L})(g+\mu)\right]^{1-\gamma}}{r-(1-\gamma)g} \tag{3.37}$$

with three scalar unknowns g, $\overline{K}$ and $\overline{L}$. The standard optimality conditions for interior solutions $\partial\Phi/\partial g = 0$, $\partial\Phi/\partial\overline{K} = 0$, $\partial\Phi/\partial\overline{L} = 0$ lead after transformation to the following equalities:

$$g = \left[A\overline{K}^{\alpha}\overline{L}^{1-\alpha}/(\overline{K}+\overline{L}) - \mu - r\right]/\gamma, \tag{3.38}$$

$$A\alpha(\overline{K}/\overline{L})^{\alpha-1} = A(1-\alpha)(\overline{K}/\overline{L})^{\alpha} = g+\mu. \tag{3.39}$$

The condition (3.39) means that the marginal product of physical capital should be equal to the marginal product of human capital. By (3.39), the optimal ratio of the stocks of physical and human capital is

$$\overline{K}/\overline{L} = \alpha/(1-\alpha). \tag{3.40}$$

Along the optimal BGP, the distribution of the investments into physical and human capital $\overline{I}/\overline{H} = \overline{K}/\overline{L} = \alpha/(1-\alpha)$ is also in accordance with their marginal products. Therefore, this distribution is similar to the golden rule of economic growth (2.32) in the Solow–Swan model of Sect. 2.2.

Substituting (3.40) into (3.38), we obtain the optimal balanced growth rate of the economy as

$$g = \left[A\alpha^{\alpha}(1-\alpha)^{(1-\alpha)} - \mu - r\right]/\gamma. \tag{3.41}$$

Let the given model parameters A, α, μ, r, and γ be such that $g > 0$ and $g < r/(1-\gamma)$. Under this choice of parameters, the economy grows endogenously with the rate g despite that the actual physical labor is constant (but the human capital $L(t)$ increases with the rate g). So the model (3.32)–(3.36) is a *model of endogenous growth*.

Dynamic Optimization

The dynamic optimization in the model (3.29)–(3.33) involves the infinite-horizon optimal control problem (3.29) with *two independent controls* $I(t)$ and $H(t)$, $t \in [0, \infty)$, under the constraints

$I(t) \geq 0$, $H(t) \geq 0$

and the initial conditions

$K(0) = K_0$, $L(0) = L_0$.

This problem is investigated using the maximum principle of Sect. 2.4 and possesses the following behavior of solutions.

Long-term dynamics

The *long-term dynamics* of the model coincides with the BGP (3.37) with the endogenous growth rate (3.41) on a certain interval $[\theta, \infty)$. The optimal trajectory approaches the BGP in a finite time.

Transition dynamics

If $K_0/L_0 > \alpha/(1-\alpha)$ (the physical capital is abundant initially), then *the transition dynamics* is $I(t) = 0$ (no investment into physical capital) over the interval $[0, \theta]$. Otherwise, at $K_0/L_0 < \alpha/(1-\alpha)$ (the human capital is abundant initially), *the transition dynamics* is $H(t) = 0$ (no investment into physical capital) on $[0, \theta]$. The transition dynamics ends at an instant θ such that the ratio $K(\theta)/L(\theta) = \alpha/(1-\alpha)$.

So the problem (3.29)–(3.33) is technically similar to the Solow–Ramsey optimization model (3.24) but provides quite different economic implications. Namely, the model (3.29)–(3.33) can generate an exponential growth of an economy under constant labor. Such models are known as *the models of endogenous growth*.

3.3.3 Two-Sector Model with Physical and Human Capital (Uzawa–Lucas Model)

In the one-sector model with physical and human capital of the previous section, the human capital is generated by the same production function as the physical capital, which is far from reality. In reality, the education industry relies on educated people as an input, and its productivity is quite different from physical production. So further research has been directed to overcome this shortage. Here we consider a simple and well-known two-sector model with physical and human capital: the *Uzawa–Lucas model*. This model is a modification of the one-sector model (3.29)–(3.33). The differences are:

- There are two sectors: *production* and *education*. The human capital $L(t)$ is produced using a different technology (in the education sector) that involves the human capital only.
- The fraction $u(t)$ of the human capital $L(t)$ is used in the production sector and $1 - u(t)$ is used in the education sector. The fraction u is a control variable of the optimization problem together with the investment I. The aggregate characteristics $Q(t)$, $C(t)$, and $K(t)$ retain the same meaning as in the one-sector model (3.29)–(3.33).

The two-sector Uzawa–Lucas model is of the form:

$$\max_{I, u} \int_0^\infty e^{-rt} \frac{C^{1-\gamma}(t) - 1}{1 - \gamma} dt, \tag{3.42}$$

$$Q(t) = AK^{\alpha}(t)[u(t)L(t)]^{1-\alpha}, \quad 0 < \alpha < 1, \tag{3.43}$$

$$Q(t) = C(t) + I(t), \tag{3.44}$$

$$K'(t) = I(t) - \mu K(t), \quad \mu = \text{const} > 0, \tag{3.45}$$

$$L'(t) = B(1 - u(t))L(t) - \mu L(t), \tag{3.46}$$

where A and B are constant technological parameters in the production and education sectors correspondingly. The structure of this model is illustrated in Fig. 3.3.

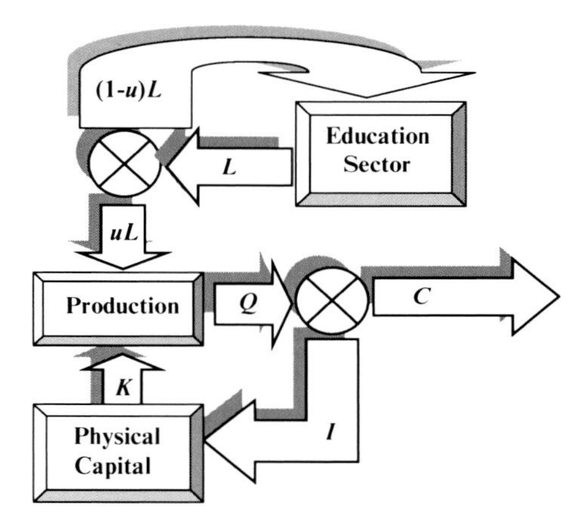

Fig. 3.3 The flow diagram of the two-sector model with physical capital K and human capital L

Steady-State Analysis

The *steady-state analysis* of model (3.42)–(3.46) is provided analogously to the previous section. It can be shown that the model has a BGP with the variables $Q(t)$, $C(t)$, $I(t)$, $K(t)$ and $L(t)$ that grow with the same constant rate $g > 0$ while the variable $u(t)$ is constant. The optimal balanced growth rate of the economy is

$$g^* = \frac{B - \mu - r}{\gamma} \tag{3.47}$$

and the optimal constant fraction u of human capital used in production is

$$u^* = \frac{r + \mu(1 - \gamma)}{B\gamma} + \frac{\gamma - 1}{\gamma}. \tag{3.48}$$

If $B > \mu + r$, then the optimal growth rate $g^* > 0$ and $u^* < 1$. The transversality condition is $g^* > B - \mu$ and ensures that $u^* > 0$.

Dynamic Optimization

The model (3.42)–(3.46) is the *infinite-horizon* optimal control problem with *two independent controls* $I(t)$ and $u(t)$, $t \in [0, \infty)$, under the constraints $I(t) \geq 0$, $0 \leq u(t) \leq 1$ and the initial conditions $K(0) = K_0$, $H(0) = H_0$. This problem is investigated using maximum principle.

The *long-term dynamics* of the model coincides with the optimal balanced growth path with the endogenous growth rate (3.47) on the infinite interval $[\theta, \infty)$, except for a certain transition dynamics period $[0, \theta]$.

The *transition dynamics* demonstrates that the optimal trajectory approaches the optimal balanced path (3.47) and (3.48) in a finite time. The transition dynamics appears to be richer and more asymmetric than in the one-sector model (3.29–3.33). Namely, if the human capital is abundant initially (K_0/L_0 is small), then the transitional growth rate of the gross output is larger than in the case if the human capital is scarce initially (K_0/L_0 is large). An applied interpretation of the transition dynamics is that an economy will recover faster from a war that from an epidemic.

Comparing the expressions for the optimal endogenous growth rate, we can say that the optimal dynamics of the two-sector model is simpler and more natural than that of the one sector model. At the same time, the two-sector model produces new nontrivial economic results. In particular, the endogenous economy growth (3.47) depends on the efficiency B of the education sector.

3.3.4 Knowledge-Based Models of Economic Growth

In modern economy, technological change requires considerable investments into R&D and learning. Correspondingly, a growing family of the endogenous growth models includes a separate R&D sector to model the economic mechanisms of R&D impact on the productivity growth. In terms of the two-sector model (3.42)–(3.46), the technology level A in the production function (3.43) becomes an endogenous variable that determines the stock of *knowledge* (or *technology*) in an economy.

The *knowledge* is understood as the accumulation of ideas that are produced in the R&D sector. Knowledge is a non-rival production factor that raises the productivity of both capital and resource inputs. However, knowledge may decrease production output: some workers have to devote their labor effort to research rather than to the production of final goods, or some of the economy outputs are used as an input (e.g., scientific equipment) in R&D. Correspondingly, the accumulation of knowledge A, i.e., the increase of $A(t)$ describes the endogenous technological change in such models. The knowledge stock $A(t)$ becomes an additional input of endogenous growth models while the total labor $L(t)$ is exogenous.

Let $1 - u(t)$ be the fraction of labor $L(t)$ used in the R&D sector. Then, a popular hypothesis of knowledge accumulation is that the rate $A'(t)$ is linearly proportional to the labor $(1 - u(t))L(t)$ in the R&D sector:

$$A'(t) = \delta(1 - u(t))L(t), \qquad (3.49)$$

where the parameter $\delta > 0$ is interpreted as the *arrival rate of new R&D ideas* and it can be governed by a Poisson process. Equation (3.49) resembles the human capital accumulation equation (3.46) in the Uzawa–Lucas model of Sect. 3.3.3 but leads to different results.

More careful description of knowledge accumulation considers the nonlinear dependence of the knowledge accumulation on R&D labor, for instance, as

$$A'(t) = \delta[(1 - u(t))L(t)]^{\lambda} A^{\phi}(t), \qquad (3.50)$$

where

The knowledge stock $A(t)$ is in line with the above definition.

The parameter λ, $0 < \lambda \leq 1$, describes possible reduction in the total number of innovation because of duplication and overlapping of research.

The *R&D complexity parameter* ϕ, $0 < \phi < 1$, captures the *decreasing returns to scale* of the R&D process.

An economy with production sector and R&D sector can be described in the social planner framework by the following nonlinear growth model [8]:

$$\max_{c,u} \int_0^{\infty} \mathrm{e}^{-rt} \frac{c^{1-\gamma}(t) - 1}{1 - \gamma} \mathrm{d}t, \quad c = C/L, \qquad (3.51)$$

$$Q(t) = K(t)^{\alpha} [A(t)u(t)L(t)]^{1-\alpha}, 0 < \alpha < 1, \qquad (3.52)$$

$$Q(t) = C(t) + K'(t), \qquad (3.53)$$

$$L'(t) = \eta L(t), \quad \eta = \text{const} > 0, \qquad (3.54)$$

where the dynamics of the technology factor $A(t)$ is described by equation (3.50).

This model is similar to the two-sector model (3.42)–(3.46) with human capital. The model assumes no deterioration of the physical capital: $\mu = 0$ in (3.45), which reduces the equations (3.44)–(3.45) to $I(t) = K'(t)$ in the equation (3.53).The major difference is that $A(t)$ is endogenous in (3.52) while $L(t)$ is endogenous in (3.43). The dynamics of these models appears to be quite different because of distinct feedbacks through inputs A and L. The structure of the model (3.51)–(3.54) with R&D sector is illustrated in Fig. 3.2.

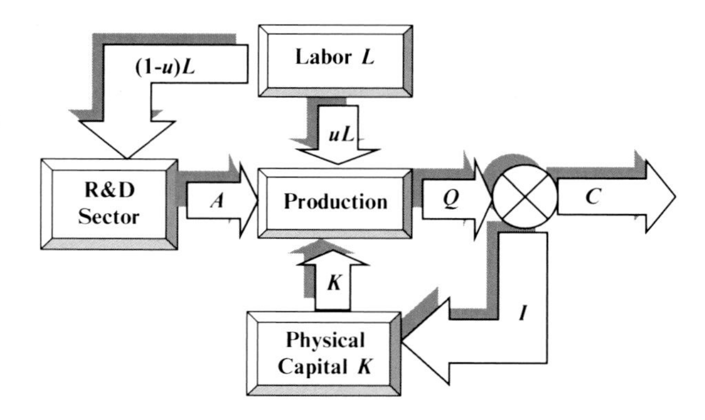

Fig. 3.4 The flow diagram of the endogenous growth model with R&D sector, production sector, and given labor L

Elements of Model Analysis

The total labor $L(t)$ is exogenous and found immediately from the equation (3.54) as $L(t) = \overline{L}e^{\eta t}$. Thus, the reduced model (3.50)–(3.53) has five unknown functions $u(t)$, $Q(t)$, $C(t)$, $A(t)$, and $K(t)$, $t \in [0, \infty)$. This problem is investigated analogously to previous sections using the steady-state analysis and maximum principle of Sect. 2.4.

Balanced Growth and Static Optimization

One can prove that the model (3.50)–(3.54) has a *balanced growth path* (BGP) along which $u(t)$ is constant, while A and per capita variables $q = Q/L$, $c = C/L$, $k = K/L$ grow with the same constant rate $g > 0$. The optimal balanced growth rate of the economy is determined by static optimization as

$$g^* = \frac{\lambda\eta}{1-\phi} \tag{3.55}$$

and the optimal u is

$$u^* = 1 - \left[1 + \frac{1}{\lambda}\left(\frac{r(1-\phi)}{\lambda\eta} + \frac{1}{\gamma} - \phi\right)\right]^{-1}. \tag{3.56}$$

Correspondingly, the unknown functions $Q(t)$, $C(t)$, and $K(t)$, $t \in [0, \infty)$, of the model (3.50)–(3.53) grow along the optimal BPG with the rate $g^* + \eta$.

The steady state dynamics of this model is different from two previous models with human capital. In particular, the economy growth rate (3.47) in the two-sector model (3.42)–(3.46) with human capital depends on the *level variable* B, which is not supported by economic reality. In (3.55), the optimal growth rate g^* is determined by the exogenous labor growth rate η, the R&D efficiency parameter λ, and the R&D complexity ϕ, which better fits empiric data. However, the economy growth is not completely endogenous: $g^* > 0$ if and only if $\eta > 0$, which requires the total labor growth for the economy growth. Such dynamics is classified as *semi-endogenous growth*.

Dynamic Optimization

The model (3.50)–(3.54) is the infinite-horizon optimal control problem with *two independent controls* $c(t)$ and $u(t)$, $t \in [0, \infty)$, under the constraints $c(t) \geq 0$, $0 \leq u(t) \leq 1$ and the initial conditions $K(0) = K_0$, $H(0) = H_0$. The *long-term dynamics* of the model coincides with the optimal BGP with the endogenous growth rate (3.55) of the per capita variables q, c, k on $[\theta, \infty)$, except for a certain transition interval $[0, \theta]$.

A specific outcome of the R&D-based models is that the endogenous growth is possible only when the effectiveness of the R&D sector is sufficient to deliver constant returns as in the equation (3.49). Then, a society that spends enough on R&D can maintain a sustainable long-run growth without relying on the population growth or exogenous technological change. However, the linear dependence (3.49) between the knowledge A accumulation rate A' and the R&D labor is not supported by current economic data.

A decentralized equilibrium model that describes microeconomics of knowledge accumulation involves three sectors [8]: the *R&D sector* produces new designs and sells them to the intermediate sector, *the intermediate sector* firms use these designs to produce a variety of products of increasing quality, and the *final good sector* uses labor and intermediate products as inputs. The resulting aggregate equations of this model are similar to its social planner's version (3.50)–(3.54) (Fig. 3.4).

An obvious limitation of the endogenous growth models is the shortage of real data about the knowledge accumulation process and its impact on the economic productivity. It limits practical applications of existing models and the development of more detailed models with endogenous technological change and vintage effects.

3.4 Modeling of Technological Innovations

The modern endogenous growth theory builds macroeconomic models on micro-economic foundations, focusing on new technologies and human capital. The description of endogenous growth ranges from simple nonlinear relations between the technological change rate and other aggregate economic parameters (as in the models of Sects. 3.2 and 3.3) to sophisticated frameworks with spillover effects, increasing numbers of goods, increasing quality of labor, and so on. In particular, models explore the sources of technological change as well as the instruments of its implementation.

3.4.1 *Inventions, Innovations, and Spillovers*

Contemporary models tell apart three stages of technological change:

- *Invention* as the first development of a new product or process.

- *Innovation* as commercialization of a new product or process, which makes it available on the market.
- *Diffusion* as a gradual adoption of the new product or process by firms or individuals.

The invention and innovation stages are carried out primarily by private firms through the R&D process. A firm can innovate without inventing. The engine of technological change is the entrepreneurial initiative that makes an available technology profitable through innovative decisions.

The *creative destruction theory* of J. Schumpeter considers innovations by entrepreneurs as the major driving force of technological progress and long-term economic growth. Schumpeterian models involve an endogenous microeconomics of innovation and employ the general equilibrium framework discussed in Sect. 2.3.3. An important aspect of Schumpeterian models is *creative destruction*: when a new or improved product or production process is developed, it replaces the old one. A variety of such economic models describe various aspects of economies with creative destruction processes. Their analysis is based on analytic tools and numeric simulation of agent-based computer models.

The *spillovers* are unintended benefits of technological change resulting from R&D activities of other firms and organizations. A *knowledge spillover* is simply an exchange of ideas among individuals. Spillovers are free for a specific firm because their costs are paid by other firms. Their importance increases in modern economy because of the explosion of globalization, networks, online social networking, and so on. Spillover effects stimulate technological improvements and are an important feature of innovation dissemination.

The dissemination of technological innovations has been investigated by means of *substitution*, *diffusion*, and *evolution models* that deal with different aspects of the innovation dissemination.

3.4.2 Substitution Models of Technological Innovations

Substitution models describe the innovation process as a *substitution operation*, which consists of the replacement of a particular product or process associated with one technology by another product associated with a new technology. Mathematically, such models are similar to those used in the modeling of biological populations in Chap. 6, but possess different interpretation. In this section, we consider some simple deterministic models of innovation substitution. The substitution is defined as the replacement of an old technology (capital asset, equipment, or machine) with a new one. Let $N_i(t)$ be the number of production units (firms, plants, enterprises) that use the technology i, $i = 1,2$, at time t.

Logistic Substitution Model

The logistic model of substitution [7] is one of the most known substitution models. Its key assumption is that the instantaneous growth rate of the fractional substitution $f = N_2/(N_1 + N_2)$ of an old technology N_1 with a new one N_2 is linearly proportional to the remaining amount $N_1/(N_1 + N_2)$ of the old technology to be substituted:

$$f'/f = kN_1/(N_1 + N_2). \quad (3.57)$$

It leads to the *Verhulst–Pearl model* (*logistic model*) of the form:

$$df/dt = kf(1 - f), \quad (3.58)$$

which is well known in the modeling of biological populations (see Sect. 6.1.3). Its solution (6.6) is known as the *S-shaped or logistic curve* and is shown in Fig. 6.2. Despite its simplicity, this model often appears to fits well the empirical results about the substitution of industrial technologies.

Lotka–Volterra Model of Substitution

The generalized Lotka–Volterra equations

$$dN_1/dt = a_1N_1\left(N_1^{max} - N_1 + b_1N_2\right) - d_1N_1, \quad (3.59)$$

$$dN_2/dt = a_2N_2\left(N_2^{max} - N_2 + b_2N_1\right) - d_2N_2, \quad (3.60)$$

describe a substitution process as the competition among technologies [3]. Here:

- N_i^{max} is the maximum number of firms that can use the technology i.
- a_i and d_i are the natural entry (fertility) and exit (mortality) coefficients of the firms using the technology i.
- b_i characterize the degree of using a common resource by both technologies. If $b_1 = b_2 = 1$, then both firms use the same resource.

The condition for the success of the new technology is found from a stability analysis of the stationary states of the equations (3.59)–(3.60). Namely, if

$$b_2\left(N_2^{max} - d_2/a_2\right) > b_1\left(N_1^{max} - d_1/a_1\right), \quad (3.61)$$

then the new technology will completely substitute the old one. In other words, starting from a state ($N_1 > 0, N_2 = 0$), the dynamic system (3.59) and (3.60) will reach a new stable state ($N_1 = 0, N_2 > 0$).

The *Lotka–Volterra equations* (3.59) and (3.60) is a classic dynamic model of competing biological populations (Sect. 6.2.1) with different interpretation of the parameters a_i and d_i as the natural fertility and mortality coefficients.

Generalized Substitution Model

A generalized model of substitution [5] is described by the equations

$$dN_i/dt = \left(E_i + B_iN_j\right)N_i - k_iN_i, \quad i = 1, 2, \quad (3.62)$$

under the assumption that the total number of firms is constant:

$$N = N_1 + N_2 = \text{const.} \tag{3.63}$$

This model combines both one-dimensional logistic substitution model (3.58) and two-dimensional Lotka–Volterra model (3.59) and (3.60) as special cases at certain choice of parameters.

Depending on the choice of parameters E_i and B_i, the model (3.62) and (3.63) produces a linear ($B_1 = B_2 = 0$) or a nonlinear (at least one $B_i > 0$) economic growth. By (3.63), the new technology can succeed only if it replaces the old one. The ratio

$$\alpha = \begin{cases} E_1/E_2 & \text{if} \quad B_1 = B_2 = 0 \\ B_1/B_2 & \text{if} \quad B_2 > 0 \end{cases} \tag{3.64}$$

can be used as an indicator of the success of a new technology. Namely, the new technology 2 is successful as compared to the old technology 1 at $\alpha > 1$ and is not successful at $\alpha < 1$.

3.4.3 Diffusion and Evolution Models of Technological Innovation

Diffusion models describe innovation in the terms of spreading out a new product or technology from its manufacturers to final users or adopters. Such models focus on diffusion and adoption aspects of the innovation process. Diffusion processes in continuous time are described by special partial differential equations, known as *diffusion equations*. Diffusion equations are used for modeling of ecological populations in Sect. 7.2 and the environmental contamination in Chaps. 8 and 9.

The majority of substitution and diffusion models is deterministic. Stochastic *evolutionary models* have been introduced to describe the random nature of innovation processes. *Self-organization evolutionary models* consider innovation as a structural fluctuation under technological change and unite the aspects of diffusion of new technologies, diversity, learning mechanisms, and age-dependent effects. They lead to interesting results such as the existence of evolutionary long waves and low-dimensional chaotic properties. The self-organization models also appeared first in the study of biological evolution. In general, a deep analogy exists between biological and technological processes, which mutually can enhance development of mathematical models for both fields.

3.4.4 General Purpose Technologies and Technological Breakthroughs

Recent economic research demonstrates that a gradual continuous technical progress does not provide a complete explanation of key aspects of economic growth. It appears that long periods of economic growth have been driven by few major technologies, called the *general purpose technologies*. Examples of such technologies are the steam engine, gasoline engine, electric power, electric motor, and semiconductor in the past and computers, networks, and cell phones.

The general purpose technologies are used as inputs by many production sectors, and the effectiveness of R&D in these sectors can sharply increase as a consequence of an innovation in such technologies. As a result, two fundamentally different modes of technical progress coexist:

- A gradual improvement (a "normal" mode) when technological improvements occur incrementally as a result of accumulated experience.
- A *technological breakthrough* (the radical improvement mode) when a radically new innovation is capable of displacing an older general purpose technology among competing technologies.

Modern economic theory considers technological breakthroughs as radical innovations caused by the substitution of one general-purpose technology by another. Such breakthroughs explain economy-wide structural changes. Many economists interpret the recent IT revolution as a major breakthrough.

Exercises

1. Prove that the Cobb–Douglas production function with autonomous technological change (3.5) is Hicks neutral.
   HINT: Replace L with $A(t)L$ in the formula $F(K, L) = A_0K^{\alpha}L^{1-\alpha}$.
2. Prove that the Cobb–Douglas production function (3.5) is Harrod neutral.
   HINT: Replace K with $A(t)K$ in the formula $F(K, L) = A_0K^{\alpha}L^{1-\alpha}$.
3. Prove that the Cobb–Douglas production function (3.6) with autonomous technological change is Hicks, Harrod, and Solow neutral at the same time.
4. Justify that the formulas (3.17) and (3.18) describe a balanced growth in the Solow–Swan model (3.7)–(3.10) with autonomous technological change.
   HINT: Substitute (3.17) into the model equations (3.7)–(3.10) and obtain formulas (3.18).
5. In the Solow–Ramsey model (3.24) and (3.25), find the exact formula for $k^*(t) = k_{tr}(t)$ on the transition interval $[0, \theta_1]$ in the case $k_0 > k^*(0)$.
   HINT: Determine $s(t)$ from (3.27), substitute it into the linear differential equation (3.25) and solve this equation with respect to $k(t)$.
6. In the Solow–Ramsey model (3.24) and (3.25), find the exact formula for $k^*(t) = k_{tr}(t)$ on the transition interval $[0, \theta_1]$ in the case $k_0 < k^*(0)$. Use the hint from previous exercise.

7. Snow that the one-sector model with physical and human capital (3.29)–(3.34) has a balanced growth path.

   HINT: Substitute the formulas (3.35) into equations (3.30)–(3.33) and obtain formulas (3.36).
8. In the one-sector model (3.29)–(3.34) with physical and human capital, derive the equalities (3.35) and (3.36) from the static optimality conditions for interior solutions $\partial\Phi/\partial g = 0, \partial\Phi/\partial\overline{K} = 0, \partial\Phi/\partial\overline{L} = 0$.
9. Snow that the Uzawa–Lucas model (3.42)–(3.46) has a balanced growth path such that the variables $Q(t)$, $C(t)$, $I(t)$, $K(t)$, and $L(t)$ grow with the same constant rate $g > 0$, while the variable $u(t)$ is constant.

   HINT: Substitute $Q(t) = \overline{Q}\,\mathrm{e}^{gt}$, $C(t) = \overline{C}\,\mathrm{e}^{gt}$, $I(t) = \overline{I}\,\mathrm{e}^{gt}$, $\overline{K}(t) = \overline{K}\,\mathrm{e}^{gt}$, $L(t) = \overline{L}\,\mathrm{e}^{gt}$, and $u(t) =$ const into equations (3.43)–(3.46).
10. Show that the knowledge accumulation equation (3.50) at constant L and u and $0 < \phi < 1$ produces less than exponential growth of the unknown $A(t)$.

    HINT: Find the solution of the differential equation $A'/A = \text{const}/A^{1-\phi}$ using the separation of variables (see Sect. 1.3.3)

References[1]

1. Arrow, K.: The economic implications of learning by doing. Rev. Econ. Stud. **29**, 155–173 (1962)
2. 📖 Barro, R.J., Sala-i-Martin, X.: Economic Growth. The MIT Press, Cambridge, MA (2003)
3. Bhargava, S.C.: Generalized Lotka-Volterra equations and the mechanism of technological substitution. Technol. Forecast. Soc. Change **35**, 319–326 (1989)
4. 📖 Black, J., Bradley, J.F.: Essential Mathematics for Economists, 2nd edn. John Wiley and Sons, New York (1980)
5. Bruckner, E., Ebeling, W., Jimenez Montano, M.A., Scharnhorst, A.: Nonlinear stochastic effects of substitution—an evolutionary approach. J. Evol. Econ. **6**, 1–30 (1996)
6. 📖 Caputo, M.R.: Foundations of Dynamic Economic Analysis: Optimal Control Theory and Applications. Cambridge University Press, New York (2005)
7. Fisher, J.C., Pry, R.H.: A simple substitution model of technological change. Technol. Forecast. Soc. Change **3**, 75–88 (1971)
8. Jones, C.I.: R&D-based models of economic growth. J. Polit. Econ. **103**, 759–784 (1995)
9. 📖 Lambert, P.J.: Advanced Mathematics for Economists, Static and Dynamic Optimization. Blackwell Publishing, New York (1985)
10. 📖 Leonard, D., Van Long, N.: Optimal Control Theory and Static Optimization in Economics. Cambridge University Press, New York (1992)
11. Shell, K.: Optimal programs of capital accumulation for an economy in which there is exogenous technical change. In: Shell, K. (ed.) Essays on Theory of Optimal Economic Growth, Chapter I, pp. 1–30. MIT Press, Cambridge (1967)
12. Solow, R.: A contribution to the theory of economic growth. Quart. J. Econ. **70**, 65–94 (1956)

[1] The book symbol 📖 means that the reference is a textbook recommended for further student reading.

Chapter 4
Models with Heterogeneous Capital

This chapter explores economic growth models with heterogeneous capital and labor described by the integral or partial differential equations. Such models are imperative in explaining economic development under embodied technological change. Section 4.1 describes the well-known macroeconomic growth models with vintage capital of R. Solow and L. Johansen and analyzes links among them. Optimization vintage capital models at a firm level are portrayed in Sect. 4.2. Section 4.3 considers models with investment into different vintages of capital. The last section discusses two fundamental replacement problems of Operations Research: the serial replacement of a single machine and the parallel replacement of several machines.

There are many reasons to distinguish heterogeneous production factors in economic models, and technological change of Chap. 3 is just one of them. In economic practice, the capital and labor are never homogeneous. Capital assets (productive capacities, equipment, machines) have different productivity, price, operating costs, and other relevant factors. The labor force also naturally varies with respect to the experience, education, and compensation. In economics, the models with heterogeneous capital and/or labor are known as the *vintage capital models* or *models with vintages*.

Vintage models with heterogeneous assets can be presented in discrete or continuous time. Discrete-time models are commonly used in Operations Research because they better suit available data and real decision making processes. The *Operations Research* (OR) is a discipline that develops advanced analytical methods to help make better decisions in complex management and engineering problems. Continuous-time models have certain mathematical advantages and are frequently employed in theoretical economic research. Mathematically, the continuous-time vintage models use integral or partial differential equations.

N. Hritonenko and Y. Yatsenko, *Mathematical Modeling in Economics, Ecology and the Environment*, Springer Optimization and Its Applications 88,
DOI 10.1007/978-1-4614-9311-2_4,

4.1 Macroeconomic Vintage Capital Models

First vintage models were developed for macroeconomic growth by R. Solow [14, 15], L. Johansen [7], and other economists in the 1960s. These models extend the concept of production functions from Chaps. 2 and 3 to the case of heterogeneous production inputs. The most known of them are the *Solow vintage models of 1960 and 1966*.

4.1.1 Solow Vintage Capital Model

According to the Solow vintage model, the technological change is embedded in the physical capital. All capital units installed at the same time belong to the same *vintage* and have identical efficiency. Because of the technological change, the vintages created recently are more effective than the vintages created at earlier times.

For better understanding, we start with a *discrete version of the Solow model*. Let us introduce the discrete time $t = 1, 2, 3, \ldots$, and the following functions:

$i(t)$—the amount of installed capital of vintage t (capital investment at time t),
$v(\tau, t)$—the amount of capital of vintage τ, that was installed at instant $\tau \le t$ and is still in operation at current time t,
$l(\tau, t)$—the amount of labor assigned to the capital $v(\tau, t)$.

The output $q(\tau, t)$ produced at time t by the capital $v(\tau, t)$ created at time τ can be described as

$$q(\tau, t) = F(\tau, v, l) = F(\tau, v(\tau, t), l(\tau, t)), \tag{4.1}$$

where $F(\tau, v, l)$ is a production function for the capital v and labor l of vintage τ. The dependence of the vintage production function $F(\tau, v, l)$ on the time τ of vintage installation is crucial and represents the embodied technological change. The function F increases in τ because the newer capital is more productive. The *Cobb–Douglas–Tinbergen* production function *with exponential technological change* (Sect. 3.4) is chosen in the Solow model [14]:

$$F(\tau, v, l) = e^{g\tau} v^{1-\alpha} l^{\alpha}, \quad 0 < \alpha < 1, \tag{4.2}$$

where $g > 0$ is *the rate of embodied technological change*.

The total production output $Q(t)$ at time t is the aggregation of the outputs from all operating vintages from the infinitum up to the current time t:

$$Q(t) = \sum_{\tau=-\infty}^{t} q(\tau, t) = \sum_{\tau=-\infty}^{t} F[\tau, v(\tau, t), l(\tau, t)] \tag{4.3}$$

and the total required labor is

$$L(t) = \sum_{\tau=-\infty}^{t} l(\tau, t). \tag{4.4}$$

Equalities (4.1)–(4.4) constitute the *discrete-time Solow vintage model.*

Let us now assume that the time t is continuous. Then, instead of (4.3), the total production output $Q(t)$ is described by the integral

$$Q(t) = \int_{-\infty}^{t} F[\tau, v(\tau, t), l(\tau, t)] \mathrm{d}\tau, \tag{4.5}$$

and the total labor $L(t)$ is

$$L(t) = \int_{-\infty}^{t} l(\tau, t) \mathrm{d}\tau. \tag{4.6}$$

It is supposed that the investment $I(t)$ goes into the newest capital vintage $v(t, t)$ and the older vintages are gradually removed from the operation due to their *physical* depreciation with a constant depreciation *rate* $\mu > 0$:

$$v(\tau, t) = I(t)\mathrm{e}^{-\mu(t-\tau)}. \tag{4.7}$$

The equalities (4.1), (4.2), (4.5)–(4.7) describe the production part of the *Solow* [14] *vintage capital model.*

In this model, the obsolescence of capital takes place through its depreciation. By (4.7), the capital is left to depreciate forever, with exponentially declining portions of capital and labor existing at any distant past timeτ. The model (4.1)–(4.7) assumes that the capital and labor can be substituted at any time. Such models are known in economics as the *putty–putty* models (their counterparts, *putty–clay* and *clay–clay* models will be discussed in Sect. 4.1.2).

Because of the constant depreciation and putty–putty assumption, the Solow vintage model can be aggregated to the standard Solow growth model of Chap. 3. Namely, the steady-state analysis of the vintage model (4.2), (4.5)–(4.7) with constant depreciation rate μ and constant technological change rate g leads to the same outcomes as for the aggregate (not integral) Solow model (3.7) with exponential autonomous technological change. R. Solow demonstrated six years later [15] that vintage models possess principally new features not inherent in other economic models in the case of nonuniform economic development. In particular, they allow us to control *the scrapping of obsolete capital vintages.*

4.1.2 *Vintage Models with Scrapping of Obsolete Capital*

The endogenous scrapping of capital means that old capital vintages are no longer removed from operations because of their physical depreciation but as a result of directed managerial decisions. First vintage models with endogenous capital lifetime were developed for macroeconomic growth [7, 15]. They can be represented as the modification of the Solow vintage model (4.5)–(4.7):

$$Q(t) = \int_{a(t)}^{t} F[\tau, v(\tau), l(\tau)]\mathrm{d}\tau, \quad L(t) = \int_{a(t)}^{t} l(\tau)\mathrm{d}\tau. \tag{4.8}$$

The principal novelty of the model (4.8) is in the new endogenous variable $a(t)$ that represents the installation time of the capital vintage to be scrapped at current time t. Then, the difference $T(t) = t - a(t)$ is the *lifetime* (*useful life*) of the oldest operating capital vintage (created at time $a(t)$). The integral in (4.8) means that only capital vintages introduced between $a(t)$ and t are used in production at time t.

In (4.8), the functions Q, F, v, L, and l have the same meaning as in the Solow [14] model. The investment I still goes into the newest vintages following (4.7). For simplicity, the capital depreciation rate is taken $\mu = 0$ in this model, therefore, $v(\tau) = I(\tau)$ by (4.7) and depends on the capital installation time τ only. The physical depreciation of capital is not essential in the model (4.8), because the obsolescence of capital takes place through its finite lifetime. In contrast to the Solow model (4.5)–(4.7) where capital vintages gradually vanish with the rate μ, the vintages disappear in (4.8) only by means of their endogenous scrapping and replacement with the newest vintages.

4.1.2.1 Solow [15] Vintage Capital Model

The next Solow vintage model of [15] assumes fixed proportions between capital and labor, so the capital and labor are not substitutable at any time. Such vintage models are known as *clay–clay* models. The vintage production function in (4.8) is taken as the Leontief production function:

$$F(\tau, I(\tau), l(\tau)) = \alpha(\tau)I(\tau) = \beta(\tau)l(\tau), \tag{4.9}$$

where the given functions α and β are the *labor–output* and *capital–output* coefficients respectively. The increase of functions $\alpha(\tau)$ and $\beta(\tau)$ reflects capital-augmenting and labor-augmenting embodied technological change. By (4.9), the capital $I(\tau)$ and labor $l(\tau)$ of vintage τ are connected *with fixed proportions* as $I(\tau) = l(\tau)\beta(\tau)/\alpha(\tau)$. Then, the model (4.8) becomes

$$Q(t) = \int_{a(t)}^{t} \alpha(\tau)I(\tau)\mathrm{d}\tau, \quad L(t) = \int_{a(t)}^{t} \frac{\alpha(\tau)}{\beta(\tau)}I(\tau)\mathrm{d}\tau. \tag{4.10}$$

Similarly to the Solow–Swan growth model with homogeneous capital of Sect. 3.2, the product output $Q(t)$ is distributed between investment and consumption: $Q(t) = I(t) + C(t)$ and the model assumes the constant *accumulation norm* (*saving ratio*) s between the investment $I(t)$ and consumption $C(t)$:

$$I(t) = sQ(t), \quad C(t) = (1 - s)Q(t), \tag{4.11}$$

The paper [15] analyzes a balanced growth with a constant endogenous lifetime of capital $t - a(t) = T$ in model (4.10) and (4.11) under the exponential labor-augmenting technological change $\beta(\tau) = \mathrm{e}^{g\tau}$, $\alpha(\tau) = 1$. An optimization version of this model with the variable unknown saving rate $s(t)$ is considered in Sect. 5.3.

4.1.2.2 Johansen Vintage Capital Model with Neoclassical Production Function

The two-factor production function $F(\tau, I, l)$ in the vintage model (4.8) means that the capital and labor can be substituted at timeτ. However, the vintage structure of the capital should be fixed for the investments already made. Such vintage models known as the *putty–clay models* were first introduced by L. Johansen [7]. In other words, putty–clay models assume different flexibility in substituting production factors before (*ex ante*) and after (*ex post*) capital is installed.

As in standard two-factor production functions of Sect. 2.1.4, we can represent the neoclassical vintage production function $F(\tau, I, l)$ in (4.8) in the *per capita variables* (Sect. 2.1.4) as

$$F(\tau, I(\tau), l(\tau)) = F(\tau, I(\tau)/l(\tau), 1)l(\tau) = \beta(\tau, k(\tau))l(\tau), \tag{4.12}$$

where

$k(\tau) = I(\tau)/l(\tau)$ is the *capital–labor ratio*,
$\beta(\tau, k(\tau)) = F(\tau, k(\tau), 1)$ is the *output per worker (or productivity)* of the vintage τ.

Then, the vintage model (4.8) and (4.9) can be written *in the intensive form* as

$$Q(t) = \int_{a(t)}^{t} \beta(\tau, k(\tau))l(\tau)\mathrm{d}\tau, \quad L(t) = \int_{a(t)}^{t} l(\tau)\mathrm{d}\tau. \tag{4.13}$$

As in the clay–clay Solow model (4.10), the output equation (4.13) is linear with respect to the endogenous function l. However, it depends on another important endogenous variable k. The intensive form (4.13) of the putty–clay vintage model is more convenient for analysis.

Let the planning horizon start at $t = 0$. The *putty–clay* assumption reflects that the model (4.13) involves an initial condition on a certain prehistory $[a(0), 0]$, where the capital structure $k(\tau)$ and $l(\tau)$ is fixed. Alternatively, the vintage model (4.8) and (4.9) can be written in the terms of the investment I as:

$$Q(t) = \int_{a(t)}^{t} I(\tau)\mathrm{d}\tau, \quad L(t) = \int_{a(t)}^{t} \frac{I(\tau)}{\beta(\tau, k(\tau))}\mathrm{d}\tau. \tag{4.14}$$

Since 1960s, numerous extensions of the first vintage models (4.10) and (4.13) have been suggested and analyzed. The modifications include economic optimization (profit maximization or expense minimization), additional balances (energy, environment contamination, etc.), learning, nonlinear adjustment costs, microeconomic equilibrium, and endogenous nature of technological change [1–4, 6, 9].

4.1.3 Two-Sector Vintage Model

The embodied technological change is investment-specific and is implemented through the production of new more efficient capital vintages. This process can be described by two-sector macroeconomic vintage models that generalize the Solow–Swan model (3.7). The first sector produces consumption goods, and the second one produces new capital goods. The capital can be used in the production of consumption goods or in the production of new capital.

A flexible two-sector vintage model of macroeconomic growth is represented by the following modification of the Solow vintage model (4.10)

$$I(t) = \int_{a(t)}^{t} \alpha(\tau, t)s(\tau, t)l(\tau)\mathrm{d}\tau, \tag{4.15}$$

$$C(t) = \int_{a(t)}^{t} \beta(\tau, t)[1 - s(\tau, t)]l(\tau)\mathrm{d}\tau, \tag{4.16}$$

$$L(t) = \int_{a(t)}^{t} l(\tau)\mathrm{d}\tau. \tag{4.17}$$

where

$I(t)$ is the output of capital goods per time unit,
$C(t)$ is the output of consumption goods,
$\alpha(\tau, t)$ and $\beta(\tau, t)$ are the productivities (output/labor coefficients) of vintage τ at time t in two sectors respectively,
$s(\tau, t)$ is a variable accumulation norm.

As the Solow vintage model (4.10), the model (4.15)–(4.17) assumes the *Leontief technology with fixed proportions* when labor and capital are not freely substitutable. The model is closed, i.e., new capital vintages cannot enter the economy from outside but are produced inside the economy. Because of the embodied technological change, newer vintages are more effective than the older ones, i.e., the functions $\alpha(\tau, t)$ and $\beta(\tau, t)$ increase in τ. The dependence of productivities $\alpha(\tau, t)$ and $\beta(\tau, t)$ on the current time t reflect learning-by-doing and disembodied technical progress that affects all vintages simultaneously.

Compared to the one-sector vintage models, the two-sector vintage model possesses the new control function $s(\tau, t)$. Different forms of this function have important economic interpretation:

- In the case of *freely substitutable machines between sectors*, the function $s = s(t)$ depends on current time t only and controls the distribution of all (new and existing) vintages between production sectors. The function $s(t)$ is referred to as the *variable accumulation norm*.
- In the case when *machines are not substitutable between sectors*, the function $s = s(\tau)$ depends on the vintage installation time τ and controls the distribution of only new vintages τ between production sectors. Then, the new capital is assigned to the production sector at the time of its creation and the distribution of older vintages is fixed after the investment has been made: the vintages cannot move from one sector to another during their lifetime $[a(t), t]$. This assumption is equivalent to the *investment irreversibility* in high-tech industries, where capital is specific to the industry. The function $s(\tau)$ is referred to as the *variable assignment coefficient*.
- The case of the *constant accumulation norm* $s \equiv \text{const}$ is equivalent to the clay–clay Solow vintage model (4.10) and (4.11) at $\alpha k \equiv \beta$, where k is the capital–labor ratio. Then, using $I(t) = k(t)m(t)$ and adding (4.15) and (4.16) gives $Q(t) = I(t) + C(t) = \int_{a(t)}^{t} \beta(\tau,t) l(\tau) d\tau$, which is the equality (4.10) in terms of l.
- The special case $s \equiv 0$ leads to the one-sector vintage model (4.10).

The vintage models (4.10), (4.13), and (4.15)–(4.17) turn out to be a foundation for many problems discussed later on. From a mathematical point of view, such models involve *nonlinear integral equations with controlled delay*. The vintage models have been also formulated in partial differential equations and discrete settings.

4.1.4 Optimization Problems in Vintage Models

Following the logic of economic growth theory, the next modeling step is the optimization in the central planner or general equilibrium framework (Chap. 2). Optimization problems in the vintage capital models use objective functions similar to the neoclassic economic models of Chaps. 2 and 3.

Commonly analyzed central planner problems in the vintage model (4.15)–(4.17) are to find the unknown functions I, a, and s that:

- maximize the present value of the consumption per capita c over the infinite planning horizon $[0, \infty)$:

$$\max \int_0^\infty e^{-rt} c(t) dt \tag{4.18}$$

- maximize the present value of the consumer utility $u(c)$ over $[0, \infty)$:

$$\max_s \int_0^\infty e^{-rt} u(c(t)) dt \tag{4.19}$$

subject to the model constraints (4.15)–(4.17) and certain initial conditions. In (4.18) and (4.19), $r > 0$ is a constant discount rate.

A smooth nonlinear *utility function* $u(c)$, $u'(c) > 0$, $u''(c) < 0$, in (4.19) describes the value of the consumption c for consumers; $u(c)$ is concave downward because c is more valuable when it is small. The common choices of the utility function are the linear utility $u(c) = c$, power utility $u(c) = c^{1-\gamma}$, or logarithmic utility $u(c) = \ln c$ (see Sect. 2.3.3).

The unknown variables I, a, and s should satisfy the following inequality-constraints

$$I(t) \geq 0, \quad c(t) \geq 0, \quad 0 \leq s(t) \leq 1, \quad a(t) < t, \quad a'(t) \geq 0. \tag{4.20}$$

The last inequality (4.20) is the *irreversibility of capital scrapping*: once vintages have been scrapped, they cannot be used again. A complete setup of optimal control problems for one- and two-sector vintage models is provided in Chap. 5.

Analogously to Chap. 3, optimization versions (4.18) and (4.19) of the two-sector vintage model (4.15)–(4.17) are often referred to as the *Ramsey vintage capital model*. The two-sector vintage model describes the simultaneous optimization of capital modernization strategies and the distribution of capital between production sectors. The strategy of capital modernization is determined by the capital lifetime $t - a(t)$ and by the investment $I(t)$ in new vintages. The optimization problem (4.15)–(4.17), (4.19), (4.20) is explored in Sect. 5.3.

The two-sector model (4.15)–(4.20) can be extended to three-, four-, and multi-sector vintage models, which describe simultaneous processes of capital modernization and the distribution of operating vintages among several industries (sectors). A multi-sector integral vintage model [6] describes an economy that uses n types of capital, produces l various kinds of consumption goods c_k, and consumes p different resources R_s (labor, capital, energy, raw materials, etc.). The structure of integral operators is the same as in the above one- or two-sector models.

4.2 Vintage Capital Models of a Firm

The vintage models with embodied technological change were initially applied to large-scale macroeconomic systems. However, capital replacement processes are similar for all levels of economic management.

4.2.1 *Malcomson Model*

The first vintage model for the optimal replacement of capital equipment in a separate firm under technological change is proposed by J. M. Malcomson in 1975 [11]. It is described essentially by the same equations (4.13) as the one-sector macroeconomic vintage model but with slightly different interpretation.

Let us consider a firm that makes a single product, invests into new machines and scraps the oldest machines. We introduce the following dynamic characteristics:

$Q(t)$—the amount of product produced and sold at time t,
$\pi(t, Q(t))$—the price of product Q sold at time t,
$\beta(\tau, t)$—the output of one machine of vintage τ at time t,
$o(\tau, t)$—the operating cost of one machine of vintage τ at time t,
$p(t)$—the price and installation cost of one new machine of vintage t,
$m(t)$—the number of new machines installed at time t.

The efficiency $\beta(\tau, t)$ of vintage τ increases and the operating cost $o(\tau, t)$ decreases in τ. Indeed, new vintages are more productive and usually require less maintenance because of embodied technological change. The output $\beta(\tau, t)$ and the operating cost $o(\tau, t)$ can change in time t depending on vintage deterioration, disembodied technological change, and learning. The dependence of the product price $\pi(t, Q)$ on Q reflects the supply–demand theory and describes the *market power* of the firm: the product price can go down, when the output Q is larger. Then, the total product output is

$$Q(t) = \int_{a(t)}^{t} \beta(\tau, t)m(\tau)\mathrm{d}\tau, \tag{4.21}$$

and the total operating cost of all machines is

$$O(t) = \int_{a(t)}^{t} o(\tau, t)m(\tau)\mathrm{d}\tau. \tag{4.22}$$

4.2.1.1 Profit Maximization Versus Expense Minimization

Common optimization problems in the vintage model (4.21) and (4.22) involve the profit maximization or expense minimization:

- A profit-maximizing firm aims to maximize the present discounted value of its net profit over the finite or infinite horizon $[0, T)$, $T \leq \infty$:

$$\int_0^T e^{-rt}[\pi(t, Q(t))Q(t) - O(t) - p(t)m(t)]dt, \tag{4.23}$$

where the constant $r > 0$ is the industry-wide discount rate.

- An expense-minimizing firm minimizes the present discounted value of its total expenses over $[0, T)$:

$$\int_0^T e^{-rt}[p(t)m(t) + O(t)]dt. \tag{4.24}$$

under a given output Q.

In both optimization problems (4.23) and (4.24), the unknown variables are m and a, which should satisfy the equality-constraints (4.21) and (4.22), the inequality-constraints $m(t) \geq 0$, $a(t) < t$, and the scrapping irreversibility constraint $a'(t) \geq 0$ (once machines have been scrapped, they cannot be used again).

4.2.1.2 Putty–Clay and Clay–Clay Vintage Models

The Malcomson model (4.21) and (4.22) is a *clay–clay* vintage model, which does not allow any substitutability between labor and capital. It can be extended to a *putty–clay* form by adding a new endogenous variable p:

$$Q(t) = \int_{a(t)}^{t} \beta(p(\tau), \tau, t)m(\tau)d\tau, \tag{4.25}$$

$$O(t) = \int_{a(t)}^{t} o(p(\tau), \tau, t)m(\tau)d\tau. \tag{4.26}$$

Here, the endogenous price $p(\tau)$ of new machines is chosen at the time τ of machine purchase and positively affects the machine productivity $\beta(\tau, t)$ and/or decreases the machine operating cost $o(\tau, t)$.

The model (4.25) and (4.26) assumes that a continuous set of new machines with productivities that depend on the price p is available on market for the firm to buy. Correspondingly, the price p becomes an additional unknown variable in the optimization problem (4.23) or (4.24). By analogy with the intensive-form production function (2.18), the integrand of (4.25) becomes a two-factor production function for vintage τ because of its dependence on endogenous m and p.

In the model (4.25) and (4.26), firms are free to choose factor proportions *ex ante* but cannot change them *ex post* after the investment has been made and the machines have been installed. The model assumes that the specific vintage capital structure (p, m) is fixed over $[a(\tau), \tau]$ after the investment $m(\tau)$ has been made. In particular, the structure of the capital is fixed for the investments already made on the pre-history $[a(0), 0]$. Such two-factor putty–clay vintage models are more flexible than the clay–clay vintage models.

4.2.1.3 Additional Balances for Energy and Environment Contamination

Traditional growth models of a firm do not consider any resource limitations, while the macroeconomic growth models typically assume that a key resource (labor, land, energy, environmental quality) is given. The classic economic theory limits the growth of a firm by the supply–demand relationship only.

However, in economic practice, the firms are often subjected to various external physical conditions and government regulations, which set up resource restrictions on the firm's growth. Some relevant examples of such resource constraints are:

- A firm cannot extend the size by buying or renting neighboring land. In this situation, the resource is the *land* (*space*).
- In the case of continuously dry markets, for instance labor markets, the firm has to cope with the *restricted labor resource*.
- A firm operates on a network (cellular phone, railway, gas, water companies, etc.) but does not own it and cannot increase the amount of network service. In this case, the resource is the *network capacity*.
- The newest example is forthcoming *quotas on* CO_2 *emissions*. Setting such quotas is already a common practice in European countries and will soon become the case in USA.

In general, any restriction put in place by the government would be an example of such constraints. So firms often face resource constraints imposed from outside. Mathematically, such constraints can be described in the form of the aggregate balance relation

$$\int_{a(t)}^{t} r(\tau, t)m(\tau)\mathrm{d}\tau \leq R(t) \tag{4.27}$$

with a given resource quota $R(t)$ and corresponding interpretation of new functions R and r.

4.2.1.4 Vintage Models and Creative Destruction Theory

Vintage capital models have resemblance to the Schumpeterian *creative destruction theory* that considers entrepreneurial innovations as the driving force of technological change (see Sect. 3.4.1). Few Schumpeterian models explicitly address the embodied technological change and capital obsolescence. Vintage capital models of a firm under exogenous technological change isolate the source of innovations from the firm's investment/replacement process. They focus on endogenous scrapping of the capital that becomes obsolete because of the smoothened embodied technological change resulting from innovations on the industry level. The market availability of new vintages in the vintage models plays the role of innovations in the Schumpeterian models.

4.2.2 Aggregate Production Functions

The distribution approach to the *aggregation of production functions* has been developed for an industry level by L. Johansen in 1972. It also leads to integral models that are at first sight different from the vintage capital models of Sects. 4.1 and 4.2.1.

Let us assume that an industry consists of a continuum of heterogeneous production units (firms) that differ by their labor input coefficient (*specific labor expenditure*) ξ, where $\hat{\xi} \leq \xi < \infty$. The last inequality means that the industry has the *industry-wide best* (smallest) value $\hat{\xi}$ of ξ and the labor expenditure ξ of any firm cannot be smaller than this value $\hat{\xi}$. The continuous *capacity distribution* $m(\xi)$ describes the known distribution of capacity output in the terms of the variable labor expenditure ξ over the interval $[,\hat{\xi},\infty)$. Then, the *aggregate production function* of the industry can be described as

$$Q = \int_{\hat{\xi}}^{\phi} m(\xi)\mathrm{d}\xi, \quad L = \int_{\hat{\xi}}^{\phi} \xi m(\xi)\mathrm{d}\xi, \tag{4.28}$$

The production function (4.28) determines the dependence of the aggregate output Q of the industry on the total labor L employed in the industry through the given capacity distribution $m(\xi)$ and the unknown integration limit ϕ. The model (4.28) assumes suggesting an efficient economic behavior, namely, that the operating capacity is filled with production units with the best available technology ξ, starting with $\hat{\xi}$ and up to an unknown threshold valueϕ, $\hat{\xi} \leq \phi < \infty$. The full capacity of the industry is $J = \int_{\hat{\xi}}^{\infty} m(\xi)\mathrm{d}\xi$.

The model (4.28) with one variable input (labor) can be extended to the case of several variable inputs. If heterogeneous production units are characterized by different input–output coefficients for several independent factors (rather than only labor), then the models involve multiple integrals with several unknown limits of integration [8].

4.2.2.1 Dynamic Aggregate Production Function

A relevant extension is the dynamic version of the model (4.28), where the capacity distribution $m(t,\xi)$ and the parameters $Q(t)$, $L(t)$, $\hat{\xi}(t)$, and $\phi(t)$ depend on the current time t. Then, the *dynamic aggregate production function* of a developing economy can be presented as

$$Q(t) = \int_{\hat{\xi}(t)}^{\phi(t)} m(t,\xi)\mathrm{d}\xi, \quad L(t) = \int_{\hat{\xi}(t)}^{\phi(t)} \xi m(t,\xi)\mathrm{d}\xi. \tag{4.29}$$

In particular, the dynamic model (4.29) can describe the embodied technological change by assuming that the industry-wide best labor input coefficient $\hat{\xi}(t)$ monotonically decreases in time t.

4.2.2.2 Relations Between Aggregate Production Functions and Vintage Models

In model (4.29), the distribution $m(t,\xi)$ of operating production units depends on the variable labor input ξ rather than on the unit installation time τ as in the previous vintage models (4.8)–(4.26). So the integrals are evaluated for the independent variable ξ and the upper integration limit $\phi(.)$ is unknown. In the presence of embodied technological change, new vintages installed at time t use the best available technology with the smallest $\hat{\xi}(t)$. Then, the substitution $\xi \rightarrow \tau$ of the variables ξ and τ transforms the model (4.29) into the vintage model (4.10) [6].

4.3 Vintage Models with Distributed Investments

In all vintage models described above, only the latest vintage can be installed at any point of time and all investments flow into capital of the latest vintage, where the efficiency of investment is the highest, and no investment goes into any earlier vintage. However, in reality, firms invest in new more efficient capital as well as investing in older capital. Old structures are renovated, old machines are repaired, and old workers are retrained. In this section, we consider vintage models [10] that allow for investment into different vintages of capital.

Let us consider a firm that invests into the new capital of vintage t and into the old capital of various vintages $\tau < t$ (of the age $t - \tau$) and introduce the following variables:

$x(t)$—the investment in new capital at time t,
$i(\tau, t)$—the investment in the old capital vintages $\tau < t$, $\tau \in (-\infty, t]$, $t \in [0, \infty)$.
$v(\tau, t)$—the amount of active capital of vintage τ at time t,
$Q(t)$—the product output at time t.

If the planning horizon starts at $t = 0$, then, all the investments made before the time $t = 0$ are fixed and we can control only investments for $t > 0$. At time $t = 0$, the age distribution of capital over past vintages $\tau < 0$ is known:

$$v(\tau, 0) = v_0(\tau), \quad \tau \in (-\infty, 0]. \tag{4.30}$$

Let the capital deteriorate with a given constant deterioration rate $\mu > 0$. Then, the amount of active capital is a result of the previous investments and can be expressed as

$$v(\tau, t) = e^{-\mu(t-\tau)} x(\tau) + \int_{\tau}^{t} e^{-\mu(t-s)} i(\tau, s) ds \tag{4.31}$$

for the vintages $\tau \geq 0$ and

$$v(\tau, t) = e^{-\mu(t-\tau)} v_0(\tau) + \int_{0}^{t} e^{-\mu(t-s)} i(\tau, s) ds \tag{4.32}$$

for the older vintages $-\infty < \tau < 0$ installed before the initial time $t = 0$. Formulas (4.31) and (4.32) have a clear and intuitive economic interpretation. In particular, (4.31) means that the investment into a vintage $\tau > 0$ was made at time τ as a new vintage and at later times s, $\tau < s < t$, as an old vintage. Formula (4.32) emphasizes that each vintage $\tau < 0$ was already old at the initial time $t = 0$ with the given capital amount $v_0(\tau)$ and all investments into this vintage were made at all times $0 < s < t$ as into an old vintage.

Finally, the product output of the firm is

$$Q(t) = \int_{-\infty}^{t} \beta(\tau) A(t - \tau) v(\tau, t) d\tau, \quad t \in [0, \infty), \tag{4.33}$$

where

$\beta(\tau)$ is the unit efficiency of the capital of vintage τ,
$A(t-\tau)$ is the *age-dependent learning curve* for the capital of vintage τ.

The function $A(t-\tau)$ in (4.33) describes learning-by-doing. The *learning-by-doing* implies that the efficiency depends on the age $t-\tau$ of the vintage. Empiric evidence at the plant level demonstrates that it takes some time for a new vintage (equipment, plant, or machine) to operate at peak efficiency.

The function $\beta(\tau)$ describes the embedded technological change, for example, as $\beta(\tau) = \exp(g\tau)$. The final efficiency can also depend on the current time t, which reflects the disembodied technological progress that affects all vintages simultaneously.

The model (4.31)–(4.33) is a step ahead compared to the vintage models of previous sections. It allows analyzing the optimal distribution of operating vintages with respect to their age, which occurs because of the joint effect of technical change, deterioration, and learning. Possible modifications of the model (4.31)–(4.33) include adding disinvestment (scrapping or selling old vintages), salvage value of the scrapped vintages, and nonlinear adjustment cost. Similar age-structured models are discussed in Sect. 6.3 for biological populations.

4.3.1 Optimization Problems

Let us discuss optimization problems in the vintage capital model (4.31)–(4.33) with distributed investments.

4.3.1.1 Utility Maximization

A common optimization problem in the central planner setup is to maximize the present value of the consumer's utility over $[0, \infty)$:

$$\max_{i,x} \int_0^\infty e^{-rt} u(C(t)) dt, \tag{4.34}$$

with a constant discount rate $r > 0$, where the consumption $C(t)$ is determined by the total investment $I(t)$ as

$$C(t) = Q(t) - I(t), \quad I(t) = x(t) + \int_{-\infty}^{t} i(v, t) dv. \tag{4.35}$$

The decision variables (independent controls) in the optimization problem (4.31)–(4.35) are the investments into the new capital $x(t)$ and into old capital $i(\tau, t)$ that satisfy the non-negativity constraints: $x(t) \geq 0$, $i(\tau, t) \geq 0$ for $t \in [0, \infty)$, $\tau \in (-\infty, t]$. The unknown capital stock $v(\tau, t)$ and product output $Q(t)$ are determined from equations (4.31)–(4.33).

4.3.1.2 Profit Maximization

At the firm level, the common objective is to maximize the discounted value of its net profit over the finite or infinite horizon $[0, T)$, $T \leq \infty$:

$$\max_{i,x} \int_0^T \mathrm{e}^{-rt} \left(\pi(t, Q(t))Q(t) - p(t,t)x(t) - \int_0^t p(\tau,t)i(\tau,t)\mathrm{d}\tau \right) \mathrm{d}t, \tag{4.36}$$

where

$\pi(t, Q)$ is the price of product Q sold at time t,
$p(t, t)$ is the purchasing and installation cost of one new machine of vintage t,
$p(\tau, t)$ is the installation cost of one old machine of vintage $\tau < t$.

4.3.1.3 Expenses Minimization

Another common objective at the firm level is to minimize total expenses to support operations at a desirable level. In such problems, the *total operating capacity* (the number of operating machines) is given as

$$N(t) = \int_{-\infty}^t v(\tau,t)\mathrm{d}\tau \tag{4.37}$$

and the problem is to minimize the present value of the future total expenses to keep the required number (4.37) of operating machines over the horizon $[0, T)$:

$$\min = \int_0^T \left(p(t,t)x(t) + \int_{-\infty}^t \left[p(\tau,t)i(\tau,t) + o(\tau,t)v(\tau,t) \right] \mathrm{d}\tau \right) \mathrm{d}t, \tag{4.38}$$

where

$p(\tau, t)$ is the installation cost of one machine of vintage $\tau \leq t$ at time t,
$o(\tau, t)$ is the specific operating and maintenance cost of one machine.

Maintaining a proper age distribution of capital is important in such problems because both the machine cost $p(\tau, t)$ and maintenance expenditure $o(\tau, t)$ usually depend on the machine age $t - \tau$.

4.3.2 Relations to Differential Models of Equipment Replacement

The evolution of heterogeneous vintage capital can also be described by means of partial differential equations. In particular, the integral vintage model (4.31) and (4.32) with distributed investments is equivalent to the following linear *partial differential equation* (PDE)

$$\frac{\partial v(t,a)}{\partial t} + \frac{\partial v(t,a)}{\partial a} = -\mu(t,a)v(t,a) + i(t,a), \tag{4.39}$$

with the boundary conditions

$$v(t,0) = x(t), \quad t \in [0,\infty), \quad v(0,a) = v_0(-a), \quad a \in [0,\infty). \tag{4.40}$$

Here, the economic meaning of all functions is the same as in the model (4.31)–(4.33) but the independent variable is the capital age $a = t - \tau$ (instead of the vintage τ).

The linear PDE (4.39) is known as the *evolutionary equation* in mathematical population ecology. Such PDE-based models have been intensively used for modeling of age-structured ecological populations. The structure and properties of the equation (4.39) are analyzed in Sect. 6.4 and Chap. 7.

4.3.2.1 Equivalence of Integral and PDE-Based Vintage Models

The integral vintage capital model (4.31) and (4.32) with distributed investments can be formally derived from the PDE vintage capital model (4.39) and (4.40) in the case of constant deterioration rate μ. Namely, switching the equation (4.39) back to the variables (t,τ), one can show (e.g., [4]) that the linear PDE boundary problem (4.39) and (4.40) is equivalent to the following Volterra integral equations of the second kind:

$$v(\tau,t) = x(\tau) + \int_\tau^t [i(\tau,s) - \mu(s-\tau)v(\tau,s)]\mathrm{d}s \\ \text{at } 0 \le \tau < t, \quad 0 \le t < \infty, \tag{4.41}$$

$$v(\tau,t) = v_0(\tau) + \int_0^t [i(\tau,s) - \mu(s-\tau)v(\tau,s)]\mathrm{d}s \\ \text{at } -\infty < \tau < 0, \quad 0 \le t < \infty. \tag{4.42}$$

This equivalence is easily verified by taking the derivatives of formulas (4.41) and (4.42). Next, solving the linear Volterra integral equations (4.41) and (4.42) in the case of constant μ leads to formulas (4.31) and (4.32).

The differential vintage models of the type (4.39) and (4.40) are effective when the constant lifetime of capital is infinite (or is fixed constant). If the capital lifetime is endogenous and depends on time, then it is more convenient to use the integral vintage models with endogenous scrapping described in Sects. 4.1 and 4.2.

4.4 Discrete and Continuous Models of Machine Replacement

The rational replacement of equipment (assets, machines) in industrial systems belongs to complex decision-making problems, in which the endogenous lifetime of production elements is important. The Operations Research distinguishes two fundamental categories of equipment replacement problems [5, 12, 13, 16]:

- *Serial replacement* of a single machine.
- *Parallel replacement* of several machines that are economically interdependent and operate in parallel.

The *economic interdependence of machines* can be caused by various economic factors:

- The requirement of keeping a prescribed number of assets in service at all times (capacity demand constraints).
- Given output demand constraints.
- Restrictions on capital expenditures (rationing budgeting constraints).
- The presence of fixed replacement costs or nonlinear adjustment costs that occur when one or more assets are replaced.

The first three factors exhibit *constant return-to-scales* with respect to the number of machines. The last assumption introduces *the economy of scale* in purchasing price, which makes the replacement problem more difficult to analyze.

The optimal replacement of machines has been modeled in deterministic and stochastic settings. This section focuses on deterministic models. In OR theory, the equipment replacement processes are usually modeled in discrete time as sequential decision problems. Mathematically, the majority of OR equipment replacement models are *discrete (or integer) programming problems*. The numeric simulation of such problems uses efficient algorithms based on dynamic programming. However, the qualitative analysis of the discrete replacement models is challenging.

4.4.1 Multi-machine Replacement Model in Discrete Time

This section explores a discrete-time model of parallel machine replacement under common assumptions of the OR theory. Let us consider a firm (factory, plant) that must keep P, $P \geq 1$, machines of a particular type at all times. Newer vintages of machines are better and require less maintenance due to technological change. The performance of operating machines (measured by maintenance costs) deteriorates as the machines become older. So the firm should consider selling old machines at a certain point of time and buying new machines. The same situation repeats with the new machines, so the firm shall make a *chain of replacement decisions*.

Let us consider a (finite or infinite) planning horizon $\{1, \ldots, T\}$ in the *discrete time* $j = \ldots, -1, 0, 1, 2, \ldots$ We assume that only one technology (type of machines) is available at time j and that the firm knows the capital and maintenance costs of new machines over the future horizon $\{1, \ldots, T\}$:

$p(j)$ is the purchasing price and installation cost of a machine bought at time j (the machine of vintage j),
$q(j, k)$ is the operating and maintenance cost for the vintage j machine during the time period $k \geq j$, and
$s(j, k)$ is the salvage value of the vintage j machine at the end of period $k \geq j$.

Since the purchased price $p(j)$ involves a certain installation (switching) cost, we assume that $p(j) > q(j, k)$ for all $k \geq j$. Because of deterioration, $q(j, k)$ increases and $s(j, k)$ decreases in the age $a = j-k$ at a fixed time j. Under improving technology, the sequence $q(j, k)$ decreases in j for any fixed machine age $a = t-k$.

A popular OR approach is to solve the optimal replacement problem tracking the replacement chain for each of P machines. The alternative approach is based on choosing the machine lifetime as a control. Let two standard rules of parallel equipment replacement hold [5]:

- *Older Cluster Replacement Rule*: an optimal replacement policy always replaces older machines first, and
- *No-Splitting Rule*: machines of the same age are either kept or replaced at the same time period.

Under these rules, the model can be expressed in the terms of machine lifetimes. Namely, let us introduce the following variables:

L_j—the lifetime (service life) of machines *replaced* in period j,
m_j—the number of new machines purchased during the period j, $1 \leq j \leq T$.

Then, a machine purchased in the period $j-L_j$ will be used during the periods $j - L_j + 1$ through j and be replaced in period j. The requirement of keeping the total number P of operating machines is expressed by the following *demand constraint*:

$$\sum_{k=j-L_j+1}^{j} m_k = P, \quad j = 1, \ldots, T. \tag{4.43}$$

Let the firm be in business for a while and have P machines at $j = 0$. Then, the number $m^0(k)$ of machines purchased in each instant k, $-L_0 \leq k \leq 0$, before the starting time $j = 0$ is known. So the initial condition of the replacement problem at time $j = 0$ is

$$\sum_{k=-L_0+1}^{0} m_k^0 = P. \tag{4.44}$$

Under the *no-splitting rule*, machines of the same age are always replaced during the same period. Thus, the equalities (4.43) and (4.44) completely determine the vector $\{L_k, 1 \le k \le j\}$ under a given $\{m_k, 1 \le k \le j\}$, and vice versa. Subtracting the equality (4.43) at $j - 1$ from itself at j gives the following expression for the number of machines replaced in period j

$$m_j = \sum_{k=j-L_{j-1}}^{j-L_j} m_k. \tag{4.45}$$

If the given L_j are known, then (4.45) determines the amount of new machines:

- $m_j = 0$ if $L_j = L_{j-1} + 1$ (no machine is replaced during the period j),
- $m_j = m_{j-L_j}$ if $L_j = L_{j-1}$ (machines of vintage $j-L_j$ are replaced at j),
- $m_j = m_{j-L_{j-1}} + \ldots + m_{j-L_j}$ if $L_j < L_{j-1}$ (machines of several vintages $j-L_{j-1}$, …, $j-L_j$ are replaced in the same period j).

Conversely, if no machine is replaced at time j: $m_j = 0$, then $L_j = L_{j-1} + 1$ (all functioning machines become one period older).

The discounted total replacement cost over the T-period horizon $[1, \ldots, T]$ can be written as

$$J(T) = \sum_{j=1}^{T} \rho^j p(j) m_j + \sum_{j=1}^{T} \rho^j \sum_{k=j-L_j}^{j-1} q(j,k) m_k - \sum_{j=1}^{T} \rho^j \sum_{k=L_{j-1}}^{L_j} s(j, j-k) m_{j-k}, \tag{4.46}$$

where the first term represents the total price of purchased machines, the second term is the total maintenance cost, and the last term stands for the total salvage value of the replaced machines. The given parameter ρ, $0 < \rho \le 1$, in (4.46) denotes the *discrete-time discount factor* during the elementary time interval. The relation of ρ to the *continuous-time discount rate* $r > 0$ will be shown below in (4.51).

Now, we can formulate the machine replacement problem as the nonlinear *optimization problem*

$$\min_{m_j, L_j, j=1,\ldots,T} J(T) \tag{4.47}$$

with $2T$ discrete unknown variables L_j and m_j, $1 \le j \le T$, subject to the constraints $L_j \in \mathbf{I}$, $m_j \in \mathbf{I}$, $L_j \ge 0$, $m_j \ge 0$, (4.43), and the initial condition (4.44) ($\mathbf{I}$ is the set of integer numbers).

4.4.1.1 Relations Between Discrete Replacement Models and Continuous-Time Vintage Capital Models

The discrete-time equipment replacement models are based on the following assumptions: the time is discrete, the lifetimes of machines are discrete, and the amounts (numbers) of machines are discrete. Relaxing all or some of these assumptions leads to continuous-time optimal control problems, similar to appeared in the vintage capital models of Sects. 4.1–4.3. Continuous-time vintage capital models describe the machine lifetime and amount of assets as continuous variables [16].

4.4.2 One-Machine Replacement in Discrete and Continuous Time

The majority of the machine replacement models under technological change analyze one-machine serial replacement. Let us consider a production shop that keeps indefinitely *one* machine of a particular type. The shop should determine a policy of selling the operating machine and buying a new one under improving technology. The one-machine replacement model can be formally obtained from the multi-machine model (4.43)–(4.47) at $P = 1$, however, the model becomes simpler because the variables m_j are not necessary in the case of one machine.

Let us consider the infinite horizon case $T = \infty$. Then, the *replacement policy* can be completely defined by the sequence $\{L_i,\ i = 1,\ 2,\ \ldots\}$ of the unknown lifetimes L_i of the consecutively replaced asset. The policy is an infinite series $\{L_k\}$, $k = 1, \ldots, \infty$, of finite lifetimes or a finite number of replacements $\{L_k\}$, $k = 1, \ldots, N$, $N \geq 0$, with the last infinite lifetime $L_N = \infty$.

Assuming that the replacement process started at $t = 0$ and the first machine was purchased at the known time $\tau_0 \leq 0$, the sequence $\{L_i, i = 1, 2, \ldots\}$ determines the sequence $\{\tau_i,\ i = 1,\ 2,\ \ldots\}$ of the endogenous replacement times

$$\tau_i = L_i + \tau_{i-1}, \quad i = 1, 2, \ldots \tag{4.48}$$

We will analyze the optimal one-machine replacement using models in continuous and discrete time. The discrete-time model is easier to construct but appears to be more complicated for analytical study.

4.4.2.1 Discrete-Time Serial Replacement Model

First, let us describe the replacement process in the *discrete time* $t = \ldots, -1, 0, 1, 2, \ldots$

The present value of the total cost of the replacement policy over the infinite horizon $[\tau_0, \infty)$ can be expressed as

$$\hat{J}(\tau_1, \tau_2, \ldots) = \sum_{i=1}^{\infty} \rho^{\tau_i} p(\tau_i)$$
$$+ \sum_{i=0}^{\infty} \left[\sum_{k=\tau_i}^{\tau_{i+1}} \rho^k q(\tau_i, \tau_k) - \sigma \rho^{\tau_{i+1}} p(\tau_i) \theta^{-(\tau_{i+1}-\tau_i)} \right], \tag{4.49}$$

where

$0 < \rho < 1$ is the *discrete-time* (annual) discount factor,
$p(t)$, $t \in [\tau_0, \infty)$, is the cost of a new asset (purchase price and installation cost) at time t;
$q(t, u)$, t, $u \in [\tau_0, \infty)$, is the operating and maintenance (O & M) cost at time u for the asset bought at time $t \le u$;
σ, $0 \le \sigma < 1$, is the annual salvage value multiplier for the new asset;
θ, $0 < \theta < 1$ is the annual decrease factor of the salvage value.

Because of deterioration, the O & M cost $q(t, u)$ increases in u at fixed t as the asset becomes older (the asset age $u - t$ increases). The technological change leads to the availability of newer assets (challengers) that require less maintenance and are less expensive, i.e., $p(t)$ and $q(t, u)$ decrease in t for any fixed asset age $u - t$.

In the expression (4.49), $\rho^{\tau_i} p(\tau_i)$ is the discounted cost of a machine purchased at τ_i, and the sum of $\rho^k q(\tau_i, \tau_k)$ over $[\tau_i, \tau_{i+1}]$ is the discounted O & M cost for this machine until next machine replacement. Correspondingly, the first sum in (4.49) represents the total cost of purchased machines, the second term is the discounted total O & M cost, and the third one is the discounted total salvage value.

Now we can formulate the replacement problem as the problem of finding the optimal replacement times $\{\tau_k^*, k = 1, 2, \ldots\}$ that *minimize* the present value of the total replacement cost (4.49):

$$J(\tau_1*, \tau_2*, \ldots) = \min_{\tau_i, i=1,\ldots,\infty} J(\tau_1, \tau_2, \ldots). \tag{4.50}$$

The optimization problem (4.49) and (4.50) is an *integer programming* problem. It assumes that the unknown τ_i are integer-valued: $\tau_j \in \mathbf{I}$. This restriction essentially complicates both qualitative analysis and numeric solution of the replacement problem.

4.4.2.2 Continuous-Time Replacement Model

Let us now consider the replacement of a single machine in the *continuous time t*, $t \in [\tau_0, \infty)$, assuming that the initial purchase time τ_0 of the machine is given.

The corresponding optimization problem is presented as the minimization of

$$J(\tau_1, \tau_2, \ldots) = \sum_{i=1}^{\infty} e^{-r\tau_i} p(\tau_i) + \sum_{i=0}^{\infty} \left[\int_{\tau_i}^{\tau_{i+1}} e^{-ru} q(\tau_i, u) du - \sigma e^{-r\tau_{i+1}} p(\tau_i) e^{-s(\tau_{i+1} - \tau_i)} \right], \quad (4.51)$$

with respect to the real-valued unknowns τ_i, $i = 1,2,\ldots$, $\tau_0 < \tau_1 < \tau_2 < \tau_3 < \ldots$. In (4.51), the parameter $r = -\ln\rho > 0$ denotes the instantaneous discount rate, $s = -\ln\theta > 0$ is the instantaneous decrease rate of the salvage value, and all other parameters have the same meaning as in model (4.49).

The continuous-time model (4.51) is not subjected to the restriction of the integer-valued unknowns τ_i of the discrete model (4.49) and (4.50). As result, it can be investigated using standard optimization techniques. Section 5.1 provides a theoretic analysis of the continuous replacement model (4.51) and establishes qualitative properties of the optimal machine lifetime, which lead to new results on the machine replacement under improving technology.

Exercises

1. Suggest a formula for the total operating capital $K(t)$ through the past investments $I(\tau)$ at $\tau \leq t$ in the Solow vintage model of Sect. 4.1.1.
   HINT: The formula will have an integral structure similar to formulas (4.10) for $Q(t)$ and $L(t)$.
2. Show that, in the absence of technological change, i.e., at constant $\alpha(\tau)$ and $\beta(\tau)$, the Solow [15] vintage model is reduced to the aggregate production function $Q(t) = \alpha K(t) = \beta L(t)$ with fixed proportion.
   HINT: Use the formula for $K(t)$ from the previous exercise.
3. Write the intensive-form vintage model (4.13) in the case when the production function F has the Cobb–Douglas–Tinbergen form (4.2).
4. Write the vintage model (4.14) for the case of the Cobb–Douglas–Tinbergen production function F.
5. Show that the two-sector vintage model (4.15)–(4.17) in the case of the *constant accumulation norm* $s \equiv$ const is reduced to the Solow vintage model (4.10) and (4.11) at $\alpha(\tau, t)k(t) = \beta(\tau, t)$, where $k(t)$ is a certain given function.
6. Show that, in the absence of technological change, i.e., at $\alpha = \alpha(t)$ and $\beta = \beta(t)$, the two-sector vintage model (4.15)–(4.17) at $s \equiv$ const is reduced to the non-vintage model $Q(t) = \beta(t)L(t)$, $I(t) = sQ(t)$, $C(t) = (1-s)Q(t)$.
7. Show that, the vintage model (4.31)–(4.33) at $i \equiv 0$ (with no investments into older vintages) is equivalent to the Solow vintage model (4.5)–(4.7) with a linear one-factor production function $F(\tau, t, v)$.

8. Show that the Volterra integral equations (4.41) and (4.42) in the case of constant $\mu > 0$ lead to the PDE problem (4.39) and (4.40).

   HINT: Take the partial derivatives of (4.41) and (4.42) and combine them to obtain equalities (4.39) and (4.40).
9. Show that the formulas (4.31) and (4.32) give the solution of the Volterra integral equations (4.41) and (4.42) in the case of a constant $\mu > 0$.

   HINT: The differentiation of (4.41) and (4.42) in t at a fixed τ leads to linear ordinary differential equations for $v(\tau, t)$ (at a fixed τ), which are solved using standard formulas.
10. Compare the discrete and continuous replacement models (4.49) and (4.51) in the special case when the function $q = q(t)$ does not depend on u. Discuss and justify the relation $r = -\ln\rho$ between the discount parameters r and ρ of these models.

References[1]

1. 🕮 Boucekkine, R., de la Croix, D., Licandro, O.: Vintage capital growth theory: three breakthroughs. In: de la Grandville, O. (ed.) Economic Growth and Development, Chapter 5, pp. 87–116. Emerald Group Publishing, Bradford, England (2011)
2. Boucekkine, R., Germain, M., Licandro, O., Magnus, A.: Creative destruction, investment volatility and the average age of capital. J. Econ. Growth **3**, 361–384 (1998)
3. Cooley, T., Greenwood, J., Yorukoglu, M.: The replacement problem. J. Monet. Econ. **40**, 457–499 (1997)
4. Goetz, R., Hritonenko, N., Yatsenko, Y.: The optimal economic lifetime of vintage capital in the presence of operating cost, technological progress, and learning. J. Econ. Dyn. Control **32**, 3032–3053 (2008)
5. 🕮 Hartman, J.: Engineering Economy and the Decision-Making Process. Pearson Prentice Hall, Upper Saddle River, NJ (2007)
6. 🕮 Hritonenko, N., Yatsenko, Y.: Modeling and Optimization of the Lifetime of Technologies. Kluwer Academic Publishers, Dordrecht (1996)
7. Johansen, L.: Substitution versus fixed production coefficients in the theory of economic growth: a synthesis. Econometrica **27**, 157–175 (1959)
8. Johansen, L.: Production Functions. North-Holland, Amsterdam (1972)
9. Jovanovic, B.: Vintage capital and inequality. Rev. Econ. Dyn. **1**, 497–530 (1998)
10. Jovanovic, B., Yatsenko, Y.: Investment in vintage capital. J. Econ. Theory **147**, 551–569 (2012)
11. 🕮 Malcomson, J.M.: Replacement and the rental value of capital equipment subject to obsolescence. J. Econ. Theory **10**, 24–41 (1975)
12. Newman, D., Eschenbach, T., Lavelle, J.: Engineering Economic Analysis, 9th edn. Oxford University Press, New York (2004)
13. Sethi, S., Chand, S.: Planning horizon procedures in machine replacement models. Manage. Sci. **25**, 140–151 (1979)

[1] The book symbol 🕮 means that the reference is a textbook recommended for further student reading.

14. Solow, R.: Investment and technical progress. In: Arrow, K., Karlin, S., Suppes, P. (eds.) Mathematical Methods in the Social Sciences, 1959. Stanford University Press, Stanford (1960)
15. Solow, R., Tobin, J., von Weizsacker, C., Yaari, M.: Neoclassical growth with fixed factor proportions. Rev. Econ. Stud. **33**, 79–115 (1966)
16. Yatsenko, Y., Hritonenko, N.: Discrete-continuous analysis of optimal equipment replacement. Int. Trans. Oper. Res. **17**, 577–593 (2010)

Chapter 5
Optimization of Economic Renovation

This chapter analyzes optimization problems in the economic models with heterogeneous capital and labor of Chap. 4. Such models are important in explaining economic modernization under improving technology. Section 5.1 provides a qualitative analysis of the continuous-time optimization problem of one-machine replacement from Sect. 4.4 using standard tools of nonlinear optimization. Section 5.2 explores the optimal modernization of vintage capital in a profit-maximizing firm under environmental constraints. Section 5.3 investigates an optimization problem with nonlinear utility in the Ramsey vintage capital model of Sect. 4.2. A balanced growth regime is established and analyzed under exponential technology and labor. It possesses new properties compared to the linear utility case. Section 5.4 contains a mathematical appendix that derives extremum conditions for vintage capital models using variation techniques and Lagrange multipliers.

5.1 Optimal Replacement of One Machine

As shown in Sect. 4.4, the replacement process of a single machine under embodied technological change can be described by the following optimization problem

$$J(\tau_1{}^*, \tau_2{}^*, \ldots) = \min_{\tau_1, \tau_2, \ldots} J(\tau_1, \tau_2, \ldots), \tag{5.1}$$

$$J(\tau_1, \tau_2, \ldots) = \sum_{i=1}^{\infty} \mathrm{e}^{-r\tau_i} p(\tau_i) + \sum_{i=0}^{\infty} \left[\int_{\tau_i}^{\tau_{i+1}} \mathrm{e}^{-ru} q(\tau_i, u) \mathrm{d}u - \sigma \mathrm{e}^{-r\tau_{i+1}} p(\tau_i) \mathrm{e}^{-s(\tau_{i+1} - \tau_i)} \right],$$

with real-valued unknown replacement times $\tau_i{}^*$, $i = 1, 2, \ldots$ The given parameters are as follows:

N. Hritonenko and Y. Yatsenko, *Mathematical Modeling in Economics, Ecology and the Environment*, Springer Optimization and Its Applications 88,
DOI 10.1007/978-1-4614-9311-2_5,

the purchase time τ_0 of the first machine,
the instantaneous discount rate $r > 0$,
salvage multiplier for the new asset $\sigma \geq 0$,
the salvage value decrease rate $s > 0$.

The given new machine cost $p(t)$ and the operating and maintenance (O & M) cost $q(t,u)$ are positive and continuously differentiable for $t,\ u \in [\tau_0, \infty)$, $t \in [\tau_0, \infty)$. The optimization problem (5.1) in the continuous time $t \in [\tau_0, \infty)$ can be investigated using standard tools of the nonlinear optimization theory.

5.1.1 Necessary Condition for an Extremum

Although the minimized function (5.1) is expressed through the unknowns τ_i, $i = 1, 2, \ldots$, the analysis appears to be simpler with using the unknowns $L_i = \tau_i - \tau_{i-1}$, $i = 1, 2, \ldots$. Assuming that the initial purchase time $\tau_0 \leq 0$ of the machine is given, the sequence of the replacement times $\{\tau_i,\ i = 1,\ 2, \ldots\}$ determines the sequence $\{L_i,\ i = 1, 2, \ldots\}$ of the endogenous lifetimes $L_i = \tau_i - \tau_{i-1}$ of sequentially replaced machines, $i = 1, 2, \ldots$.

Optimality Condition: If an optimal sequence $\{L_i{}^*,\ i = 1, 2, \ldots\}$ exists, then each unknown value $L_i{}^*$, $0 < L_i{}^* < \infty$, satisfies the condition

$$-rp(\tau_i) + p'(\tau_i)\left[1 - \sigma e^{-(r+s)L_{i+1}}\right] + \sigma\left[(r+s)p(\tau_{i-1})e^{-sL_i} - sp(\tau_i)e^{-(r+s)L_{i+1}}\right] + \\ + q(\tau_{i-1}, \tau_i) - q(\tau_i, \tau_i) + \int_{\tau_i}^{\tau_{i+1}} e^{-r(u-\tau_i)} \frac{\partial q(\tau_i, u)}{\partial \tau_i} du = 0, \qquad i = 1, 2, \ldots \tag{5.2}$$

The formula (5.2) can be proven using elementary extremum conditions [7]. Indeed, if a differentiable function $f(x)$ has an extremum at a point x^*, then $df(x^*)/dx = 0$. Similarly, if an optimal policy $\tau_i{}^*$, $i = 1, 2, \ldots$, exists, then it should satisfy the necessary condition for an extremum of the function $J(\tau_1, \tau_2, \tau_3, \ldots)$:

$$\partial J / \partial \tau_i = 0, \quad i = 1, 2, \ldots. \tag{5.3}$$

To find the derivative $\partial J / \partial \tau_i$ for a fixed number i, we notice that the unknown variable τ_i appears in five terms of the infinite sum in the objective function (5.1). So at a fixed i we can write (5.1) as a sum of five terms with τ_i plus an infinite sum that does not depend on τ_i. Differentiating (5.1) in τ_i, we obtain

$$\frac{\partial J}{\partial \tau_i} = -re^{-r\tau_i}p(\tau_i) + e^{-r\tau_i}p'(\tau_i) + e^{-r\tau_i}q(\tau_{i-1}, \tau_i) - e^{-r\tau_i}q(\tau_i, \tau_i) \\ + \int_{\tau_i}^{\tau_{i+1}} e^{-ru} \frac{\partial q(\tau_i, u)}{\partial \tau_i} du - \sigma p'(\tau_i) e^{-r\tau_{i+1} - s(\tau_{i+1} - \tau_i)} \\ + \sigma(r+s)p(\tau_{i-1})e^{-r\tau_i - s(\tau_i - \tau_{i-1})} - \sigma sp(\tau_i)e^{-r\tau_{i+1} - s(\tau_{i+1} - \tau_i)}, \tag{5.4}$$

which after transformation leads to the equality (5.2).

Finding a sufficient condition for an extremum is more difficult and imposes more restrictions on the given functions.

5.1.2 Qualitative Analysis of Optimal Replacement Policy

The optimality condition of previous section leads to the following conclusions.

5.1.2.1 Case with No Technological Change and Deterioration

In the case $q(t,u)$ = const, $p(t)$ = const with no technological change and deterioration, the derivative $\partial J/\partial \tau_1 > 0$ by (5.2). Therefore, the optimal time of the first replacement is $\tau_1{}^* = \infty$ and the optimal policy is "*no replacement policy*."

5.1.2.2 Exponential Technological Change and Deterioration

To illustrate a simple solution with replacement in the optimization problem (5.1), we impose specific assumptions on the given functions. Namely, let us assume that both technological change and deterioration are *exponential*:

$$q(t,u) = q_0 e^{c_d(u-t)} e^{-c_q t}, \quad p(t) = p_0 e^{-c_p t}, c_q + c_d > 0, \quad 0 \le c_p + c_d < r. \quad (5.5)$$

Case (5.5) often occurs in replacement models. The *exponential technological change* means that the O & M cost (at a fixed age) and the new machine price drop by constant factors after each time period. The *exponential deterioration* means that the O & M costs increase by a constant factor when the machine age increases. Usually, both new machine and O & M costs decrease: $c_p > 0$, $c_q > 0$, because of technological change. The technological change impact on these costs can be different: $c_p \neq c_q$. By (5.5), the O & M and machine costs can even increase ($c_p < 0$ and/or $c_q < 0$) but slower than the deterioration rate c_d.

Under the condition (5.5), the extremum condition (5.2) has the form:

$$\left[e^{(c_d+c_q)L_i} - 1\right] + \frac{c_d + c_q}{r - c_d}\left[e^{-(r-c_d)L_{i+1}} - 1\right] + \\ + \sigma \frac{p_0}{q_0} e^{(c_q-c_p)\tau_i}\left[(c_p - s)e^{-(r+s)L_{i+1}} + (r+s)e^{(c_p-s)L_i}\right] = \frac{p_0}{q_0}(r + c_p)e^{(c_q-c_p)\tau_i}. \quad (5.6)$$

It is of great interest for the economic theory to identify possible cases of a constant optimal lifetime (which corresponds to the *balanced growth* concept).

5.1.2.3 Constant Lifetime at Proportional Technological Change

The technological change *with equal rates* c_q *and* c_p is referred to as *the proportional technological change*. In the case $c_q = c_p$, the equality (5.6) does not depend on τ_i explicitly. So we can try the constant optimal lifetime $L_i \equiv L > 0$ as a solution to (5.6).

Indeed, substituting $L_i \equiv L$ to (5.6), we obtain the equation

$$\begin{aligned} & e^{(c+c_d)L} - 1 + \frac{c+c_d}{r-c_d}\left[e^{-(r-c_d)L} - 1\right] + \sigma\frac{p_0}{q_0}\left[(c-s)e^{-(r+s)L} + (r+s)e^{(c-s)L}\right] \\ & -\frac{p_0}{q_0}(r+c) = 0 \end{aligned} \tag{5.7}$$

for the unknown L. To prove that (5.7) has a solution $L^* > 0$, let $F(L)$ denote the left side of (5.7). Then, $F(0) = (\sigma - 1)p_0(c + r)/q_0 < 0$. The behavior of $F(L)$ at large L is determined by two exponents with the positive coefficients $(c + c_d)L$ and $(c - s)L$, hence, $F(L)\rightarrow\infty$ at $L\rightarrow\infty$. Since $F(L)$ is continuous, (5.7) has a solution $L^* > 0$.

Finally, the derivative

$$\begin{aligned} \frac{dF}{dL} &= (c+c_d)\left[e^{(c+c_d)L} - e^{-(r-c_d)L}\right] \\ &+ \sigma\frac{p_0}{q_0}(r+s)(c-s)\left[e^{(c-s)L} - e^{-(r+s)L}\right] \end{aligned} \tag{5.8}$$

is positive under (5.5), hence, the solution L^* is unique. Thus, we have proved the following property.

Property 5.1: ***Optimal constant lifetime*** in the case of *the proportional technological change*. If the rates c_q and c_p are equal: $c_q = c_p = c$ in (5.5), then the optimal machine lifetime is constant $L_k^* = L^*$, $k = 1, 2,\ldots$, where the constant $L^* > 0$ is uniquely determined from the nonlinear equation (5.7).

5.1.2.4 Impact of Technological Change on Optimal Lifetime

Using the simple replacement policy at proportional technological change $c_q = c_p = c$ from Property 5.1, we can analyze how the intensity of technological change affects the optimal lifetime.

Let us denote the left-hand side of (5.7) as $F(L, c)$. Then, the equality (5.7) can be treated as the implicit function $F(L, c) = 0$. The derivative

$$\frac{\partial F}{\partial c} = Le^{(c+c_d)L} - \frac{p_0}{q_0} + \frac{1}{r - c_d}\left[e^{(c+c_d)L} - 1\right] + \sigma\frac{p_0}{q_0}\left[e^{-(r+s)L} + (r+s)Le^{(c-s)L}\right]$$

is positive:

$$Le^{(c+c_d)L} - \left[e^{(c+c_d)L} - 1\right] \Big/ (c + c_d) + \frac{p_0}{q_0}\frac{r - c_d}{c + c_d} > 0 \text{ at } L^* > 0 \text{ and } \sigma = 0.$$

Next, $\frac{\partial F}{\partial L} > 0$ by (5.8) and, therefore, $\frac{\partial L}{\partial c} = -\frac{\partial F}{\partial c} / \frac{\partial F}{\partial L} < 0$ by the implicit function theorem (see Sect. 1.3.1). It means that L decreases when c increases, i.e., we have

Property 5.2: Under the proportional technological change $c_q = c_p = c$ and $\sigma = 0$, the (constant) optimal lifetime L^* is shorter when the technological change rate c is larger.

5.1.2.5 Approximate Formula for the Optimal Lifetime

In the case of small rates r, c_d, and $c = c_q = c_p$, such that $c \ll 1$, $c_d \ll 1$, and $r \ll 1$, we can find an approximate solution of the nonlinear equation (5.7) for L. For simplicity, let the salvage value $\sigma = 0$. Then (5.7) is

$$e^{(c+c_d)L} - 1 + \frac{c + c_d}{r - c_d}\left[e^{-(r-c_d)L} - 1\right] - \frac{p_0}{q_0}(r + c) = 0. \tag{5.9}$$

Expanding the exponential function e^x into the Taylor series (see Sect. 1.3) and neglecting the cubic and higher terms of x, we obtain the approximate formula $e^x \approx 1 + x + x^2/2$, which holds for $x \ll 1$. Next, applying this formula to two exponential functions in (5.9), we obtain

$$\frac{(c + c_d)^2 L^2}{2} + \frac{c + c_d}{r - c_d}\frac{(r - c_d)^2 L^2}{2} - \frac{p_0}{q_0}(r + c) \approx 0 \text{ or}$$

$$\frac{L^2}{2}(c + c_d)(r + c) = \frac{p_0}{q_0}(r + c). \tag{5.10}$$

Solving (5.10), we have the approximate formula for L:

$$L \approx \sqrt{\frac{2p_0}{(c + c_d)q_0}}. \tag{5.11}$$

The formula (5.11) is well known as the *Terborgh formula for the optimal asset lifetime* [10].

Notations $\ll$***and***$\gg$: The notation $\ll$ ("*much smaller than*") is commonly used to indicate that one value is smaller than another by two or more orders of magnitude (by a factor of 100 or larger). Orders of magnitude are generally used to make approximate comparisons, and reflect large differences. It is common among scientists to say that a parameter is "of the order of" some value. If two numbers differ by one order of magnitude, then one is about ten times larger than the other. If two numbers differ by two orders of magnitude, then they differ by a factor of about 100.

5.1.2.6 Case of Variable Optimal Asset Lifetime

The case $c_q \neq c_p$ is more difficult to analyze. For brevity, let the salvage value $\sigma = 0$. Then, the extremum condition (5.6) has the form:

$$\left[e^{\left(c_d+c_q\right)L_i} - 1\right] + \frac{c_d + c_q}{r - c_d}\left[e^{-(r-c_d)L_{i+1}} - 1\right] = \frac{p_0}{q_0}\left(r + c_p\right)e^{\left(c_q-c_p\right)\tau_i}, \quad i = 1, 2, \ldots \tag{5.12}$$

The left-hand side of (5.12) for L_i increases in τ_i (and in i) at $c_q > c_p$, and decreases in τ_i if $c_q < c_p$. Therefore, the solution L_i should be different for different values i. Moreover, if the optimal policy L_i^*, $i = 1, 2, \ldots$, exists, then the following property holds.

Property 5.3: ***Optimal variable lifetime***. If $c_q < c_p$, i.e., the O & M cost $q(t,u)$ at fixed age $u - t$ decreases in t slower than the machine price $p(t)$, then the optimal machine lifetime decreases, $L_k^* > L_{k+1}^*$ for $k = 1, 2, \ldots$(and converse).

The formal proof of Property 5.3 is out of scope of this chapter. It requires converting the nonlinear equation (5.12) to a special nonlinear integral equation similar to (5.23) of the next section.

5.1.2.7 Economic Interpretation

Properties 5.1–5.3 provide interesting insight into the optimal replacement policies under improving technology depending on the changing machine price and O & M cost. They highlight the qualitatively different impact of the machine price (capital cost) and O & M cost on the dynamics of optimal machine service life. Property 5.1 identifies cases when the optimal lifetime of assets is constant and shows that it happens *if and only if* the decrease rate is the same for both O & M cost and machine price (the proportional technological change). By Property 5.2, the optimal lifetime is always shorter for more intense proportional change.

However, the impact of technological change may be different if it affects mainly one of these costs [7, 9]. By Property 5.3, the optimal lifetime decreases if the machine price decreases faster than the O & M expenses (and increases otherwise). In particular, the optimal lifetimes are *shorter* when the technological change in new machine price is more intense (the rate c_p is larger) but the rate c_q in

O & M cost remains the same. Hence, an acceleration of the change of the new machine price speeds up the introduction of new technologies. For the same rate c_p, the optimal lifetimes are *longer* when the O & M cost rate c_q increases. Hence, a more intense technological change in the O & M cost delays the introduction of new assets.

5.2 Profit-Maximizing Firm Under Resource Restrictions

In this section, we analyze an optimal capital modernization strategy of a firm under improving technology and restrictions on available resource. The restricted resource may be space, land, labor, finance, or environmental pollution. An essential practical issue is whether the presence of such constraints encourages or discourages technological modernization of the firm's productive assets [3].

We consider a firm that produces some output, uses a limited resource, invests into new more productive capital vintages, and scraps older obsolete vintages with the goal of maximizing the total net profit over time [5]. Following Sect. 4.3, the profit maximizing strategy of the firm is described by following optimal control problem with respect to the unknown functions Q, m, and a:

$$\max_{m,a,Q,C} I = \max \int_0^\infty \mathrm{e}^{-rt} \frac{C^{1-\gamma}}{1-\gamma} \mathrm{d}t \tag{5.13}$$

$$C(t) = Q(t) - p(t)m(t), \tag{5.14}$$

$$Q(t) = \int_{a(t)}^{t} \beta(\tau)m(\tau)\mathrm{d}\tau, \tag{5.15}$$

$$L = \int_{a(t)}^{t} m(\tau)\mathrm{d}\tau, \tag{5.16}$$

the inequality-constraints:

$$0 \le m(t) \le M(t), \tag{5.17}$$

$$a'(t) \ge 0, \quad a(t) < t, \quad t \in [0, \infty), \tag{5.18}$$

and the initial conditions:

$$a(t_0) = a_0, \quad m(\tau) \equiv m_0(\tau), \quad \tau \in [a_0, 0]. \tag{5.19}$$

Following Sect. 4.3,

$C(t)$ can be interpreted as the net profit,
$Q(t)$ is the product output per period,
$m(t)$ is the investment into new capital (measured in resource consumption units),

$p(t)$ is the given price of new capital (in resource consumption units),
$\beta(\tau)$ is the efficiency of the vintage introduced at instant τ,
$a(t)$ is the time of installing the oldest vintage in use,
L is the per period consumption of a key resource (energy, labor, etc.),
$r > 0$ is the discounting rate.

The investment constraint (5.17) sets the upper limit $M(t)$ on possible investment. Also, $\beta_\tau > 0$ because of the embodied technological change and the condition

$$\int_{t_0}^{\infty} \mathrm{e}^{-rt}\beta(t)\mathrm{d}t < \infty \tag{5.20}$$

is imposed to guarantee the convergence of the improper integral in (5.13).

5.2.1 Necessary Condition for an Extremum

A detailed treatment of a more general optimal control problem (5.51)–(5.56) is provided in Appendix (Sect. 5.4). The problem (5.13)–(5.19) coincides with (5.51)–(5.56) at $\gamma = 0$. Correspondingly, applying optimality conditions (5.65)–(5.66) from Appendix, a solution $m^*(t)$, $t \in [0, \infty)$, of the problem (5.13)–(5.19) (if it exists) satisfies the conditions

$$\begin{aligned} &I'(t) \le 0 \text{ at } m^*(t) = 0, \\ &I'(t) \ge 0 \text{ at } m^*(t) = M(t), \\ &I'(t) \equiv 0 \text{ at } 0 < m^*(t) < M(t), t \in [0, \infty), \end{aligned} \tag{5.21}$$

where the gradient of the functional I is

$$I'(t) = \int_{t}^{a^{-1}(t)} \mathrm{e}^{-ru}[\beta(t) - \beta(a(u))]\mathrm{d}u - \mathrm{e}^{-rt}p(t), \tag{5.22}$$

$a^{-1}(t)$ is the inverse of the function $a(t)$.

The gradient (5.22) depends on a and does not depend on m and Q. This fact allows us to characterize the complete dynamics of the model (5.13)–(5.19) with linear utility. By (5.22), an interior optimal control $\widetilde{a}$ can be determined from the integral-functional equation

$$\int_{t}^{\widetilde{a}^{-1}(t)} \mathrm{e}^{-r\tau}[\beta(t) - \beta(\widetilde{a}(\tau), \tau)]\mathrm{d}\tau = \mathrm{e}^{-rt}p(t), \quad t \in [0, \infty). \tag{5.23}$$

We will refer to the solution $\widetilde{a}$ of (5.23), if it exists, as the *turnpike* of the model (5.13)–(5.19). We will return to turnpikes later in this chapter.

5.2.1.1 Case of Exponential Technological Change and Capital Price

In order to obtain a complete dynamic picture of optimal solutions, let us consider the exponential technological change and exponential capital price:

$$\beta(\tau) = \overline{\beta}\mathrm{e}^{g\tau}, \quad p(t) = \overline{p}\mathrm{e}^{g_p t}, \quad 0 < g < r,\ g_p \leq g. \tag{5.24}$$

Then, the gradient (5.22) is

$$I_m{}'(t) = \overline{\beta} \int_t^{a^{-1}(t)} \mathrm{e}^{-r\tau}\left[\mathrm{e}^{gt} - \mathrm{e}^{ga(\tau)}\right]\mathrm{d}\tau - \overline{p}\mathrm{e}^{(g_p - r)t}. \tag{5.25}$$

and nonlinear integral equation $I'_m(a;t) \equiv 0$ possesses a unique solution (the turnpike) $\widetilde{a}(t),\ t \in [0, \infty)$ [5]. The notation $I'_m(a;t)$ emphasizes that the gradient depends on the unknown a only.

5.2.1.2 Interior Turnpike for Capital Scrapping Time

It can be shown [5] that (5.23) in the case (5.24) has a unique solution $\widetilde{a}(t) < t$, $\mathrm{d}\widetilde{a}(t)/\mathrm{d}t > 0, t \in [0, \infty)$, such that:

(a) if $g > g_p$, then $t - \widetilde{a}(t) \to 0$ at $t \to \infty$;
(b) if $g = g_p$, then $\widetilde{a}(t) \equiv t - T, t \in [0, \infty)$, where the constant T is found from the nonlinear equation

$$r\mathrm{e}^{-gT} - g\mathrm{e}^{-rT} = (r - g)\left(1 - r\overline{p}/\overline{\beta}\right). \tag{5.26}$$

If $0 < g < r << 1$, then

$$T \approx \sqrt{2\overline{p}/\left(\overline{\beta}\,g\right)}. \tag{5.27}$$

If $g < g_p$, then (5.23) does not have a solution on the entire $[0, \infty)$.

The case (a) has a clear resemblance to Property 5.1 about the constant optimal lifetime of machine in the case of equal rates $c_q = c_p$ technical change in (5.5). Then, the nonlinear equation (5.26) has a structure similar to the nonlinear equation (5 . 7) for the optimal constant machine lifetime. Finally, the formula (5.27) is very similar to the *Terborgh formula* (5.11).

5.2.2 Structure of Optimal Trajectories

Let us analyze the structure of solutions to the problem (5.13)–(5.19). As in the neoclassic growth economic problems of Chap. 3, the solutions will involve *transition dynamics* (with a corner solution) and *long-term dynamics*. The turnpike trajectory $\widetilde{a}$ plays an important role in the structure of optimal trajectories. Namely, under the exponential technological change and capital price (5.24), the solution (m^*, a^*) of the optimization problem (5.13)–(5.19) exists and has the following structure:

A. **The *transition dynamics* period** $[0, \mu)$.

The optimal control m^* is boundary on $[0, \mu)$ and depends on the difference between a_0 and $\widetilde{a}(0)$:

Case $a_0 > \widetilde{a}(0)$: Then $m^*(t) = 0$ and $a^*(t) = a_0$ on $[0, \mu)$, where the instant $\mu > 0$ is determined from the condition $a_0 = \widetilde{a}(\mu)$.

Case $a_0 < \widetilde{a}(0)$: Then $m^*(t) = M(t)$, $a^*(t) = a_{\max}(t)$ is determined from (5.16) and increases fast, and the instant $\mu > 0$ is determined by $a_{\max}(\mu) = \widetilde{a}(\mu)$.

B. **The *long-term dynamics* interval** $[\mu, \infty)$.

Both optimal control m^* and trajectory a^* are interior on $[\mu, \infty)$:

$$a^*(t) = \widetilde{a}(t),\ m^*(t) = m^*(\widetilde{a}(t))\widetilde{a}'(t),\ t \in [\mu, \infty). \tag{5.28}$$

The length μ of the transition period is defined by the difference $|\widetilde{a}(0) - a_0|$. If $\widetilde{a}(0) = a_0$, then $\mu = 0$ (*no transition dynamics*).

Proof of the optimal structure A-B: A direct check shows that the constructed functions m^*, a^* satisfy all restrictions (5.14)–(5.18). Next, we prove that the functions m^*, a^* satisfy the necessary extremum condition (5.21) and, therefore, represent a solution of the problem (5.13)–(5.19).

Let us consider the case $a_0 > \widetilde{a}(0)$. Since $a(t)$ is continuous, then $a^*(t) = a_0 > \widetilde{a}(t)$ on a certain interval $[0, \mu)$ such that $a^*(\mu) = \widetilde{a}(\mu)$, i.e., μ is the intersection point of the curves $a = a_0$ and $\widetilde{a}$. Then, $a^{-1}(t) > \widetilde{a}^{-1}(t)$ for $t \in [0, \mu)$, by the definition of the inverse a^{-1}, and the gradient (5.22) can be written and estimated as

$$\begin{aligned}
I'(a^*, t) = &\int_t^{\mu} e^{-r\tau}\overline{\beta}\left[e^{gt} - e^{ga_0}\right]d\tau + \\
&+ \int_{\mu}^{a_{\lim}^{-1}(t)} e^{-r\tau}\overline{\beta}\left[e^{gt} - e^{g\widetilde{a}(\tau)}\right]d\tau + \overline{p}e^{-rt}e^{g_p t} = \\
= &\int_t^{\mu} e^{-r\tau}\overline{\beta}\left[e^{g\widetilde{a}(\tau)} - e^{ga_0}\right]d\tau + \\
&+ \int_t^{a_{\lim}^{-1}(t)} e^{-r\tau}\overline{\beta}\left[e^{gt} - e^{g\widetilde{a}(\tau)}\right]d\tau + \overline{p}e^{-rt}e^{g_p t} < \\
< &\int_t^{a_{\lim}^{-1}(t)} e^{-r\tau}\overline{\beta}\left[e^{gt} - e^{g\widetilde{a}(\tau)}\right]d\tau + \overline{p}e^{-rt}e^{g_p t} < \\
< &\int_t^{\widetilde{a}^{-1}(t)} e^{-r\tau}\overline{\beta}\left[e^{gt} - e^{g\widetilde{a}(\tau)}\right]d\tau + \overline{p}e^{-rt}e^{g_p t} = 0
\end{aligned}$$

for $t \in [t_0, \mu)$. So $I'(a^*, t)$ is negative at $t \in [0, \mu)$ for any m^*. Then, $m^*(t)$ should be minimum possible at $t \in [0, \mu)$ by the optimality condition (5.21) and, therefore, $m^*(t) = 0$ is optimal. The optimal $a^*(t) = a_0$ is constant on $[0, \mu)$, while $\widetilde{a}(t)$ increases. Therefore, they will intersect at a certain point $t = \mu$. Next, at $t \in [\mu, \infty)$, an interior trajectory (m^*, a^*) exists such that $a^*(t) = \widetilde{a}(t)$ and $I'(a^*, t) = I'(, \widetilde{a}, t) = 0$ for $t \in [\mu, \infty)$. Therefore, the extremum condition (5.21) holds for $t \in [0, \infty)$ and (m^*, a^*) is a solution of the problem. The case $a_0 < \widetilde{a}(0)$ is investigated similarly.

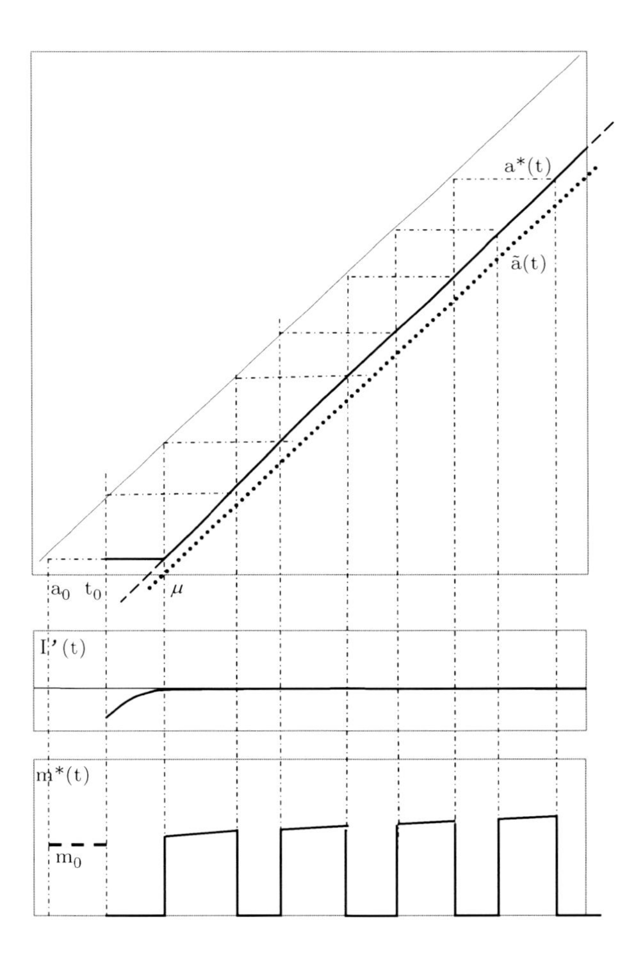

Fig. 5.1 The optimal capital scrapping time a^*, investment m^*, and gradient I' in the optimal control problem (5.13)–(5.19). The optimal m^* possesses *replacement echoes*

The optimal trajectory (m^*, a^*) is illustrated in Fig. 5.1 for the case $a_0 > \widetilde{a}(0)$ and $g > g_p$. For simplicity, here $m_0 =$ const. The turnpike trajectory $\widetilde{a}$ is indicated with the dash line. For comparison, the dotted line shows the constant lifetime trajectory $t - T$, which is optimal in the case $g = g_p$.

5.2.3 Economic Interpretation

The obtained structure of the solution (m^*, a^*) demonstrates significant differences of the vintage capital models with heterogeneous capital of Chap. 4 from the traditional growth models of Chaps. 2 and 3. The vintage model (5.13)–(5.19) involves constant returns and a single-factor production function, so corner solutions with maximum or zero investment naturally appear during the transition dynamics.

5.2.3.1 Investment Replacement Echoes

Figure 5.1 demonstrates a clear repetition pattern in the *optimal investment* trajectory m^* on $[0, \infty)$. This effect is known as the *replacement echoes* in the theory of vintage capital models [1, 4, 6].

The *replacement echoes* appear because of the impact of the initial condition $a(0) = a_0$ imposed on the unknown a. In the general case $a_0 \neq \widetilde{a}(0)$, the solution $m^*(t)$ is boundary by (5.17) (the minimal $m_{\min}(t) = 0$ or maximal $M(t)$) at the beginning part $[0, \mu]$ of the transition dynamics. After that, $m^*(t) = m(\widetilde{a}(t))\widetilde{a}'(t)$ is found on $[\mu, \infty)$ from (5.16). The last formula demonstrates that the initial shape $\begin{cases} m_0(t), & t \in [a_0, 0] \\ M(t) \text{ or } 0, & t \in (0, \mu] \end{cases}$ of m on $[a_0, \mu]$ is approximately replicated throughout the entire horizon $[0, \infty)$. Namely, when we reach the part $[\widetilde{a}^{-1}(0), \widetilde{a}^{-1}(\mu)]$ of $[0, \infty)$, then $m(t)$ is similar to $m(\widetilde{a}(t))$ at $[0, \mu]$, at least, when $\widetilde{a}' \approx 1$. The same pattern appears on the interval $[\widetilde{a}^{-1}(\widetilde{a}^{-1}(0)), \widetilde{a}^{-1}(\widetilde{a}^{-1}(\mu))]$, and so on.

So the replacement echoes disseminate the transition dynamics of the optimal investment m to the future infinite period. These echoes do not decline. On the other side, the optimal capital lifetime $a^*(t)$ does not have any irregularities after the instant μ. If the initial condition is "perfect": $a_0 = \widetilde{a}(0)$, then the replacement echoes are *absent* and the optimal a^* coincides with the turnpike $\widetilde{a}$ indicated with the dash line in Fig. 5.1 from the very beginning $\mu = 0$.

In the finite-horizon case of the problem (5.13)–(5.19), there exists even a more powerful echo pattern: the so-called *anticipation echoes* [5, 6]. It is caused by the existence of "zero-investment period" $(\theta, T_{\max}]$ at the right end of the finite horizon $[0, T_{\max}]$ such that $m^*(t) \equiv 0$ for $t \in (\theta, T_{\max}]$.

5.2.3.2 Qualitative Properties of Optimal Capital Replacement

The problem (5.13)–(5.19) may involve different dynamic patterns on the following intervals:

- *Extensive transition growth on* $[0, \mu]$, where $m^*(t) = 0$, $a^*(t) = a_0$, and the optimal lifetime $t - a^*(t) = t - a_0$ increases.

- *Intensive transition growth on* $[0, \mu]$, where $m^*(t) = M(t)$, $a^*(t)$ increases faster than t and the optimal lifetime $t - a^*(t)$ decreases.
- *Long-term growth on* $[\mu, \infty)$, where $a^*(t) = \widetilde{a}(t)$ and $m^*(t)$ is determined from (5.16) and has a cyclic behavior.

The optimal capital lifetime $T(t)$ may be finite only if the vintage productivity $\beta(t)$ increases, i.e., if $g > 0$. Otherwise, no replacement policy is optimal (in the absence of deterioration).

During the long-term growth, the optimal capital lifetime $T(t) = t - \widetilde{a}(t)$ does not depend on the production scale and the initial capital structure. It is determined only by the rates of technological change g, capital cost g_p, and discount r. The balance between the productivity $\beta(t)$ and capital price $p(t)$ determines the dynamics of the optimal capital lifetime $T(t)$:

- If $g > g_p$ (i.e., the output per new capital cost increases), then the capital lifetime decreases.
- If $g < g_p$, then the capital lifetime increases and becomes infinite at some finite instant. The replacements become less frequent and finally stop.
- If $g = g_p$ (*proportional* technological change), the optimal capital lifetime is constant. Then, the constant lifetime depends only on the discount rate r and the constant ratio $\beta(t)/p(t)$ between the productivity and capital price.

5.2.3.3 Turnpike Properties of Capital Lifetime

The optimal structure of the problem (5.13)–(5.19) can be interpreted as *the turnpike theorem in the strongest form* for the optimal capital scrapping time a^* in the infinite-horizon case. This theorem states that, starting from some instant $\mu \geq 0$, the optimal trajectory $a^*(t)$ coincides with the turnpike $\widetilde{a}(t)$ on the infinite interval $[\mu, \infty)$. The turnpike $\widetilde{a}(t)$ does not necessarily satisfy the initial condition $a(0) = a_0$. The optimal investment m^* does not possess any turnpike property.

5.3 Nonlinear Utility Optimization in Ramsey Vintage Model

This section investigates an optimization problem with nonlinear utility in the Ramsey vintage capital model (4.15)–(4.17). The necessary condition for an extremum is derived and a qualitative analysis is provided under the exponential dynamics of technological change, discount, and labor resource. A balanced growth regime is found and a turnpike theorem is established for the optimal lifetime of capital. The obtained results demonstrate new qualitative behavior of optimal trajectories in the cases of linear and nonlinear utility.

As in Sect. 4.1.4, the economy in the Ramsey vintage model produces the new capital $m(t)$ and consumption good $C(t)$ and uses the given labor resource $L(t)$:

$$m(t) = \int_{a(t)}^{t} \alpha(\tau, t)s(t)m(\tau)\mathrm{d}\tau, \tag{5.29}$$

$$C(t) = \int_{a(t)}^{t} \beta(\tau, t)[1 - s(t)]m(\tau)\mathrm{d}\tau, \tag{5.30}$$

$$L(t) = \int_{a(t)}^{t} m(\tau)\mathrm{d}\tau, \quad t \in [0, \infty). \tag{5.31}$$

For mathematical consistency of this chapter, we use the notation $m(t)$ for the output of capital goods (in capacity per worker units) instead of $l(t)$ in Sect. 4.1. The rest of notations are the same as in Sect. 4.1:

$C(t)$ is the output of consumption goods,
$\alpha(\tau,t)$ and $\beta(\tau,t)$ are productivities of vintageτ at time t in two sectors,
$s(t)$ is a variable accumulation norm (*saving rate*),
$a(t)$ is the *scrapping time* of the obsolete capital.

Then, t-$a(t)$ is the *lifetime of capital*. The functions $\alpha(\tau,t)$ and $\beta(\tau,t)$ increase in τ because newer machines are more productive under embodied technical change.

By Sect. 4.1.4, the optimization problem in the Ramsey vintage model consists of finding the unknown functions $s(t)$, $m(t)$, $a(t)$, and $C(t)$, $t \in [0, \infty)$, that maximize

$$I = \int_{0}^{\infty} \mathrm{e}^{-rt} \frac{C^{1-\gamma}}{1-\gamma} \mathrm{d}t \tag{5.32}$$

subject to (5.29)–(5.31) under the given L. In (5.32), $r > 0$ is the discount rate. The unknowns s, m, and a should satisfy the following restrictions

$$0 \le s(t) \le 1, \quad m(t) \ge 0, \quad a(t) < t, \quad a'(t) \ge 0, \quad t \in [0, \infty), \tag{5.33}$$

and the initial conditions

$$a(0) = a_0 < 0, \quad m(\tau) = m_0(\tau), \quad \tau \in [a_0, 0]. \tag{5.34}$$

The given functions α, β, L, and m_0 are positive and satisfy (5.29)–(5.34) at $t = 0$.

5.3.1 Reduction to One-Sector Optimization Problem

The optimization problem (5.29)–(5.34) in the Ramsey vintage model, can be separated into two consecutive problems of determining the optimal renovation intensity (m, a) and finding the corresponding optimal saving rate s. Namely, assuming one-sector economy, the productivity dynamics is the same in production of consumption and capital goods. It means that the *relative rates of technological change* are the same in both sectors:

$$[\partial\beta(\tau,t)/\partial\tau]/\beta(\tau,t) \equiv [\partial\alpha(\tau,t)/\partial\tau]/\alpha(\tau,t). \tag{5.35}$$

The assumption (5.35) is equivalent to

$$\beta(\tau,t) = k(t)\alpha(\tau,t), \tag{5.36}$$

where $k(t)$ is a given function. Following Sect. 4.1, $k(t)$ can be interpreted as the given *capital–labor ratio*. Then $i(t) = k(t)m(t)$ is the output of capital goods in the consumption (monetary) units.

Substituting (5.36) and (5.30) into (5.32) and using (5.29), we obtain

$$\begin{aligned} I &= \int_0^\infty \mathrm{e}^{-rt}\left(\int_{a(t)}^t \beta(\tau,t)m(\tau)\mathrm{d}\tau - s(t)\int_{a(t)}^t \beta(\tau,t)m(\tau)\mathrm{d}\tau\right)^{1-\gamma}\mathrm{d}t/(1-\gamma) \\ &= \int_0^\infty \mathrm{e}^{-rt}(Q(t) - k(t)m(t))^{1-\gamma}\mathrm{d}t/(1-\gamma), \end{aligned} \tag{5.37}$$

where

$Q(t) = \int_{a(t)}^t \beta(\tau,t)m(\tau)\mathrm{d}\tau$ is the *total product output*.

The cost functional (5.37) does not depend on s, and therefore, the maximization problem (5.30)–(5.34) with unknowns s, C, m, a is reduced to the maximization of (5.37) in Q, m, and a. So we obtain two following optimization problems:

Optimization of Renovation Intensity:
To find the unknown functions $m(t)$, $a(t)$ and $Q(t)$, $t \in [0, \infty)$, which maximize

$$I = \int_0^\infty \mathrm{e}^{-rt}\frac{(Q(t) - k(t)m(t))^{1-\gamma}}{1-\gamma}\mathrm{d}t \tag{5.38}$$

subject to the equality-constraints:

$$Q(t) = \int_{a(t)}^t \beta(\tau)m(\tau)d\tau, \tag{5.39}$$

$$L(t) = \int_{a(t)}^t m(\tau)\mathrm{d}\tau, \tag{5.40}$$

the inequality-constraints $0 \le m(t) \le M(t)$, $a'(t) \ge 0$, $a(t) < t$, and the initial conditions (5.34).

Optimization of Saving Rate:
Under known optimal y, m, and a, the optimal saving rate s and the consumption C are explicitly determined from (5.29) and (5.30) as

$$s(t) = m(t)/Q(t), \quad C(t) = [1 - s(t)]Q(t), \quad t \in [0, \infty). \tag{5.41}$$

5.3.2 Interior Solutions

It is easy to see that the nonlinear-utility maximization problem (5.38)–(5.40) is equivalent to the problem (5.51)–(5.56) of Sect. 5.4, so we can use the optimality conditions (5.66) to study the problem (5.38)–(5.40). To analyze the structure of its solutions, we restrict ourselves to the exponential technological change, capital price, and labor resource:

$$\beta(\tau) = \overline{\beta}\mathrm{e}^{g\tau}, k(t) = \overline{k}\mathrm{e}^{g_k t}, L(t) = \overline{L}e^{\eta t}, g, g_k, \eta > 0, \eta < g. \tag{5.42}$$

Assumptions (5.42) are common in related research [2, 8]. In many economic problems, the long-term dynamics of the solutions is close to interior trajectories. Possible interior trajectories of the problem (5.38)–(5.40) are determined from the condition $I'(t) = 0$, $t \in [0, \infty)$, or by (5.66), from

$$\int_t^{a^{-1}(t)} \mathrm{e}^{-ru} \frac{\beta(t) - \beta(a(u))}{(Q(u) - k(u)m(u))^{\gamma}} \mathrm{d}u = \frac{\mathrm{e}^{-rt}\lambda(t)}{(Q(t) - k(t)m(t))^{\gamma}}, \quad t \in [0, \infty). \tag{5.43}$$

Expression (5.43) together with (5.39)–(5.40) forms a system of three nonlinear integral-functional equations in the unknowns $a(t)$, $m(t)$, $Q(t)$, $t \in [0, \infty)$, with the initial conditions (5.34). Let us denote its solution on the infinite interval $[0, \infty)$ (if it exists) as $\left(\widetilde{m}, \widetilde{a}, \widetilde{Q}\right)$.

In the linear-utility case of Sect. 5.2, the gradient $I'(t)$ depends only on the unknown variable a and does not depend on m and Q. This fact has been used in Sect. 5.3, to establish the structure of optimal capital renovation with the linear utility. It is not true in the case of the nonlinear utility $C^{1-\gamma}$.

5.3.3 Balanced Growth

Let us start with the analysis of a possible balanced growth in the model. In economics, the *balanced growth* means the constant increase rates of all endogenous characteristics with fixed proportions among these rates. In vintage capital models, the balanced growth regime also requires the capital lifetime $t - \widetilde{a}(t)$ to be constant. Substituting $\widetilde{a}(t) = t - T$ to the system (5.39), (5.40), (5.43), one can see that the balanced growth is possible only at $g = g_k$.

Property 5.4: ***Balanced Growth***. In the case of (5.42) and

$$g = g_k, g < r/\gamma, r - \gamma(g+\eta) < \overline{\beta}/\overline{k}, \tag{5.44}$$

the system of nonlinear equations (5.39), (5.40), (5.43) has the following solution

$$\widetilde{a}(t) \equiv t - T, \quad \widetilde{m}(t) = \overline{m}\mathrm{e}^{\eta t}, \quad \widetilde{Q}(t) = \overline{Q}\,\mathrm{e}^{(g+\eta)t}, \quad t \in [0, \infty), \tag{5.45}$$

if $a_0 = -T$ and $m_0 \equiv \widetilde{m}$ in (15). The constant T is uniquely determined from the nonlinear equation

$$v\left(1 - \mathrm{e}^{-(v+g)T}\right) - (v+g)\mathrm{e}^{-gT}\left(1 - \mathrm{e}^{-vT}\right) = \frac{\overline{k}}{\overline{\beta}} v(v+g), \tag{5.46}$$

$v = r - \gamma(g + \eta) - g$, and the constants $\overline{m}$ and $\overline{Q}$ are

$$\overline{m} = \begin{cases} \dfrac{\eta \overline{L}}{(1 - \mathrm{e}^{-\eta T})}, & \eta \neq 0, \\ \dfrac{\overline{L}}{T}, & \eta = 0, \end{cases} \qquad \overline{Q} = \frac{\overline{\beta}\,\overline{m}\left[1 - \mathrm{e}^{-(g+\eta)T}\right]}{g+\eta}.$$

In particular,

$$T \approx \sqrt{\frac{2\overline{k}}{g\overline{\beta}}} \text{ for small } g, r << 1. \tag{5.47}$$

Formulas (5.45)–(5.47) can be verified by the substitution of (5.45) into the system (5.39), (5.40), (5.43). To show that the obtained nonlinear equation (5.46) has a unique solution, let us denote its left-hand part as $F(T)$. Since $F(T) = 0$ and $F(\infty) = v > 0$, (5.46) has a solution $T^* > 0$ if $\overline{k}(v+g) < \overline{\beta}$ (that leads to (5.44)). Next, $F'(T) = g(g + v)\mathrm{e}^{-gT}(1 - \mathrm{e}^{-vT}) > 0$ hence, the solution T^* is unique.

By (5.46), the optimal constant capital lifetime T in (5.45) depends on g, r, γ, η, $\overline{k}, \overline{\beta}$, and does not depend on the "economic scale parameter" $\overline{L}$. Therefore, the optimal capital lifetime $T = t - \widetilde{a}(t)$ is scale-independent as in the corresponding linear-utility problem. However, it can be implemented only for the specified scale of operations $\left(\widetilde{m}, \widetilde{Q}\right)$, whereas it is possible for any scale in the linear-utility optimization of Sect. 5.3. The dependence of T on g, r, γ, η produces interesting conclusions, for example, for small g and $r \ll 1$, the optimal lifetime T does not depend on the nonlinear utility parameter γ and is the same as in the linear-utility problem.

When the given prehistory $m_0(\tau)$, $\tau \in [a_0, 0]$, perfectly matches the turnpike trajectory $\widetilde{a}(t), t \in [0, \infty)$, at the initial instant t_0, i.e., $a_0 = a(0)$, then the solution a^* coincides with the trajectory $\widetilde{a}$.

Property 5.5: If (5.42) and (5.44) hold and

$$-a_0 = T, \quad m_0(\tau) = \overline{m}e^{\eta\tau}, \quad \tau \in [a_0, 0], \quad \overline{L}\eta = \overline{m}\left[1 - e^{-\eta T}\right], \tag{5.48}$$

then $\left(\widetilde{m}, \widetilde{a}, \widetilde{Q}\right)$ determined by (5.45) is a unique solution to the problem (5.38)–(5.40).

Indeed, the substitution of (5.45) into (5.39)–(5.40) verifies that the trajectory $\left(\widetilde{m}, \widetilde{a}, \widetilde{Q}\right)$, is admissible. By (5.48), it satisfies the initial conditions (5.34): $\widetilde{a}(0) = -A = a_0$ and $\widetilde{a}'(0) = 1$. It also satisfies the optimality condition (5.43) and, therefore, is a solution. The uniqueness of the solution of $\widetilde{m}$ can be proven using the concavity of the problem [5].

5.3.3.1 Balanced Growth in Ramsey Vintage Model

Under conditions (5.42) of exponential technological change, price, and labor, a balanced growth trajectory $\left(\widetilde{m}, \widetilde{a}, \widetilde{Q}, \widetilde{s}, \widetilde{C}\right)$ also exists in the general Ramsey vintage capital model (5.29)–(5.34). Moreover, it is determined by the same expressions (5.45)–(5.47) (up to the constants). The corresponding optimal saving rate $\widetilde{s}$ is constant, whereas the consumption good output $\widetilde{C}$ increases with the rate $g + \eta$:

$$\widetilde{s} = \widetilde{m}/\widetilde{Q}, \quad \widetilde{C}(t) = [1 - \widetilde{s}]\widetilde{Q}(t), \quad t \in [0, \infty). \tag{5.49}$$

5.3.4 *Economic Interpretation: Turnpike Properties*

In general case, the initial conditions (5.34) do not coincide with conditions (5.48), hence, the optimal solution can not coincide with the balanced growth trajectory but may approach it. A *turnpike property* means the existence of an interior trajectory (a turnpike) that attracts optimal trajectories [4, 6]. In the Ramsey vintage capital model (5.29)–(5.34), a turnpike property can be established for the optimal capital lifetime a^*:

Turnpike Property 5.6: If conditions (5.42) and (5.44) hold, the parameter $(g + \eta)\gamma$ is close to the discount rate r, and the problem (5.29)–(5.34) has a solution (m^*, a^*, Q^*), then $a^*(t)$ is close to the trajectory $\widetilde{a}(t)$ on an infinitely large part of the interval $[0, \infty)$. Specifically, for any $\varepsilon > 0$ there is the value $\delta > 0$ such that for any

$r-(g+\mu)\gamma \leq \delta$ the inequality $\left|a^*(t)-\widetilde{a}(t)\right|<\varepsilon$ holds at $t\in[0,\infty)-\Delta$, where $\Delta\subset[0,\infty)$, $\mathrm{mes}(\Delta)<\infty$.

The proof of this property consists of two steps [5]: (1) constructing a special admissible solution that coincides with $\widetilde{a}(t)$ on $[0,\infty)$ except for an initial interval $[0,\eta]$; (2) proving that any other admissible solution is not optimal.

The optimal investment m^*, the corresponding saving rate s^* and consumption C^*

$$s^*(t)=m^*(t)/Q^*(t),\quad C^*(t)=[1-s^*(t)]Q^*(t),\quad t\in[0,\infty), \tag{5.50}$$

in the model (5.29)–(5.34) are of *bang-bang* type and do not possess turnpike properties.

Thus, only the capital lifetime possesses turnpike properties in the Ramsey vintage capital model (5.29)–(5.34). The *turnpike trajectory* $\tilde{a}$ for the capital lifetime a^* (which is not always a balanced growth) shows the best possible capital renovation strategy. Other optimal controls (investment m^* and distribution s^*) serve the goal of reaching this ideal pattern $\tilde{a}$ of capital renovation when it is possible.

The influence of a *nonlinear utility* is less essential than technological change. In particular:

- When the optimal capital lifetime is constant in the case of linear utility, it will remain constant for a nonlinear utility function.
- In both cases of linear and nonlinear utility, the optimal capital lifetime does not depend on the scale of operations.
- At small rates of technological change and discount, the optimal capital lifetime for *any* nonlinear utility is approximately the same as in the linear-utility case, see formula (5.47).

5.4 Appendix: Optimal Control in Vintage Capital Models

Vintage capital models with endogenous capital scrapping lead to the *optimal control problems* of nonlinear Volterra integral equations (Sect. 2.2). Investigation methods for the optimal control of integral equations are based on the abstract optimization theory. Specific optimal control problems for dynamic systems governed by standard Volterra integral equations have been investigated by different authors and several versions of the maximum principle have been developed [11]. A new feature of the *optimal control* in vintage capital models is the presence of unknown function (endogenous delay) in the integration limits. There are no general results for such dynamic systems with endogenous delay, so we need to use special techniques to derive extremum conditions.

5.4.1 Statement of Optimization Problem

In this section, we explain the investigation technique for an optimal control problem that includes the vintage capital models of Sect. 5.2 and 5.3. Namely, we analyze the problem of finding the unknown functions $Q(t)$, $m(t)$, and $a(t)$, $t \in [0, \infty)$, that maximize the following functional:

$$I = \int_0^{\infty} \mathrm{e}^{-rt} \frac{(Q - pm)^{1-\gamma}}{1-\gamma} \mathrm{d}t, \tag{5.51}$$

subject to the equality-constraints :

$$Q(t) = \int_{a(t)}^{t} \beta(\tau) m(\tau) \mathrm{d}\tau, \tag{5.52}$$

$$L = \int_{a(t)}^{t} m(\tau) \mathrm{d}\tau, \tag{5.53}$$

the inequality-constraints :

$$0 \le m(t) \le M(t), \tag{5.54}$$

$$a'(t) \ge 0, a(t) < t, \ t \in [0, \infty), \tag{5.55}$$

and the initial conditions:

$$a(t_0) = a_0, m(\tau) \equiv m_0(\tau), \tau \in [a_0, 0]. \tag{5.56}$$

The model (5.51)–(5.56) describes an economic system that does not produce new capital but acquires it from outside market by spending a certain part pm of its product Q. Possible interpretation of the functions Q, m, a, L, β, and p is discussed in Sect. 5.2 and 5.3. For clarity, we assume L in (5.53) to be constant, which is a common assumption in macroeconomics.

The given functions β, p, and M of the problem are assumed to be continuously differentiable, m_0 is piecewise continuous, and all these functions are positive and satisfy (5.52)–(5.56) at $t = 0$.

As opposed to the neoclassic optimal control problems of Chap. 3 that are handled by the Pontrjagin maximum principle from Sect. 2.4, there are no general extremum conditions for the dynamic systems with endogenous delay of type (5.51)–(5.56). So we need to use more general variational (perturbation) techniques to derive extremum conditions.

5.4.1.1 Conversion to Problem Without State Constraints

Let us choose $m(t)$, $t \in [0, \infty)$, to be the *independent control* of the problem (5.51)–(5.56), then $a(t)$ and $Q(t)$ are dependent (*state*) variables. The optimal control problem involves the constraint (5.55) on the state variable a. The optimization problems with state constraints may have many challenges and exotic behavior of solutions even in simple optimal control of linear ODEs. They cannot be handled by standard optimization procedures such as the maximum principle of Sect. 2.4 or similar. To proceed with the analysis of problem (5.51)–(5.56), we need to handle these constraints. It appears to be possible because of the special structure of the problem.

The constraint $a(t) < t$ is always valid by (5.53) because m is nonnegative by (5.54). Next, under the assumption L = const, the derivative of the equality (5.53) is

$$0 = m(t) - m(a(t))a'(t). \tag{5.57}$$

Because m is nonnegative by (5.54), (5.57) means that $a'(t) \geq 0$ if $m(t) \geq 0$. Moreover, by (5.57), $a'(t) = 0$ if $m(t) = 0$. The "no-investment" regime $m(t) = 0$ plays an important role in applications (Chap. 4).

So the state constraints (5.55) are automatically satisfied when the inequality (5.54) holds and can be removed from the problem statement. Here and thereafter we consider the problem (5.51)–(5.54), (5.56).

The above technique works in more general cases of a variable function $L(t)$. Then, the differential constraint $a'(t) \geq 0$ is replaced with a stricter constraint $m(t) \geq m_{\min}(t) = \max\{0, L'(t)\}$ on the control m only.

5.4.2 *Variational Techniques*

Let the unknown control m be measurable on $[0, \infty)$. We define $\mathbf{U}$ as the set of the *measurable* controls m that satisfy (5.54) *almost everywhere* (a.e.) on $[0, \infty)$. For any measurable control $m \in \mathbf{U}$, a unique a.e. continuous function a is determined from the state equation (5.53), which a.e. has $a'(t)$ and satisfies (5.54). Next, the unique continuous state variable y is determined from (5.52).

The control m is called *admissible* if m and the corresponding state trajectory (a, Q) satisfy (5.52)–(5.54), (5.56). We assume that the problem (5.51)–(5.54), (5.56) has at least one admissible solution.

We will use the variational (perturbation) method and the method of Lagrange multipliers, which are common techniques to derive extremum conditions for various optimal control problems.

Let us add an arbitrarily small variation (increment) $\delta m(t)$ to the independent control $m(t)$, $t \in [0, \infty)$. Then, we obtain the *perturbed* control $m + \delta m$. The small variation can be defined in different ways. In particular, it is often introduced using the so-called needle-like variation $\delta m(t) = \begin{cases} 0, & t\in[0,\infty)-\Delta, \\ v, & t\in\Delta, \end{cases}$ where $v \in \mathbf{U}$ and $\Delta\subset[0,\infty)$ is a set of intervals of a small measure $0 < \mu\Delta \ll 1$. Hence, the needle-like variation $\delta m(t) \neq 0$ on Δ only.

The perturbed control $v = m + \delta m$ leads to a new state trajectory $(a + \delta a, Q + \delta Q)$. The variation δm is called *admissible* if both m and $m + \delta m$ are admissible. Then the functional I undergoes the variation

$$I = I(m + \delta m, a + \delta a, Q + \delta Q) - I(m, a, Q). \tag{5.58}$$

caused by δm. Since the domain $[0, M(t)]$ is closed, a general necessary condition for an extremum is $\delta I \geq 0$ for all admissible variations δm. Further development of this conditions leads to a maximum principle for the problem (5.51)–(5.56).

5.4.3 Method of Lagrange Multipliers

The method of Lagrange multipliers is a powerful general technique to handle optimization problems with state constraints. It uses a modified objective functional and additional unknown *dual variables* and works well for many nonstandard optimization problems, including the problem (5.51)–(5.56). In particular, it can be used to derive the maximum principle of Sect. 2.4. Here we demonstrate the use of Lagrange multipliers to determine optimality conditions for the problem (5.51)–(5.56). For clarity, we omit certain minor details.

Let us introduce the Lagrange multipliers $\lambda_1(t)$ for the constraint (5.52) and $\lambda_2(t)$ for the constraint (5.53), $t \in [0, \infty)$, and construct the following functional

$$\Lambda = \int_0^\infty \left\{ \frac{(Q - pm)^{1-\gamma}}{1-\gamma} + \lambda_1 \left[Q - \int_{a(t)}^t \beta(\tau)m(\tau)\mathrm{d}\tau \right] + \lambda_2 \left[L - \int_{a(t)}^t m(\tau)\mathrm{d}\tau \right] \right\} \mathrm{e}^{-rt}\mathrm{d}t \tag{5.59}$$

known as the *Lagrange functional* (*Lagrangian*) of the optimal control problem (5.51)–(5.56). The functions $\lambda_1(t)$ and $\lambda_2(t)$ are arbitrary at this point and will be determined below to simplify the treatment.

Since the Lagrangian (5.59) does not involve inequality-constraints, it may be maximized instead of the original functional (5.51) under natural assumptions. Indeed, for any choice of λ_1 and λ_2, the value of $\Lambda(m)$ is the same that of $I(m)$ for any admissible m that satisfies (5.52) and (5.53). So we can consider the problem of maximizing functional Λ instead of I.

As above, we give an admissible variation $\delta m(t)$, $t \in (0, \infty)$, to the control m and determine the corresponding variations $\delta a(t)$ and $\delta Q(t)$, $t \in (0, \infty)$, of the state variables Q and a. Now let us determine the variation $\delta\Lambda = \Lambda(m + \delta m) - \Lambda(m)$ of the Lagrangian (5.59) $\Lambda(m)$. Substituting $m + \delta m$, $a + \delta a$, and $Q + \delta Q$ into (5.57), we obtain that

$$\delta\Lambda = \Lambda(m+\delta m) - \Lambda(m) = \int_0^\infty \left\{ \mathrm{e}^{-rt} \frac{(Q+\delta Q - p(m+\delta m))^{1-\gamma}}{1-\gamma} - \frac{(Q-pm)^{1-\gamma}}{1-\gamma} \right.$$
$$+ \lambda_1(t)\left[\delta Q(t) - \int_{a(t)+\delta a(t)}^{t} \beta(\tau)\delta m(\tau)\mathrm{d}\tau - \int_{a(t)}^{a(t)+\delta a(t)} \beta(\tau)m(\tau)\mathrm{d}\tau \right]$$
$$\left. +\lambda_2(t)\left[-\int_{a(t)+\delta a(t)}^{t} \delta m(\tau)\mathrm{d}\tau - \int_{a(t)}^{a(t)+\delta a(t)} m(\tau)\mathrm{d}\tau \right] \right\} \mathrm{d}t. \tag{5.60}$$

The integral in (5.60) includes the function $x^{1-\gamma}/(1-\gamma)$, where $x = Q$–pm. Using the Taylor's expansion

$$(x+\delta x)^{1-\gamma}/(1-\gamma) = (x)^{1-\gamma}/(1-\gamma) + x^{-\gamma}\delta x + o(\delta x)$$

for $x^{1-\gamma}/(1-\gamma)$ up to the second order with respect to δx, we obtain that

$$\frac{(Q+\delta Q - p(m+\delta m))^{1-\gamma}}{1-\gamma} - \frac{(Q-pm)^{1-\gamma}}{1-\gamma} \approx \frac{(\delta Q - p\delta m)}{(Q-pm)^{-\gamma}}.$$

Next, using the well-known approximation of the integral

$$\int_{a(t)}^{a(t)+\delta a(t)} \beta(\tau)m(\tau)\mathrm{d}\tau = \beta(a(t))m(a(t))\delta a(t) + o(\delta a(t))$$

and disregarding the small values of the second and upper orders with respect to δm, δa, δQ, we obtain from (5.60) that

$$\delta L \approx \int_0^\infty \left\{ \mathrm{e}^{-rt} \frac{\delta Q(t) - p(t)\delta m(t)}{(Q(t) - p(t)m(t))^{1-\gamma}} \right.$$
$$+ \lambda_1(t)\left[\delta Q(t) - \int_{a(t)}^{t} \beta(\tau)\delta m(\tau))\mathrm{d}\tau - \beta(a(t))m(a(t))\delta a(t) \right]$$
$$\left. + \lambda_2(t)\left[-\int_{a(t)}^{t} \delta m(\tau))\mathrm{d}\tau - m(a(t))\delta a(t) \right] \right\} \mathrm{d}t. \tag{5.61}$$

Next, we take into account that $\delta m(\tau) \equiv 0$ at $\tau \leq 0$, and interchange the order of integration as

$$\int_0^{\infty} \mathrm{e}^{-rt} \int_{\max(a(t),0)}^{t} \beta(\tau)\delta m(\tau)\mathrm{d}\tau \mathrm{d}t = \int_0^{\infty} \beta(\tau)\delta m(\tau) \int_{\tau}^{a^{-1}(\tau)} \mathrm{e}^{-rt}\mathrm{d}t\mathrm{d}\tau,$$

where $a^{-1}(t)$ is the inverse of $a(t)$. Then, (5.61) can be represented as

$$\begin{aligned}\delta\Lambda \approx \int_0^{\infty} \Bigg\{ & \mathrm{e}^{-rt}\left[\frac{1}{(Q(t)-p(t)m(t))^{\gamma}} - \lambda_1(t)\right]\delta y(t) \\ & + \left[-\frac{\mathrm{e}^{-rt}p(t)}{(Q(t)-p(t)m(t))^{\gamma}} + \beta(t)\int_t^{a^{-1}(t)} \lambda_1(u)\mathrm{e}^{-ru}\mathrm{d}u - \int_t^{a^{-1}(t)} \lambda_2(u)\mathrm{e}^{-ru}\mathrm{d}u\right]\delta m(t) \\ & + \left[\lambda_1(t)\beta(a(t))) - \lambda_2(t)\right] m(a(t))\delta a(t)\Bigg\}\mathrm{d}t.\end{aligned} \tag{5.62}$$

Finally, we recall that the Lagrange multipliers $\lambda_1(t)$ and $\lambda_2(t)$ may be taken arbitrary. So we select the Lagrange multipliers $\lambda_1(t)$ and $\lambda_2(t)$ to set the coefficients at the variations δa and δQ in (5.62) be equal to zero:

$$\begin{aligned}\lambda_1(t)\beta(a(t))) &= \lambda_2(t), \\ \frac{1}{(Q(t)-p(t)m(t))^{\gamma}} &= \lambda_1(t).\end{aligned} \tag{5.63}$$

The equalities (5.63) are called the *dual* (*adjoint*) equations for the problem (5.51)–(5.56) with respect to the dual variables λ_1 and λ_2. Solution of the dual problem can be challenging. Fortunately, in our problem, the dual system (5.63) has an explicit solution. Substituting the solution (λ_1, λ_2) of (5.63) to (5.62), the increment of the functional caused by a small perturbation of the control function m is represented as

$$\delta I = I(m+\delta m) - I(m) = \int_0^{\infty} \mathrm{e}^{-rt} I'(t)\delta m(t)\mathrm{d}t + o\left(\|\delta m\|^2\right), \tag{5.64}$$

where the function

$$I'(t) = \int_t^{a^{-1}(t)} \mathrm{e}^{-ru} \frac{\beta(t) - \beta(a(u))}{(Q(u)-p(u)m(u))^{\gamma}}\mathrm{d}u - \frac{\mathrm{e}^{-rt}p(t)}{(Q(t)-p(t)m(t))^{\gamma}}. \tag{5.65}$$

is called *the gradient* of the functional I in m or the *functional derivative* (*Frechét derivative*) of the functional I with respect to m.

5.4.4 Extremum Conditions

Let us denote the problem solution as m^* (if it exists). In order for a function $m^*(t)$, $t \in [0, T)$, to be a solution of the problem (5.29)–(5 . 34), it is necessary that

$$\begin{aligned} &I'(t) \le 0 \quad \text{at} \quad m^*(t) = 0, \\ &I'(t) \ge 0 \quad \text{at} \quad m^*(t) = M(t), \\ &I'(t) \equiv 0 \quad \text{at} \quad 0 < m^*(t) < M(t), \quad t \in [0, T). \end{aligned} \tag{5.66}$$

The proof of conditions (5.66) follows from the general necessary extremum condition: $\delta I = I(m^* + \delta m) - I(m^*) \ge 0$ for any admissible variation δm.

Using more sophisticated mathematical transformations, it is possible to prove that the extremum condition (5.66) is also sufficient in the nonlinear utility case $\gamma > 0$ [5]. The proof is based on the of concavity of the functional $I(m)$ and required carrying similar transformations with keeping the small values of the second order in (5.60).

Exercises

1. Prove that the equalities (5.4) lead to the optimality condition (5.2).
2. Prove that the function $a^{-1}(t) = t + T$ is the inverse to the function $a(t) = t - T$ if $T = \text{const} > 0$.
   HINT: Apply the definition of the inverse.
3. Substitute the function $a(t) = t - T$ and exponential functions (5.24) at $g = g_p$ into the nonlinear integral equation (5.23) and estimate the integrals in order to obtain the nonlinear integral equation (5.26) with respect to the constant unknown T.
   HINT: Use the fact that the inverse is $a^{-1}(t) = t + T$.
4. Rewrite the model (5.13)–(5.19) in the case of exponential technological change and capital price (5.24) and obtain the necessary condition for the optimality of a solution.
5. Show that at small rates $g < r \ll 1$, the nonlinear equation (5.26) has a solution T described by the approximate formula (5.27).
   HINT: Expand two exponential functions in (5.26) into the Taylor series and neglect the cubic and higher terms (similarly to the approximate solution (5.11) of the nonlinear equation (5.9)).
6. Write a few statements and draw a picture of how you understand the transition dynamics period, the long-term dynamics interval, and turnpike properties.
7. Prove that possible interior solutions to the problem (5.38)–(5.40), (5.42) are determined from (5.43).
8. Verify that the nonlinear equation (5.46) has a unique positive solution T^*.

9. Using the Taylor series for the exponential function, find a solution to the nonlinear equation (5.46) for small g and r.
10. Provide all steps to obtain the variational formula (5.60) from Lagrangian (5.59).

References[1]

1. Boucekkine, R., Germain, M., Licandro, O.: Replacement echoes in the vintage capital growth model. J. Econ. Theory **74**, 333–348 (1997)
2. Cooley, T., Greenwood, J., Yorukoglu, M.: The replacement problem. J. Monet. Econ. **40**, 457–499 (1997)
3. Hartman, J.: The parallel replacement problem with demand and capital budgeting constraints. Nav. Res. Logist. **47**, 40–56 (2000)
4. 📖Hritonenko, N., Yatsenko, Y.: Modeling and optimization of the lifetime of technologies. Kluwer Academic, Dordrecht (1996)
5. Hritonenko, N., Yatsenko, Y.: Optimal control of Solow vintage capital model with nonlinear utility. Optimization **36**, 581–590 (2008)
6. Hritonenko, N., Yatsenko, Y.: Turnpike and optimal trajectories in integral dynamic models with endogenous delay. J. Optim. Theory Appl. **127**, 109–127 (2005)
7. Hritonenko, N., Yatsenko, Y.: The dynamics of asset lifetime under technological change. Oper. Res. Lett. **36**, 565–568 (2008)
8. Jovanovic, B.: Vintage capital and inequality. Rev. Econ. Dyn. **1**, 497–530 (1998)
9. Regnier, E., Sharp, G., Tovey, C.: Replacement under ongoing technological progress. IIE Trans. **36**, 497–508 (2004)
10. Terborgh, G.: Dynamic equipment policy. McGraw-Hill, New York (1949)
11. Yatsenko, Y.: Maximum principle for Volterra integral equations with controlled delay time. Optimization **53**, 177–187 (2004)

[1] The book symbol 📖 means that the reference is a textbook recommended for further student reading.

Part II
Models in Ecology and Environment

Chapter 6
Mathematical Models of Biological Populations

The study of development and interaction of biological species is an important direction of modern research. As one of the central parts of this study, mathematical modeling assists in understanding the behavior of populations and provides reliable forecasts and recommendations for sustainable policies and management. A variety of mathematical models of biological populations and their investigation techniques have been pretty well developed, but practice requires new models that consider, for instance, aftereffect and joint influence of different exogenous and endogenous factors. This chapter explores well-known population models that have become a foundation to contemporary models widely used in practice. Section 6.1 presents population models based on ordinary differential equations and basic elements of their analysis. Section 6.2 explores different types of interaction among species and offers a detailed analysis of predator–prey models. Section 6.3 discusses partial differential and integral models of population dynamics.

6.1 Models of Single Species Dynamics

In biology and ecology, a *population* is defined as a community of organisms of the same species (animals, plants, humans) that live in the same geographical area. An important feature of a population is that its members can mate and produce offspring. This section considers four celebrated classic models that describe the dynamics of one-species populations.

N. Hritonenko and Y. Yatsenko, *Mathematical Modeling in Economics, Ecology and the Environment*, Springer Optimization and Its Applications 88,
DOI 10.1007/978-1-4614-9311-2_6,

6.1.1 Malthusian Growth Model

Let us introduce the following characteristics of a population:

$N(t)$—the *population size*: the number of individuals in a population at time t,
μ—the coefficient of population intrinsic growth: the difference between constant birth rate and death rates.

Then, the *average rate of change* in the population size over the time interval Δ is $(N(t + \Delta) - N(t))/\Delta$, and the *instantaneous rate of change* is

$$\lim_{\Delta \to 0} \frac{N(t+\Delta) - N(t)}{\Delta} = \frac{dN(t)}{dt}.$$

If the instantaneous rate is proportional to the population size N, then

$$dN/dt = \mu N. \tag{6.1}$$

Equation (6.1) is the first mathematical description of a population and is called the *Malthus* or *Malthusian growth model* after the British scholar *Thomas Robert Malthus* (1766–1834). Malthus authored a book *An Essay on the Principle of Population*, one of the most influential books about populations. The book had six editions between 1798 and 1826. It is interesting to mention that the first edition was published anonymously and Malthus did not use any mathematical models to present his ideas. The solution to the linear differential equation (6.1)

$$N(t) = N(0)\exp(\mu t) \tag{6.2}$$

shows that the dynamics of the population depends on the sign of the parameter μ: there is the exponential decay if $\mu < 0$, no change if $\mu = 0$, and the exponential growth if $\mu > 0$. The last case produces the *unlimited* growth in a population, which is possible only for some populations over a limited time period. Malthus was puzzled with such a result and tried to find different explanations and interpretations.

The modern view is that the model (6.1) is valid in an ideal environment with unlimited food and space resources, absence of competition, and small population size. More realistic models consider the population growth to be dependent on population density, resources, and other factors [3]. Below we discuss two models that present different ways of modeling the restricted growth.

6.1.2 Von Bertalanffy Model

The *von Bertalanffy growth equation*

$$\frac{\mathrm{d}l(t)}{\mathrm{d}t} = \mu(L - l(t)), \tag{6.3}$$

is the simplest population model of a restricted growth. In (6.3),

t—time,
$l(t)$—the length or size of population individuals,
L—the maximum length of the individuals,
μ—the *von Bertalanffy growth rate,* specific for each population.

The model (6.3) is named after the Austrian born biologist Karl Ludwig von Bertalanffy (1901–1972) who suggested it in 1934. It can be also found in earlier publications, for instance, by Verhulst in 1847 and Putter in 1920 [2]. The model (6.3) can be used for the description of trees, fish, and other populations where size or length plays an important role. The analytic solution of the linear differential equation (6.3)

$$l(t) = L - (L - l(0))\mathrm{e}^{-\mu t}, \tag{6.4}$$

approaches the maximum length L as t becomes large, $t \to \infty$. Thus, L can be interpreted as the asymptotic length for it will never be reached. The solution (6.4) is depicted in Fig. 6.1. Its first derivative $l'(t) = \mu(L - l(0))\mathrm{e}^{-\mu t}$ is positive and the second one is negative. Hence, the growth rate, which is proportional to the difference between the maximum length L and the current length l, is positive and decreases over time. Indeed, population individuals grow throughout their lives, do not exceed their maximum length, and newborns and juveniles grow faster than adults.

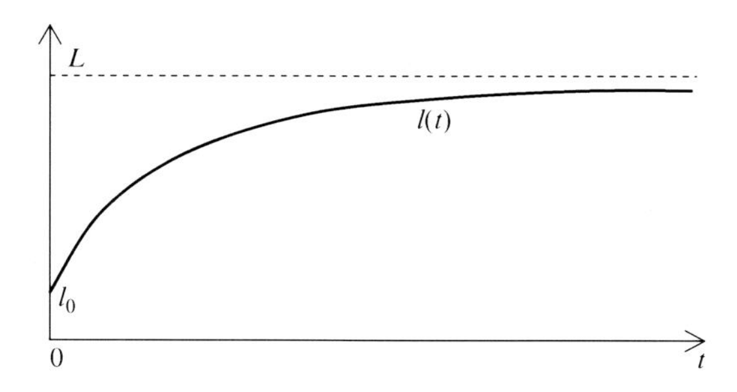

Fig. 6.1 The solution of the von Bertalanffy population model (6.3)

6.1.3 Verhulst–Pearl Model

A population model with the growth rate proportional to both population density and available resource can be described by the nonlinear differential equation:

$$\frac{\mathrm{d}N(t)}{\mathrm{d}t} = \mu N(t)\left(1 - \frac{N(t)}{K}\right), \tag{6.5}$$

where K is a positive constant and other parameters are as in the Malthus model (6.1). The multiplier $(1-N/K)$ reflects the environmental resistance to a population increase. When the population size is small, then $N/K << 1$ and the model (6.5) suggests the fast exponential growth of the population as the Malthusian model (6.1). As $N(t)$ gets closer to K, the growth rate decreases and becomes negative if $N(t)$ exceeds K.

The model (6.5) was proposed in 1838 by the Belgian mathematician Pierre-Francois Verhulst (1804–1849). He compared analytical outcomes of the model with the data on populations in France, Belgium, and Russia and found a good correspondence between the model and actual data and, thus, referred to (6.5) as the "*logistic equation*" in his papers of 1845. This model was rediscovered independently several times by Robertson in 1908, McKendrick in 1911, Pearl in 1920, Reed in 1920, and others. Fortunately, R. Pearl noticed the work of Verhulst in 1922 and returned the name "logistic" to the model. The model (6.5) provides a good mathematical description for many biological populations of microorganisms, plants, and animals. It is also widely used in statistics, economics, medicine, physics, chemistry, and other applications.

Elements of Model Analysis:

1. ***Equilibrium. Stationary trajectories*** are solutions independent of time t and, thus, $\mathrm{d}N/\mathrm{d}t = 0$. Stationary trajectories $N = \text{const}$ of the Verhulst–Pearl model (6.5) are found from the equation $\mu N(1 - N/K) = 0$, which has two solutions $N_0 = 0$ and $N^* = K$, also called the *equilibrium states*.
2. ***Stability Analysis*** aims to investigate the behavior of a solution around an equilibrium and figure out whether the solution returns to the equilibrium after a small perturbation. Because of challenges and, at the same time, importance of the stability analysis, it can be performed in different ways. One of the ways is to give a small perturbation δN to the equilibrium and investigate the behavior of the resulting system. Another way will be demonstrated in Sect. 6.2.1.

   Two simple techniques of the stability analysis for (6.5) are graphic and analytic ones. To illustrate the graphic technique, let us plot the dependence dN/dt on N using (6.5)—see Fig. 6.2. The two x-intercepts (0, 0) and $(K, 0)$ are the *points of equilibrium*. The derivative $\mathrm{d}N/\mathrm{d}t > 0$ at $N > 0$ in some neighborhood of (0, 0), which means that the trivial equilibrium $N = 0$ is *unstable*. Conversely, $\mathrm{d}N/\mathrm{d}t$ is positive at $N < K$ and negative at $N > K$ around $(K, 0)$, which means that the positive equilibrium $N^* = K$ is *stable*. Checking the sign of the first derivative of the left side of (6.5) around each equilibrium, we approach the same conclusion in the analytic way.

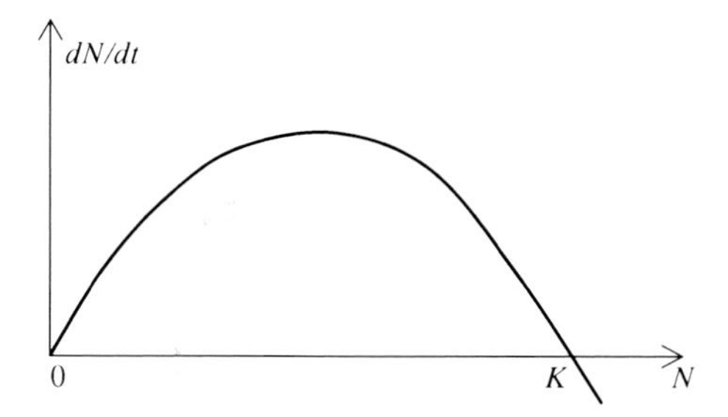

Fig. 6.2 The dependence of dN/dt on N in the Verhulst–Pearl model (6.5)

3. ***Analytic Solution***. Dividing both sides of the equation (6.5) by N^2 and rewriting it in a new variable $v = 1/N - 1/K$ as $dv/dt = -\mu v$, we obtain that the last equation has a solution $v = v(0)\exp(-\mu t)$, which, in the original notation $N(t)$, produces the exact solution to (6.5):

$$N(t) = \frac{N(0)e^{\mu t}}{1 + N(0)(e^{\mu t} - 1)/K}, \tag{6.6}$$

known as the *logistic curve*, *S-shaped curve*, *Verhulst curve*, and *sigmoid curve*. The graph of the solution $N(t)$ is depicted in Fig. 6.3. In general, the logistic curve can represent a solution not only to the Verhulst equation (6.5) but also to Gompertz and other equations.

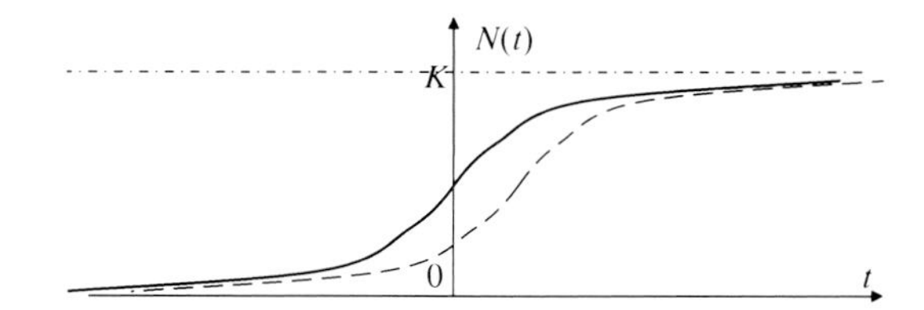

Fig. 6.3 The solution of the Verhulst–Pearl population model (6.5) known as the logistic curve

4. ***Qualitative Analysis***. Since

$$\lim_{t\to\infty} N(t) = \lim_{t\to\infty} \frac{N(0)K}{Ke^{-\mu t} + N(0)(1 - e^{-\mu t})} = K,$$

the solution $N(t)$ approaches K at large t. For this reason, the parameter K is called the *carrying capacity* and determines the maximal population size supported by the environment. If $N(0) > K$, the population size will decrease down to K, otherwise, it will increase up to K. That is, independently of its initial size, the population will approach its capacity K if $N(0) > 0$. Thus, if vital resources are reduced because of population growth, then the population growth slows down and the population size approaches its limiting environment capacity K.

The first derivative of the solution (6.6) is always positive and its second derivative is positive if $N < K/2$ and negative if $N > K/2$. Therefore, the solution of (6.5) increases and concave upward if $N < K/2$ and downward if $N > K/2$, see Fig. 6.3.

6.1.4 Controlled Version of Verhulst–Pearl Model

Control functions are introduced to population models to change the development of a population [4, 6, 8, 12, 13]. For instance, the model (6.5) can be used in medicine for modeling tumor growth and determining the time for starting chemotherapy. Then the control is a *therapy-induced death rate of tumors* $u(t)$, and a controlled version of the model (6.5) takes the form:

$$\frac{dN(t)}{dt} = \mu N(t)\left(1 - \frac{N(t)}{K}\right) - u(t)N(t). \tag{6.7}$$

In the new notations $w = \mu - u$ and $K_1 = K(\mu - u)/\mu$, the model (6.7) is written as

$$\frac{dN(t)}{dt} = wN(t)\left(1 - \frac{N(t)}{K_1}\right),$$

which is the *logistic equation* (6.5). So we have just justified a remarkable property of the model (6.5): if a population follows the logistic equation in the absence of controls, then it continues to follow a logistic equation after a linear control has been implemented but with a lower growth rate and capacity. All other properties of the logistic equation remain to be valid for (6.7).

The model (6.7) can be also interpreted as a *harvesting problem* with u as the harvesting effort. Harvesting is an important class of applied biological and economic problems: humans harvest crops, animals, trees, fossil fuel to survive. Unreasonable and irrational harvesting may lead to exhaustion and, even, disappearance of some species and nonrenewable natural resources. Chapter 7 discusses several more advanced harvesting models.

6.1.5 Verhulst–Volterra Model with Hereditary Effects

Various applications of the Verhulst model (6.5) require its modification to fit specifics of real problems. One modification, a controlled version (6.7) of the model (6.5), is considered in the previous section. Another possible extension is related to yearly seasonal changes in the environment that can be described by a periodic carrying capacity $K(t) = K(t + T)$, where T is 1 year. Other essential modifications of the Verhulst model incorporate time delays and consider the carrying capacity $K(t)$ as a function of a population prehistory, or introduce the dependence of the population dynamics on certain processes in the population. Such generalizations lead to *logistic equations with delay*, which have remarkable behavior and properties.

6.1.5.1 A Logistic Equation with Distributed Delay

Vito Volterra (1860–1940), an Italian mathematician and physicist, incorporated distributed time delays (hereditary or historical actions) into the Verhulst model (6.5) as an integral term. His equation

$$\mathrm{d}N/\mathrm{d}t = \mu N\left(1 - N/K - \int_0^t f(t-\tau)N(\tau)\mathrm{d}\tau\right), \tag{6.8}$$

describes the self-intoxication process in a population by the waste products of its own metabolism. It is a modification of the Verhulst–Pearl model (6.5), in which the additional integral *delay* (*hereditary*) term represents the decrease of the growth rate μ due to catabolic effects.

The function $f(t - \tau)$ in (6.8) is called the *hereditary function* and determines the influence of a prehistory interval on the population dynamics at the current time t. The model (6.8) considers the *distributed delay* (or *aftereffect*) on the interval $[0, t)$. Under certain assumptions, (6.8) has a unique solution $N(t)$ that approaches the *equilibrium state* $\hat{N} = K\left(1 + K\int_0^\infty f(t-\tau)N(\tau)\mathrm{d}\tau\right)^{-1}$ at $t \to \infty$.

6.1.5.2 A Logistic Equation with Point Delay

The *lumped* (*point*) *delay* introduced into the Verhulst–Pearl model (6.5) produces the following *delay* (*lag*) *differential equation*:

$$\mathrm{d}N(t)/\mathrm{d}t = \mu N(t)(1 - N(t-T)/K). \tag{6.9}$$

The model (6.9) considers the influence of only one past instant whose distance from the current time t is T. Such a behavior is characteristic for populations with a single seasonal reproduction. For some values of the delay parameter $T > 0$, the solution of (6.9) oscillates around the *environment capacity* K.

6.2 Models of Two Species Dynamics

Biological species do not live in isolation. In ecology, an interacting group of two or more different species in a common location is referred to as a *biological community* (or *biocoenosis*) [10]. For example, a forest of trees and undergrowth plants, inhabited by animals, birds, and insects, constitutes a biological community. This section considers models of two-species communities with intraspecies and interspecies interaction.

6.2.1 Lotka–Volterra Model of Two Interacting Species

In the notations:

$N_i\ (t)$—the size of the i-th species at time t,
ε_i—the growth rate of the i-th species,
γ_i—the coefficients of the interaction of the i-th species with another, $i = 1, 2$,

the model of an interaction between two species can be presented as

$$\begin{aligned} \mathrm{d}N_1(t)/\mathrm{d}t &= N_1(t)\left(\varepsilon_1 + \gamma_1 N_2(t)\right), \\ \mathrm{d}N_2(t)/\mathrm{d}t &= N_2(t)\left(\varepsilon_2 + \gamma_2 N_1(t)\right). \end{aligned} \tag{6.10}$$

Depending on the sign of ε_i and γ_i, the system (6.10) describes different relations between two species. For instance, if both γ_1 and γ_2 are positive, then both species benefit each other (*mutualism*). If γ_1 and γ_2 are negative, then the species compete for a common resource. The growth of the first species decreases in the presence of the second population if $\gamma_1 < 0$. If, in addition, $\gamma_2 > 0$, then the second population benefits from the first population and the system (6.10) depicts a *predator–prey* relation discussed in the next section.

The model (6.10) is referred to as *Lotka–Volterra model of two interacting species*. Alfred James Lotka (1880–1949), an American chemist, mathematician, and statistician, published the paper *Analytical note on certain rhythmic relations on organic systems* in 1920, where he introduced the predator–prey model (6.12) for a population of herbivores feeding on plants. The model analysis was later developed in his other papers and book *Elements of Mathematical Biology* of 1925. Unfortunately, these publications did not get much attention in the scientific community and Vito Volterra suggested simultaneously but independently the model (6.12) while working on a study on the population of cartilaginous fish, proposed, by the way, by his future son-in-law. Volterra published his results in Italian in 1926 and later in the notes of his lectures of 1928–1929.

6.2.1.1 Equilibrium

The system (6.10) has two *equilibrium states* N_1^*, N_2^*: the trivial (0,0) and nontrivial $(-\varepsilon_2/\gamma_2, -\varepsilon_1/\gamma_1)$, which is positive if ε_i and γ_i, $i = 1, 2$, have opposite signs. It is obvious that biological applications consider only nonnegative solutions.

6.2.1.2 Community Matrix

Different investigation techniques have been proposed to analyze and describe a complex interaction among species. One of them employs a *community matrix*. The *community matrix* is the Jacobi matrix at equilibrium points. Its *ij*-element $\partial f_i/\partial N_j$ shows how the growth of *i*-species changes with the size change in the *j*-species, or, in other words, it reflects the effect of *j*-species on *i*-species at equilibrium. If the *ij*-element of the matrix is negative then the growth rate of *i*-species decreases if the size of *j*-species increases, i.e., *j*-species has a negative effect on *i*-species. If it is zero, then changes in *j*-species do not affect the growth rate of *i*-species. Analogously, under a positive *ij*-element, *j*-species benefits *i*-species.

The *eigenvalues* of the community matrix determine the stability of the equilibrium point: an eigenvalue with a positive real part shows that the equilibrium is unstable. If all eigenvalues have negative real parts, then the equilibrium is stable.

The community matrix for the two-species model (6.10) at the equilibrium (N_1^*, N_2^*) is

$$D(f(N^*)) = \begin{bmatrix} \varepsilon_1 + \gamma_1 N_2^* & \gamma_1 N_1^* \\ \gamma_2 N_2^* & \varepsilon_2 + \gamma_2 N_1^* \end{bmatrix}. \tag{6.11}$$

The community matrix is used to analyze the stability of a predator–prey model in Sect. 6.2.2.

6.2.1.3 Types of Interspecies Interaction

Depending on the signs of parameters, the model (6.10) describes various types of interaction between two species:

- $\varepsilon_1 > 0, \varepsilon_2 > 0, \gamma_1 < 0, \gamma_2 < 0$—*competition*: the interaction is harmful for both species;
- $\varepsilon_1 > 0, \varepsilon_2 < 0, \gamma_1 < 0, \gamma_2 > 0$—*antagonism*: one species benefits from another, for instance, "predator–prey" or "parasite–host" relationship is negative for one species and positive for another;
- $\gamma_1 > 0, \gamma_2 > 0$—*mutualism*: beneficial for both species;
- $\varepsilon_2 > 0$, $\gamma_1 = 0$, $\gamma_2 < 0$—*amensalism*: a harmful relation for one species and neutral for the other;

- $\varepsilon_2 < 0$, $\gamma_1 = 0$, $\gamma_2 > 0$—*commensalism*: beneficial for one species, neutral for another.
- $\gamma_1 = 0$ or $\gamma_2 = 0$—*neutralism*: species develop simultaneously but independently.

A more detailed analysis of different types of interaction leads to extensions of the model (6.10) that can be also modified to include more species. Although widely used, the model (6.10) represents simplified interactions between two species and has several drawbacks. In particular, the growth of each species in absence of other species is subject to the Malthus law (6.1). To consider more realistic situations, the model (6.10) can be modified to include the logistic or other density-dependent growth.

The Lotka–Volterra model (6.10) that initially appeared in ecology has been also used in other applications, such as chemical oxidation reactions, opponent interaction in military problems, economics, innovation propagation under technological change, and others. Section 3.4 discusses some economic applications of this model.

6.2.2 Lotka–Volterra Predator–Prey Model

One of the most studied models of two-species interaction is the *Lotka–Volterra predator–prey model*. To reflect a predator–prey interaction, the model (6.10) can be rewritten as the following system with positive coefficients $\varepsilon_1, \varepsilon_2, \gamma_1, \gamma_2$:

$$\begin{aligned} \mathrm{d}N_1/\mathrm{d}t &= N_1(\varepsilon_1 - \gamma_1 N_2), \\ \mathrm{d}N_2/\mathrm{d}t &= N_2(-\varepsilon_2 + \gamma_2 N_1), \end{aligned} \tag{6.12}$$

where

N_1 is the size of a *prey population*, which positive growth rate ε_1 is negatively impacted by the presence of predators,
N_2 is the *size of predators* whose growth rate ε_2 is increased as a result of the prey consumption.

6.2.2.1 Elements of Model Analysis

1. ***Equilibrium States***. The model has two stationary *equilibrium states*:

   - *The trivial equilibrium* $N_1 = N_2 = 0$: no prey and no predators;
   - *The nontrivial equilibrium* $N_1{}^* = \varepsilon_2/\gamma_2$, $N_2{}^* = \varepsilon_1/\gamma_1$: a balance between prey and predators.

   Two special nonequilibrium trajectories of the model are as follows:

   - $N_2(t) = N_2(0)\exp(-\varepsilon_2 t)$ at $N_1 \equiv 0$: in the absence of prey, the number of predators decreases;

- $N_1(t) = N_1(0)\exp(\varepsilon_1 t)$ at $N_2 \equiv 0$: prey reproduce and grow exponentially fast in the absence of predators.

2. ***Stability of Equilibrium***. The community matrix (6.11) for the predator–prey model (6.12)

$$D\big(f(N^*)\big) = \begin{bmatrix} \varepsilon_1 + \gamma_1 N_2^* & -\gamma_1 N_1^* \\ \gamma_2 N_2^* & -\varepsilon_2 + \gamma_2 N_1^* \end{bmatrix},$$

at the trivial equilibrium (0,0) becomes

$$D(f(0,0)) = \begin{bmatrix} \varepsilon_1 & 0 \\ 0 & -\varepsilon_2 \end{bmatrix}.$$

Because the community matrix is diagonal, its eigenvalues are given by the diagonal elements $\lambda_1 = \varepsilon_1 > 0$, $\lambda_2 = -\varepsilon_2 < 0$. Thus, the trivial equilibrium (0,0) is unstable. This result has an important interpretation: if the trivial equilibrium (0,0) were stable, non-zero populations might be attracted by it, and the system would lead toward the extinction of both species. Fortunately, (0,0) is unstable, so the extinction of both species is unlikely.

The eigenvalues for the community matrix at the positive equilibrium $(\varepsilon_2/\gamma_2, \varepsilon_1/\gamma_1)$

$$D(f(N^*)) = \begin{bmatrix} 0 & -\dfrac{\varepsilon_2\gamma_1}{\gamma_2} \\ \dfrac{\varepsilon_1\gamma_2}{\gamma_1} & 0 \end{bmatrix},$$

Are found from the equation $\det(D - \lambda I) = 0$ as $\lambda_1 = i\sqrt{\varepsilon_1\varepsilon_2}$, $\lambda_1 = -i\sqrt{\varepsilon_1\varepsilon_2}$. The fact that both of them are imaginary does not allow for determining stability of the nontrivial equilibrium. Thus, a more detailed stability analysis is necessary. In the case of the system (6.12), we can find its exact solution and then look at the stability.

3. ***Analytic Solution***. Let us divide the first equation of the system (6.12) by the second one and integrate the resulting system after separation of variables:

$$\frac{\mathrm{d}N_1/\mathrm{d}t}{\mathrm{d}N_2/\mathrm{d}t} = \frac{N_1(\varepsilon_1 - \gamma_1 N_2)}{N_2(-\varepsilon_2 + \gamma_2 N_1)} \Rightarrow \int \left(-\frac{\varepsilon_2}{N_1} + \gamma_2\right)\mathrm{d}N_1 = \int \left(\frac{\varepsilon_1}{N_2} - \gamma_1\right)\mathrm{d}N_2,$$

which yields a *solution* to (6.12) in the implicit form

$$\varepsilon_2 \ln N_1 - \gamma_2 N_1 + \varepsilon_1 \ln N_2 - \gamma_1 N_2 = C, \tag{6.13}$$

where C is a constant of integration. The exponent of both sides of (6.13) produces another implicit form of the solution

$$\left(N_1^{\varepsilon_2} e^{-\gamma_2 N_1}\right)\left(N_2^{\varepsilon_1} e^{-\gamma_1 N_2}\right) = e^C = C_1, \tag{6.14}$$

where C_1 is a new constant of integration. A convenient tool for representing implicit solutions similar to (6.13) is the phase portrait.

4. ***Phase Portrait***. The phase portrait plots the solution (6.13) as a graph on the *phase plane* with coordinates N_1 and N_2 and t as a parameter. The solution curves are referred to as *trajectories*. The sign of dN_1/dt and dN_2/dt provides the direction of the motion. The properties of trajectories in the phase plane are:

   - There is at most one trajectory through any point in the phase plane.
   - A trajectory not starting at an equilibrium point cannot reach it in a finite time.
   - A trajectory cannot intersect itself, unless it is a closed curve.
   - A closed curve for a trajectory corresponds to a periodic solution in time.

   The phase trajectories of the model (6.12) can be found by numerical simulation of the implicit solution (6.14). The phase portrait is given in the Fig. 6.4 for different values of the parameter $C > 0$. Simulation shows that the shape of trajectories tends to an ellipse for $C << 1$ and to a triangle for $C >> 1$. The closed curves correspond to the undamped oscillations of the solutions $N_1(t)$ and $N_2(t)$ with respect to the positive equilibrium (N_1^*, N_2^*). They represent the periodic solutions with periods dependent on the initial conditions. Closed trajectories are not stable under perturbations because they will follow a different trajectory after a small disturbance. Thus, the positive equilibrium of (6.12) is also unstable.

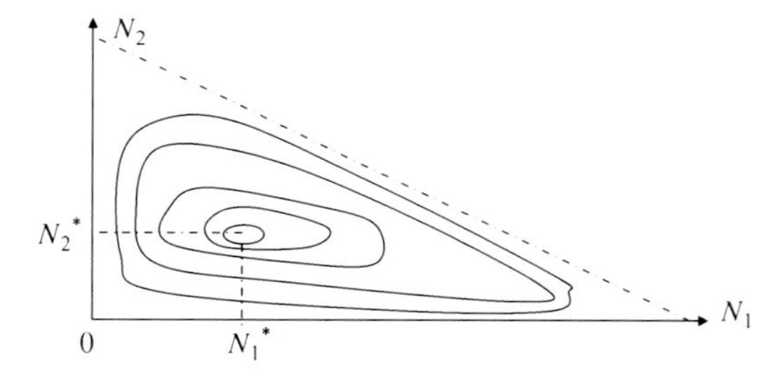

Fig. 6.4 The phase portrait of the Lotka–Volterra predator–prey model (6.12)

   The equilibrium state (N_1^*, N_2^*) is known as the "*centre*" and corresponds to undamped oscillations of $N_1(t)$ and $N_2(t)$ with respect to the values N_1^*, N_2^*.

5. ***Asymptotic of solutions***. The populations of prey and predators oscillate periodically. It can be shown that each oscillation passes through four stages illustrated in Fig. 6.5:

   - Interval T_1: the abundance of prey leads to an increase of predators, causing a decrease in a prey population. The mass of the prey transforms into the mass of the predators, $N_1(t) + N_2(t) \approx C$.

- Interval T_2: the size of prey becomes insufficient to feed predators and decreases, $N_1(t) \approx 0$, $N_2(t) \to 0$.
- Interval T_3: the prey and predators rarely meet, $N_1(t) \approx 0$, $N_2(t) \approx 0$.
- Interval T_4: the prey population grows again because of the small number of predators, $N_2(t) \approx 0$, $N_1(t) \to C$.

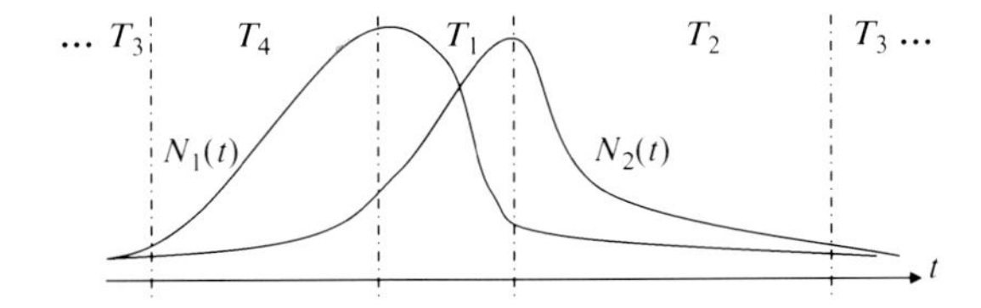

Fig. 6.5 The trajectories of the Lotka–Volterra predator–prey model at $C >> 1$

As seen in Fig. 6.5, the durations of these stages are different. A numerical analysis shows that $T_1 \approx 0.46$, $T_2 \approx 2.3$, $T_3 \approx 10$, $T_4 \approx 2.3$ for $\varepsilon_1/\gamma_1 = \varepsilon_2/\gamma_2 = 1$, $C = 10$.

6.2.3 Control in Predator–Prey Model

Let us consider a simple example of control problems in the model (6.12). Suppose that it is necessary to control prey and/or predator populations to change their nontrivial equilibrium state (N_1^*, N_2^*). This control problem can be considered as an extension of the model (6.12) with the modified coefficients $\overline{\varepsilon}_1 = \varepsilon_1 - u_1$ and $\overline{\varepsilon}_1 = -\varepsilon_2 - u_2$, where the constant values u_1 and u_2 are nonnegative control parameters. For $u_1 < \varepsilon_1$, the controlled predator–prey community has the nontrivial equilibrium state

$$N_1^* = (\varepsilon_2 + u_2)/\gamma_2, \quad N_2^* = (\varepsilon_1 - u_1)/\gamma_1.$$

Hence, the value N_1^* is bigger for larger u_2, while N_2^* is smaller for larger u_1. For $u_1 > \varepsilon_1$, the nontrivial equilibrium state (N_1^*, N_2^*) is absent and $N_1(t)$ and $N_2(t)$ decrease indefinitely. As a result, both populations die out at $u_1 > \varepsilon_1$.

All equilibria in the model (6.12) are unstable, which imposes some limitations on the model applications. A locally stable equilibrium can be obtained by including additional nonlinearities into the model, such as a nonlinear prey response to shortage of nutrition, light, or space resources or predator response to prey shortage. Realistic models of two species interaction take into account more detailed mechanisms of this interaction. For instance, the *Holling model* considers an attack threshold, search times, pursuit, prey eating, and the time between two successful hunting. Some extensions of the predator–prey model (6.12) are provided in the next sections.

6.2.4 Generalized Predator–Prey Models

In the *Kolmogorov predator–prey model*:

$$\begin{aligned} dN_1/dt &= N_1 g_1(N_1) - N_2 L(N_1), \\ dN_2/dt &= N_2 g_2(N_1, N_2), \end{aligned} \tag{6.15}$$

$L(N_1)$ is the *predator traffic function* that reflects the consumption of prey by one predator per time unit,

g_1 and g_2 are the *intrinsic growth coefficients* of prey and predators.

The model (6.15), named after Andrey Nikolaevich Kolmogorov (1903–1987), a Russian mathematician, includes a variety of predator–prey models, in particular,

- $g_1(N_1) = \varepsilon_1$, $L(N_1) = -\gamma_1 N_1$, $g_2(N_1, N_2) = -\varepsilon_1 + \gamma_2 N_1$—the Lotka–Volterra model (6.12);
- $g_1(N_1) = \varepsilon_1 - \gamma N_1$—the intraspecies competition in a population of prey, where the prey are subject to the logistic equation (6.5) in the absence of predators;
- $g_2(N_1, N_2) = \varepsilon_2 - \alpha N_2/N_1$ or $g_2(N_1, N_2) = \varepsilon_2 - \alpha N_2 N_1$ or $g_2(N_1, N_2) = \varepsilon_2[1 - \exp(-\alpha N_1)]$—a model with restrictions on the growth of predators;
- $L(N_1) = \alpha N_1/(1 + \beta N_1)$ or $L(N_1) = \alpha[1 - \exp(-\beta N_1)]$—a model with limited size of the predator population; and others.

The model (6.15) has been proposed to explain different dynamic patterns in a predator–prey community. It has very rich behavior and can produce equilibrium regimes that are stable under some conditions and unstable under others. Let us take a look at the *predator–prey model with intraspecies competition* among the prey:

$$\begin{aligned} dN_1/dt &= N_1(\varepsilon_1 - \gamma_1 N_2 - \gamma N_1), \\ dN_2/dt &= N_2(-\varepsilon_2 + \gamma_2 N_1). \end{aligned} \tag{6.16}$$

In contrast to the model (6.12), its solution $N_1(t) < \varepsilon_1/\gamma$, i.e., the prey population is bounded in the absence of predators ($N_2 \equiv 0$) and is subject to the logistic equation with the carrying capacity ε_1/γ. The model (6.16) has three equilibria: the trivial one (0,0), "no predators" (ε_1/γ, 0), and the nontrivial (N_1^*, N_2^*),

$$N_1^* = \varepsilon_2/\gamma_2, \quad N_2^* = (\varepsilon_1\gamma_2 - \varepsilon_2\gamma)/\gamma_1\gamma_2. \tag{6.17}$$

An analysis shows that the positive equilibrium $(\varepsilon_2/\gamma_2, (\varepsilon_1\gamma_2 - \varepsilon_2\gamma)/\gamma_1\gamma_2)$ is asymptotically stable for $\gamma > 0$. In contrast to (7.17), the equilibrium state $(\varepsilon_2/\gamma_2, \varepsilon_1/\gamma_1)$ in the Volterra model (6.12) is of the "centre" type and is not is asymptotically stable. The oscillations around the "centre" equilibrium are undamped and the phase portrait of such equilibrium state is shown in Fig. 6.5.

Depending on the relation among the parameters, the Kolmogorov model (6.15) can possess different phase portraits of model trajectories. They are illustrated in

Fig. 6.6, where the nontrivial equilibrium state (N_1*, N_2*) is one of the following type:

- *Stable node* (non-oscillating trajectories converge to (N_1*, N_2*)) in Fig. 6.6a
- *Stable node* (N_1*, 0) with dying out predators in Fig. 6.6b
- *Stable focus* (damped oscillations converge to (N_1*, N_2*)) in Fig. 6.6c
- *Unstable focus* and a *stable limit cycle* in Fig. 6.6d

The last case corresponds to asymptotically stable undamped oscillations of the prey $N_1(t)$ and predator $N_2(t)$ population sizes, which converge to a periodic solution (the limit cycle) shown by a closed curve in Fig. 6.6d.

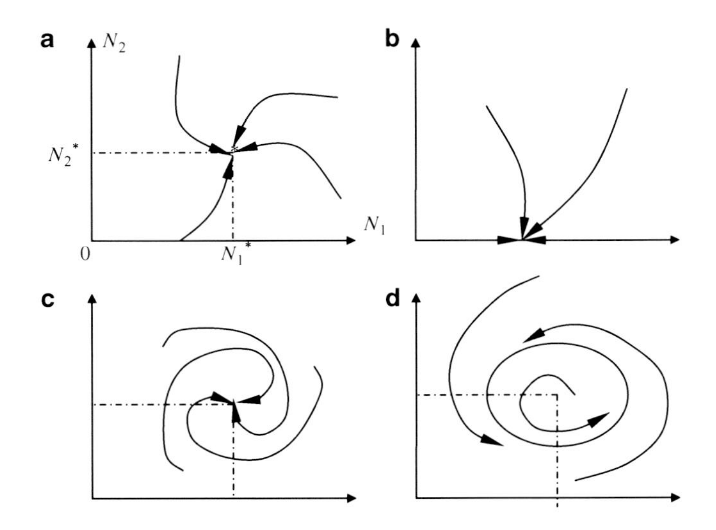

Fig. 6.6 Phase portraits of the Kolmogorov predator–prey model (6.15)

6.2.5 Predator–Prey Model with Individual Migration

The random migration of individuals can be described by a special partial differential equation, known as the diffusion equation. The derivation of this equation will be considered in Sect. 8.1 for pollution propagation in the atmosphere.

To construct a predator–prey model with individual migration, let us introduce the notations:

(x,y)—a point of the two-dimension area S of a predator–prey population habitat, t—a continuous time, $\varphi_1(x,y,t)$ and $\varphi_2(x,y,t)$—the densities of the prey and predator populations.

Then the model is of the form:

$$\begin{aligned} \partial\varphi_1/\partial t &= \kappa_1\Delta\varphi_1 + \varepsilon_1\varphi_1 - \gamma_1\varphi_1\varphi_2, \\ \partial\varphi_2/\partial t &= \kappa_2\Delta\varphi_2 + \gamma_2\varphi_1\varphi_2 - \varepsilon_2\varphi_2, \end{aligned} \tag{6.18}$$

where $\Delta = \partial^2/\partial x^2 + \partial^2/\partial x^2$ is the two-dimensional Laplace operator and κ_1, κ_2 are the *diffusion coefficients* of prey and predators. In space-distributed population models, the population density $\varphi_1(x,y,t)$ or $\varphi_2(x,y,t)$ reflects the number of individuals per space unit. The meaning of the parameters $\varepsilon_1, \varepsilon_2, \gamma_1, \gamma_2$ is the same as in the Lotka–Volterra predator–prey model (6.12).

The dynamics of *partial differential equations* is more complex compared to the ordinary differential equations. To investigate the model (6.18), we need to know initial and boundary conditions to the system (6.18). Standard *initial conditions* are of the form:

$$\varphi_i(x, y, 0) = f_i(x, y), \quad i = 1, 2. \tag{6.19}$$

If the region S is closed, then natural *boundary conditions* reflect the condition that individuals cannot migrate across the boundary δS:

$$\partial\varphi_i/\partial n_S = 0, \quad i = 1, 2, \tag{6.20}$$

where n_S is the normal to the boundary δS of the region S. An analysis shows that (6.18) with conditions (6.19)–(6.20) have *spatial periodic solutions*, with the prey and predator densities oscillating at different points of the region S.

If the region S is unbounded, then (6.18) can have solutions in the form of *travelling waves*. Figure 6.7 illustrates such solutions in a special one-dimensional case of the problem (6.18) for the initial prey distribution $f_1(x) = A\exp(-b|x|)$, $x \in R^1$. Then, introducing the new independent variable $z = x - \nu t$ (where ν is an unknown wave velocity), one can find the asymptotic solution $\varphi_i(x - \nu t)$, $i = 1, 2$, of (6.18). This solution is shown in Fig. 6.7 in the coordinates (φ_i, x) for a fixed value t. The figure demonstrates how a "wave" of predators pursues a "wave" of prey.

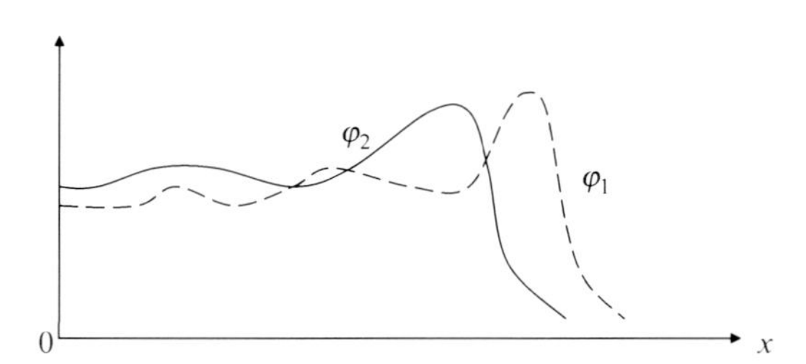

Fig. 6.7 The waves of prey and predators in the one-dimensional population model (6.18) with individual migration

6.3 Age-Structured Models of Population Dynamics

Partial differential and integral equations are used to reflect a dependence of vital parameters of a population, such as fertility and mortality, on age, size, or stage of individuals [1, 5, 7]. This section considers age-structured population models.

6.3.1 *McKendrick Linear Population Model*

Let us consider a single species population and introduce the following notations for its parameters (see Fig. 6.8):

T—time,
$0 \leq t < \infty$,
τ—the age of individuals in a population,
$x(t,\tau)$—the population density of age distribution that represents the number of population individuals of age τ at time t,
$m(t,\tau)$—the age-specific fertility rate that shows the average number of offspring from one individual of age τ at time t,
$\mu(t,\tau)$—the age-specific mortality rate,
$\phi(\tau)$—the population distribution of individuals with respect to their age τ at the initial moment $t = 0$,
$N(t)$—the population size,
A—the maximum age of individuals.

The total number of all individuals (the size) $N(t)$ of the population is

$$N(t) = \int_0^A x(t,\tau)\mathrm{d}\tau, \tag{6.21}$$

and the total number of newborns at time t is

$$x(t,0) = \int_0^A m(t,\tau)x(t,\tau)\mathrm{d}\tau, \quad A \leq t < \infty. \tag{6.22}$$

The initial population age-distribution $\phi(\tau)$ is assumed to be known at the initial time $t = 0$:

$$x(0,\tau) = \phi(t), \quad \tau \in [0,A]. \tag{6.23}$$

The equation (6.22) is called the *fertility* or *renewal equation*. It holds when $t > A$ because then all individuals are born during $[t-A, t]$ which is a part of the considered interval $0 \leq t < \infty$. The extension of (6.22) for $0 < t < A$

$$x(t,0) = \int_{t-A}^0 m(t,t-\tau)\varphi(t-\tau)\mathrm{d}\tau + \int_0^t m(t,\tau)x(t,\tau)\mathrm{d}\tau, \quad 0 \leq t \leq A, \tag{6.24}$$

splits the individuals that produce offspring at time t into two groups: those that come from the prehistory period $[t-A, 0]$ with their distribution (6.23) and those that are born during $[0, t]$.

Let us derive a differential equation for the dynamics of the age-structured population. During the time interval from t to $t + \Delta$, the age of individuals τ increases by the value Δ, so the *average rate of change* of the function $x(\tau,t)$ at time t is $(x(\tau + \Delta,t + \Delta)-x(\tau,t))/\Delta$ and the *instantaneous rate of change* is

$$\lim_{\Delta\to 0}\frac{x(t+\Delta,\tau+\Delta)-x(t,\tau)}{\Delta}$$
$$=\lim_{\Delta\to 0}\frac{x(t+\Delta,\tau+\Delta)-x(t+\Delta,\tau)+x(t+\Delta,\tau)-x(t,\tau)}{\Delta}$$
$$=\frac{\partial x(t,\tau)}{\partial t}+\frac{\partial x(t,\tau)}{\partial \tau}.$$

If the natural dearth is the only factor that changes the population size, then the population dynamics can be described by the following partial differential equation:

$$\frac{\partial x(t,\tau)}{\partial t}+\frac{\partial x(t,\tau)}{\partial \tau}=-\mu(t,\tau)x(t,\tau),\quad t\in[0,\infty),\tau\in[0,A]. \tag{6.25}$$

The equation (6.25) is called the *evolutionary equation*, because it determines the evolution of a population in time. The evolutionary equation (6.25), the fertility equation (6.22), and the initial distribution (6.23) are known as the *Lotka–von Foerster age-structured model* or *McKendrick age-structured model* of population dynamics [9].

6.3.2 MacCamy Nonlinear Population Model

There are numerous linear and nonlinear generalizations of the linear McKendrick model (6.22)–(6.25) to fit applied needs, for instance, to control the mortality rate (decrease it when it is necessary to protect endangered species, while the opposite is desired toward mortality of invasion species). If a control $u(t)$ affects the mortality, then the mortality becomes $\mu(u(t),\tau)$ and the model (6.22) is modified to

$$\frac{\partial x(t,\tau)}{\partial t}+\frac{\partial x(t,\tau)}{\partial \tau}=-\mu(u(t),\tau)x(t,\tau). \tag{6.26}$$

If the endogenous function u depends on the population size $N(t)$, then the resulting nonlinear population model (6.26) is known as *MacCamy model* or *Gurtin–MacCamy population model*.

6.3.2.1 Economic-Demographic Applications

Age-structured population models are widely used in a variety of applications. For example, the nonlinear model (6.26) can be the part of a nonlinear economic-demographic model, in which accumulated medical spending affects the expected lifespan of individuals. Then the function $\mu(u,\tau)$ in (6.26) describes effects of medical expenses per person on the longevity of human life. It depends on the individual age τ and the accumulated medical capital $u(t)$ at time t. The mortality rate $\mu(u,\tau)$ is smaller for a larger amount of medical capital u, that is, $\partial\mu/\partial u < 0$.

In applied demographic problems, μ can be chosen in more specific forms such as the Gompertz law or Coale–Demeny model life tables. Investigation techniques vary depending on the problem specifics.

6.3.3 *Euler–Lotka Linear Integral Model of Population Dynamics*

In a stationary environment with the time-independent fertility $m(t,\tau) = m(\tau)$ and mortality $\mu(t,\tau) = \mu(\tau)$, the evolutionary equation (6.25) becomes

$$\frac{\partial x(t,\tau)}{\partial t} + \frac{\partial x(t,\tau)}{\partial \tau} = -\mu(\tau)x(t,\tau). \tag{6.27}$$

Representative dynamics of the stationary age-dependent $m(\tau)$, $\mu(\tau)$, and the survival rate $l(\tau)$ from (6.31) is illustrated in Fig. 6.8.

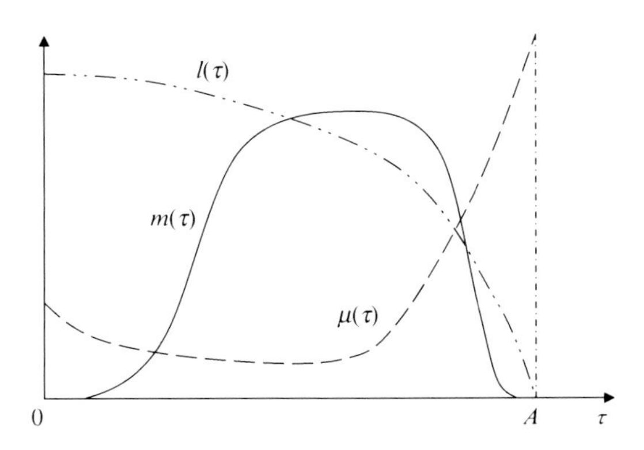

Fig. 6.8 Typical dependence of the fertility $m(\tau)$, mortality $\mu(\tau)$, and survival rate $l(\tau)$ on the individual age τ

In the new notations

$$\xi = t - \tau, \quad \overline{x}(t,\xi) = x(t,t-\xi), \ \overline{x}(t,t-\xi) = x(t,\tau), \tag{6.28}$$

the partial derivative of $\overline{x}(t,\xi)$ with respect to t is

$$\frac{d\overline{x}(t,\xi)}{dt}=\frac{\partial x(t,t-\xi)}{\partial t}+\frac{\partial x(t,t-\xi)}{\partial(t-\xi)}\frac{\partial(t-\xi)}{\partial t}=\frac{\partial x(t,\tau)}{\partial t}+\frac{\partial x(t,\tau)}{\partial\tau}.$$

Therefore, (6.27) can be rewritten for a fixed ξ as the linear *ordinary differential equation*

$$\frac{d\overline{x}(t,\xi)}{dt}=-\mu(t-\xi)\overline{x}(t,\xi). \tag{6.29}$$

By Sect. 1.3, the analytical solution of the linear equation (6.29) is

$$\overline{x}(t,\xi)=\overline{x}(\xi,\xi)\exp\left(-\int_{\xi}^{t}\mu(v-\xi)\mathrm{d}v\right). \tag{6.30}$$

The unknown function $x(t,\tau)$ in the model (6.22)–(6.25) depends on the individual age τ and the current time t, while the function $\overline{x}(t,\xi)$ in new notations (6.28) depends on the year $\xi=t-\tau$ of individual birth, which allows considering the ordinary differential equation (6.29) instead of the partial differential equation (6.25). It is worth to mention that the substitution $\xi=t-\tau$ is a part of the more advanced technique based on *characteristic curves* of partial differential equations.

Let us introduce the new notations:

$X(t)$—the *birth-rate intensity*: the number of births at time t,

$l(\tau)$—the *survival rate*: the fraction of individuals of age τ surviving to time t, that are related to the functions $x(t,\tau)$ and $\mu\ x(\tau)$ as

$$X(t)=x(t,0)=\overline{x}(t,t),\quad l(\tau)=\exp\left(-\int_{0}^{\tau}\mu(\xi)d\xi\right). \tag{6.31}$$

Substituting (6.30) and (6.31) to (6.25), we obtain the following *linear integral model of population dynamics* in new notations

$$X(t)=\int_{t-A}^{t}m(t-\xi)l(t-\xi)X(\xi)d\xi \tag{6.32}$$

with respect to the new unknown one-dimensional function $X(t)$, subject to the initial condition

$$X(\xi)=\phi(-\xi),\quad \xi\in[-A,0]. \tag{6.33}$$

The integral population model (6.32)–(6.35) is equivalent to the differential model (6.23)–(6.25). The integrand of (6.32) describes the number of all surviving individuals at time t, which is the product of all individuals born in the past and their survival rate, times their reproduction rate. The integration over all possible ages gives the total births at time t.

The change of variables $t - \xi \to \tau$ transforms the model (6.32) to the linear *integral renewal equation*

$$\begin{cases} X(t) = X^{-}(t) + \int_{0}^{t} m(\tau)l(\tau)X(t-\tau)\mathrm{d}\tau, & 0 \le t < A, \\ X(t) = \int_{t-A}^{t} m(\tau)l(\tau)X(t-\tau)\mathrm{d}\tau, & A \le t < \infty, \end{cases} \tag{6.34}$$

where

$$X^{-}(t) = \int_{t}^{T} m(\tau)l(\tau)\phi(\tau - t)\mathrm{d}\tau, \quad 0 \le t \le A.$$

Applying the contraction mapping theorem, it can be shown that the Volterra integral equation of the second kind (6.34) has a unique solution. The model (6.34) is called *Lotka equation* or *Euler–Lotka equation*, to emphasize the contribution of both Leonard Euler who suggested its special form in 1760, and Alfred James Lotka who proposed its more general version in 1911 to track females in a human population [14].

The model (6.34) describes the dynamics of an age-structured population in a stationary environment under unlimited food resources. An advantage of the integral model over the differential one is the *reduction of the dimension of unknown variables* that facilitates the theoretical and numerical analysis.

Exercises

1. The population of Texas (www.census.gov) from 1850 to 2010 is provided in the table below.

Year	Population (mln)
1850	212,592
1860	604,215
1870	818,579
1880	1,591,749
1890	2,235,527
1900	3,048,710
1910	3,896,542
1920	4,663,228
1930	5,824,715
1940	6,414,824
1950	7,711,194
1960	9,579,677

(continued)

Year	Population (mln)
1970	11,196,730
1980	14,229,191
1990	16,986,510
2000	20,851,820
2010	25,145,561

(a) Sketch the graph that represents the data in the table.
(b) In your opinion, which model better represents the data: the Malthusian model (6.2) or the Verhulst–Pearl model (6.5)?
(c) Take $N(0) = 212{,}6$ mln and estimate the parameter μ for the Malthusian model (6.2). Compare your model outcomes with the actual data. What conclusion can you make?

2. The population of the USA from 1850 to 2010 (source: www.census.gov) is presented in the table below.

Year	Population (mln)
1790	3.93
1800	5.31
1810	7.24
1820	9.64
1830	12.86
1840	17.06
1850	23.19
1860	31.44
1870	38.56
1880	50.19
1890	62.98
1900	76.09
1910	92.41
1920	106.46
1930	123.08
1940	132.12
1950	152.27
1960	180.67
1970	205.05
1980	227.22
1990	249.62
2000	282.16
2010	309.33

(a) Sketch the graph that represents the data in the table.
(b) In your opinion which model better represents the data: the Malthusian model (6.2) or the Verhulst–Pearl model (6.5)?

(c) Estimate the parameter μ for the Malthus model (6.2) and the Verhulst–Pearl model (6.5). Take $K = 350$ million. Add two more columns to the table, one with outcomes followed from the Malthusian model and another with outcomes calculated using the Verhulst–Pearl model. What conclusions can you make?

(d) Based on your calculations, what was the size of the US population in 1955 year and what is it expected to be in 2020? Compare your result with the actual size of the US population of 165.93 million in 1955 year

3. The *generalized von Bertalanffy growth model*

$$\frac{dl}{dt} = \alpha l^{2/3} - \beta l \tag{6.35}$$

is used in international fishing agreements to avoid overfishing and impose fishing restrictions. The first term in (6.35) is proportional to the fish surface area, l is the weight of fish, and the second term reflects the loss of weight and, as in (6.3), it is proportional to the weight. Investigate the model (6.35).

4. Show that the solution (6.6) to the Verhulst equation (6.5) with the initial condition $N(0) = K/2$ can be written in hyperbolic functions as $N(t) = K/2$ $(\tanh(\mu t/2) + 1)$.

5. The Gompertz model

$$\frac{dN(t)}{dt} = \mu N(t) \ln \frac{K}{N(t)} \tag{6.36}$$

is used in modeling the growth of tumors. It can be considered as a limited case of the generalized logistic equation. Find the analytic form of the solution to the Gompertz equation (6.36) and sketch its graph.

6. The interspecies and intraspecies competition between two species can be modeled as:

$$\begin{aligned} dN_1(t)/dt &= N_1(t)\left(\varepsilon_1 - \varepsilon_3 N_1(t) - \gamma_1 N_2(t)\right), \\ dN_2(t)/dt &= N_2(t)\left(\varepsilon_2 - \varepsilon_4 N_2(t) + \gamma_2 N_1(t)\right), \end{aligned} \tag{6.37}$$

where all positive parameters are as in Sect. 6.2. Analyze the model, find its equilibrium points, and investigate their stability.

7. Show that the maximal value of the integration constant C_1 in (6.12) is reached at the nontrivial equilibrium state $(\varepsilon_2/\gamma_2, \varepsilon_1/\gamma_1)$, i.e., when $N_1 = \varepsilon_2/\gamma_2$ and $N_2 = \varepsilon_1/\gamma_1$.

8. The predator–prey model (6.12) represents the interaction between two species, predators and prey, with each species growing in accordance with the Malthusian model (6.1) in absence of the other species. It can be modified to consider the intraspecies competition of both species:

$$\begin{aligned} dN_1(t)/dt &= \varepsilon_1 N_1(t)\left(1 - N_1(t)/K_1 - \gamma_1 N_2(t)\right), \\ dN_2(t)/dt &= \varepsilon_2 N_2(t)\left(1 - N_2(t)/K_2 + \gamma_2 N_1(t)\right), \end{aligned} \tag{6.38}$$

where K_1 and K_2 are carrying capacities of the two populations. The rest of parameters are as in the model (6.12). Find equilibrium points of the predator–prey model (6.38) and investigate their stability.

9. Obtain the integral version of the model (6.36).
10. Solve the linear ordinary differential equation (6.29).

References[1]

1. Anita, S.: Analysis and control of age-dependent population dynamics. Springer-Verlag, New York (2000)
2. Bacaer, N.: A short history of mathematical population dynamics. Springer-Verlag, London (2011)
3. Britton, N.F.: Essential Mathematical Biology. Springer Science and Med., New York (2003)
4. 🕮 Burghes, D.N., Borrie, M.S.: Modelling with Differential Equations. Ellis Horwood Ltd, Chichester (1981)
5. Gurtin, M., MacCamy, R.: Nonlinear age-dependent population dynamics. Arch. Rat. Mech. Anal. **54**, 281–300 (1974)
6. Hoppenstadt, F.C., Peskin, C.S.: Mathematics in medicine and the life sciences, 2nd edn. Springer-Verlag, New York (2002)
7. Hritonenko, N., Yatsenko, Y.: Age-structured PDEs in economics, ecology, and demography: optimal control and sustainability. Math. Popul. Stud. **17**, 191–214 (2010)
8. 🕮 Jones, D.S. Sleeman, B.D.: Differential equations and mathematical biology. Hall/CRC Press, Boca Raton, Florida (2003)
9. Keyfitz, B., Keyfitz, N.: The McKendrick partial differential equation and its uses in epidemiology and population study. Math. Comput. Model. **26**, 1–9 (1997)
10. 🕮 Kot, M.: Elements of mathematical ecology. Cambridge University Press, Cambridge (2001)
11. McKendrick, A.G.: Applications of mathematics to medical problems. Proc. Edinburgh Math. Soc. **44**, 98–130 (1926)
12. 🕮 Murray, J.D.: Mathematical Biology: I. An Introduction, Third Edition. Springer (2003)
13. 🕮 Taubes, C.: Modeling differential equations in biology. Upper Saddle River, NJ, Prentice Hall (2001)
14. Sharpe, F.R., Lotka, A.J.: A problem in age-distribution. Philos. Mag. **21**, 435–438 (1911)
15. 🕮 Webb, G.F.: Theory of nonlinear age-dependent population dynamics. New York, M. Dekker (1985)

[1] The book symbol 🕮 means that the reference is a textbook recommended for further student reading.

Chapter 7
Modeling of Heterogeneous and Controlled Populations

Age-structured population models, considered in Chap. 6, have become a traditional tool in biological modeling and are widely used in other disciplines. They possess well-developed investigation techniques. However, the size of individuals is a more important parameter than their age for some species. Section 7.1 considers two size-structured models that describe a population with natural reproduction and a fully managed population and explores links between size- and age-structured models. Nonlinear models of heterogeneous populations with intraspecies competition and their investigation techniques (steady-state and bifurcation analyses) are discussed in Sect. 7.2. An endogenous control is introduced into the model to address management problems in farming, fishery, forestry, and other applications. Controlled age- and size-dependent models are considered in Sect. 7.3.

7.1 Linear Size-Structured Population Models

Fertility and mortality of biological populations vary during the lifetime of individuals and depend on age, size, or stage of species development [10]. Age-dependent population models discussed in Chap. 6 are widely used in applications. In some other populations, such as trees or fish, the size of individuals is more important than their age and is significantly affected by availability of resources and changes in environmental conditions. Relevant economic parameters of such populations also depend more on size than on age: when buying fish or a Christmas tree, a customer prefers size to age. Since a relation between the age and size is rather weak, models based on size should be considered to capture the behavior and dynamics of such populations [5]. The first size-structured population models appeared three decades ago.

N. Hritonenko and Y. Yatsenko, *Mathematical Modeling in Economics, Ecology and the Environment*, Springer Optimization and Its Applications 88,
DOI 10.1007/978-1-4614-9311-2_7,

This section introduces two size-structured models for a managed population and a population with natural reproduction. An example of the first population is a tree farm where all trees are planted versus the wild forest as an example of the second population. A link between the two models is discussed in Sects. 7.1.2 and 7.2.4.

7.1.1 Model of Managed Size-Structured Population

In the notations

L—the *size of individuals* in a population, $0 \leq l_0 \leq l \leq l_m$,
l_0—the size of the smallest individual,
l_m—the size of the largest individual,
$x(t,l)$—the *population density*,
$g(t,l)$—the *growth function* that describes changes in the size over time,
$\mu(t,l)$—the size-specific *instantaneous mortality rate* that determines the probability of the natural death of an l-sized individual at time t,
$p(t)$—the *flux of new individuals* of size l_0 introduced into population at time t,
$\phi(l)$—the initial distribution of individuals by their size at time $t = 0$,

the model of a size-structured population can be presented by the following partial deferential equation

$$\frac{\partial x(t,l)}{\partial t} + \frac{\partial[g(t,l)x(t,l)]}{\partial l} = -\mu(t,l)x(t,l), \quad t \in [0,\infty), \quad l \in [l_0, l_m], \tag{7.1}$$

with respect to the *unknown* population density $x(t,l)$. The initial condition is

$$x(0,l) = \phi(l), \quad l \in [l_0, l_m], \tag{7.2}$$

and the boundary condition is

$$g(t,l_0)x(t,l_0) = p(t), \quad t \in [0,\infty). \tag{7.3}$$

The model (7.1)–(7.3) describes a managed population without natural reproduction, in which all small individuals of size l_0 are introduced into the population.

7.1.2 Connection Between Age- and Size-Structured Models

The age-structured population model (6.22) of Sect. 6.3,

$$\frac{\partial x(t,\tau)}{\partial t} + \frac{\partial x(t,\tau)}{\partial \tau} = -\mu(t,\tau)x(t,\tau), \quad t \in [0,\infty) \quad t \in [0,A], \tag{7.4}$$

with the individual age τ and the maximum age A, and the size-structured model (7.1)–(7.3) have many common features. Both of them are described by partial differential equations. The total population size in the population model (7.1)–(7.3) $N(t) = \int_{l_0}^{l_{max}} x(t, l)dl$ is similar to (6.18): $N(t) = \int_0^A x(t,\tau)d\tau$. Moreover, it is possible to establish a formal relation between the maximum age and maximum size in many cases.

The equations for the population dynamics (7.4) and (7.1) and their boundary conditions are also connected. Indeed, the size of a new individual increases by $\Delta l \approx g(t,l_0)\Delta t$ during a small interval Δt. On the other side, new individuals are brought into the population and, as a result, the density will increase by $\Delta x \approx p(t)\Delta t$ during the same time interval Δt. A combination of the last two formulas leads to the boundary condition (7.3) and explains the presence of the product gx in the second term of (7.1).

7.1.3 *Model of Size-Structured Population with Natural Reproduction*

A model of a population with natural reproduction is described by (7.1)–(7.3) and the size-structured fertility equation

$$p(t) = \int_{l_0}^{l_m} m(t, l)x(t, l)dl, \tag{7.5}$$

where

$m(t,l)$ is the *size-specific fertility rate*.

Equation (7.5) is similar to the fertility equation (6.22) in the age-structured model. The function $p(t)$ in (7.3) is now endogenous and is interpreted as the *total number of offspring* with the initial size l_0. A link between the size-structured models with and without natural reproduction is discussed in Sect. 7.2.4.

7.2 Nonlinear Population Models

The competition for nutrition, space, light, and other resources affects both mortality and fertility of a population [2]. The *intraspecies* (or *intraspecific*) *competition* can be introduced in different ways and leads to a nonlinear population model, even if the original model is linear. In this section, we include a competition component into the mortality and growth rates and consider extensions of the age-structured models of Sect. 6.3 and size-structured models of Sect. 7.1. The bifurcation analysis is illustrated for the age-structured model and the steady-state analysis is shown for the size-structured model.

7.2.1 Age-Structured Model with Intraspecies Competition

Assuming that the intraspecies competition increases the mortality rate, the linear age-structured model (7.4) can be modified to the following *nonlinear evolutionary equation*:

$$\partial x/\partial \tau + \partial x/\partial t = -\left[\mu(t,\tau) + \int_0^A b(\tau,\xi)x(t,\xi)d\xi\right]x(t,\tau), \tag{7.6}$$

where, as in Chap. 6,

t—the time,
τ—the individual age,
$A \le \infty$—the maximum age,
$x(t,\tau)$—the population age density,
$\mu\ (t,\tau)$—the mortality rate,
$b(\tau,\xi)$—an increase in the mortality of individuals of age τ caused by the individuals of age ξ (the intensity of *intraspecies competition*).

The evolutionary equation (7.6) together with the age-structured fertility equation (6.22) constitutes the nonlinear *integral-differential model* of an age-structured population with intraspecies competition and natural reproduction. The nonlinear integral equation

$$X(t) = \int_{t-A}^{t} m(t-\xi)\mathrm{e}^{-\mu(t-\xi)-\int_{t-A}^{t} b(t-\xi,i-\theta)X(\theta)d\theta} X(\xi)d\xi, \tag{7.7}$$

for the birth-intensity X describes the same population and is qualitatively equivalent to the model (7.6). The integral population model (7.7) is a nonlinear extension of the linear integral Lotka population model (6.29).

7.2.2 Bifurcation Analysis

Finding time-independent *stationary solutions* is a subject of the steady-state analysis and an important part of qualitative analysis. The *stationary solutions* are described as

- $x(t,\tau) \equiv x_S(\tau)$ *in differential models*, then $\partial x/\partial t \equiv 0$ and a partial differential equation is transformed to an ordinary differential equation with respect to $x_S(\tau)$;
- $X(t) \equiv X_S = \text{const} \ge 0$ *in integral models*, then an integral equation is reduced to a nonlinear equation for X_S.

The bifurcation analysis is another important tool in the analysis of nonlinear ecological models. It investigates *bifurcation values* of model parameters, at which the behavior of the stationary solutions changes and/or new stationary solutions appear.

This section illustrates main ideas of the steady-state and bifurcation analyses for the nonlinear integral model (7.7). We need to indentify a *bifurcation parameter* in the model first. The fertility rate is one of the most important characteristics in many biological populations. It is reasonable to take the fertility rate as $m(\tau) = \lambda \overline{m}(\tau)$, where λ is a parameter that affects the fertility intensity but does not change its structure. Then the nonlinear model (7.7) can be rewritten as:

$$X(t) = \lambda \int_0^T \overline{m}(\tau) e^{-\mu(\tau) - \int_0^T b(\tau,\xi) X(t,\xi) d\xi} X(t-\tau)\, \mathrm{d}\tau. \tag{7.8}$$

Substituting $X(t) \equiv X_S$ into (7.8), we obtain the nonlinear equation for possible stationary solutions X_S

$$X_S \left[1 - \lambda \int_0^T \overline{m}(\tau) e^{-\mu(\tau) - X_S \int_0^T b(\tau,\xi)\, \mathrm{d}\xi} \mathrm{d}\tau \right] = 0, \tag{7.9}$$

that always has one *trivial stationary state* $X^0{}_S = 0$, which is the only stationary state for small values of $\lambda > 0$. Another stationary state appears in (7.9) if λ is greater than the threshold value λ^* determined from the following equality

$$R(\lambda) = \lambda \int_0^T \overline{m}(\tau) e^{-\mu(\tau)}\, \mathrm{d}\tau = 1. \tag{7.10}$$

The function $R(\lambda)$ is called the *reproduction number* or *biological potential* of an age-structured population and describes the average number of offspring from one individual during its lifetime. For stable real populations, $R \approx 1$.

Thus, the value λ^* of the parameter λ turns to be a *bifurcation value*: a new nontrivial stationary state $X^*{}_S$ appears as $\lambda > \lambda^*$. The state $X^*{}_S$ is found as the second solution of the nonlinear equation (7.9).

The point $(\lambda^*,0)$ of the plane (λ,X) is called a *ramification point* of the nonlinear equation (7.9). Three solution branches $X(\lambda)$ exist in the neighborhood of λ^*:

- at $\lambda < \lambda^* \Rightarrow$ one branch: the stable trivial solution $X^0{}_S(\lambda) = 0$,
- at $\lambda > \lambda^* \Rightarrow$ two branches: the unstable trivial solution $X^0{}_S(\lambda) = 0$ and a stable positive solution $X^*{}_S(\lambda)$.

These solutions are illustrated in Fig. 7.1. The steady-state analysis of the nonlinear population model (7.8) becomes more complicated when the bifurcation parameter λ increases. Then, in addition to the unstable stationary solutions $X^0{}_S = 0$ and $X^*{}_S > 0$, the model allows for various periodic regimes and, even, chaos.

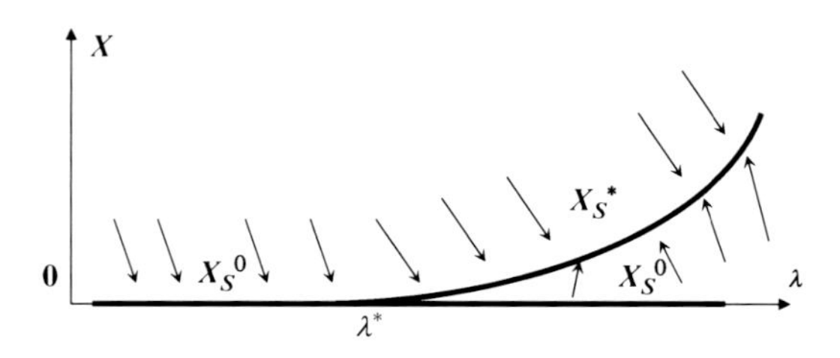

Fig. 7.1 Stationary solutions of the nonlinear population model (7.8) near the bifurcation value λ^*

The bifurcation analysis leads to important recommendations for rational exploitation of a population. In particular, if the biological potential $R(\lambda)$ of a population becomes less than one because of random environmental changes, natural or other disturbances, then the trivial solution is the only stable solution, that is, an irreversible deterioration process may start and lead to the population extinction. Hence, the condition $R > 1$ should be maintained for stable populations. Thus, the rational management of biological populations should involve special control tools for taking measures against random perturbations.

7.2.3 *Nonlinear Size-Structured Model*

The nonlinear model of a managed size-structured population with intraspecies competition can be written as

$$\frac{\partial x(t,l)}{\partial t} + \frac{\partial[g(E(t),l)x(t,l)]}{\partial l} = -\mu(E(t),l)x(t,l), \quad t \in [0,\infty), \quad l \in [l_0, l_m], \tag{7.11}$$

$$E(t) = \int_{l_0}^{l_m} \chi(l)x(t,l)\,\mathrm{d}l, \tag{7.12}$$

with the boundary conditions

$$g(E(t),l_0)x(t,l_0) = p(t), \quad t \in [0,\infty), \quad x(0,l) = x_0(l), \quad l \in [l_0, l_m]. \tag{7.13}$$

The model (7.11)–(7.13) is a nonlinear extension of the linear size-structured model (7.1)–(7.3) of population without natural reproduction with some modifications in parameters. The new parameters are as follows:

$E(t)$—an aggregated parameter that reflects the crowdedness of the population and determines the *intensity of the intraspecies competition*,
$\chi(l)$—a parameter specific to the population category ($\chi(l) = al^2$ in forestry),
$g(E(t),l)$—the growth function that depends on the intraspecies competition,
$\mu(E(t),l)$—the instantaneous mortality rate affected by the intraspecies competition.

The model (7.11)–(7.13) includes the dependence of the mortality $\mu(E(t),l)$ and growth $g(E(t),l)$ rates on the intensity of the intraspecies competition $E(t)$. The unknown variables in the *nonlinear integral-differential equations* (7.11)–(7.13) are the intensity $E(t)$ and the population density $x(t,l)$. The unknown $E(t)$ depends on the dynamics $x(t,l)$ of the entire population (since $l_0 \le l \le l_m$) at time t. Such models are known as *population models with global (nonlocal) nonlinearities*.

7.2.4 Steady-State Analysis

The *sustainable development* of biological systems is of an increasing interest for applications. It is relevant in fishery (as related to possible overfishing), forestry (e.g., the disappearance of Amazon rainforests), environmental sciences (global warming), and other areas. Well-known fishery models (e.g., the Gordon-Schaefer model) address the sustainable harvesting. Mathematically, sustainable regimes correspond to time-independent *steady states* of a population model. If a stable steady state does not satisfy initial conditions, then it can be achieved after certain time period, called a *transition period* or *start-up period*. The steady-state analysis lessens mathematical complexity of population models by decreasing their dimension and reduces two-dimensional partial differential equations to ordinary differential equations. The steady state analysis is widely used in other chapters of this textbook. In this section, we analyze possible steady states of the *nonlinear integral-differential* model (7.11)–(7.13).

The stationary solution of the size-structured model (7.11)–(7.13) is independent of the current time t:

$$x(t,l) = x(l), \quad E(t) = E, \quad l \in [l_0, l_m], \quad t \in [0, \infty). \tag{7.14}$$

Substitution of (7.14) to (7.13) demonstrates that a necessary condition for the existence of stationary solutions in the population model (7.11)–(7.13) is

$$p(t) = p = \text{const}, \quad t \in [0, \infty). \tag{7.15}$$

Combining (7.14), (7.15), and (7.11)–(7.13), one can see that a possible stationary regime $(x(l), E)$ should satisfy the following *integral-differential equations*

$$\frac{\mathrm{d}(g(E,l)x(l))}{\mathrm{d}l} = -\mu(E,l)x(l), \tag{7.16}$$

$$E = \chi \int_{l_0}^{l_m} l^2 x(l)\, \mathrm{d}l, \quad l \in [l_0, l_m], \tag{7.17}$$

$$g(E, l_0) x(l_0) = p. \tag{7.18}$$

Assuming for a moment that the function E is known, the initial problem for the ordinary differential equation (7.16), (7.18) has the analytic solution

$$x(l) = \frac{p}{g(E,l)} e^{-\int_{l_0}^{l} \frac{\mu(E,\xi)}{g(E,\xi)} d\xi}, \quad l \in [l_0,\ l_m]. \tag{7.19}$$

Therefore, E is found from the nonlinear equation

$$E - \chi \int_{l_0}^{l_m} \frac{pl^2}{g(E,l)} e^{-\int_{l_0}^{l} \frac{\mu(E,\xi)}{g(E,\xi)} \mathrm{d}\xi} \mathrm{d}l = 0, \tag{7.20}$$

which comes from the combination of (7.17) and (7.19). It has a unique solution under natural conditions. Indeed, the continuous function $F(E)$, that represents the left side of (7.20), has the properties of $F(E) < 0$ at $E = 0$ and $F(E) > 0$ for large E that proves the existence of at least one $E > 0$ such that $F(E) = 0$. The value E^* is unique when the derivative $F'(E) > 0$.

The stationary solution (7.19), (7.20) also holds for populations with natural reproduction. Namely, the *stationary* solution of this model is the same as for the model (7.11)–(7.13) at $p = g(E,l_0)x(l_0)$. The substitution of (7.19) into equality (7.5) leads to an important link between two size-structured models, (7.1)–(7.5) and (7.1)–(7.3).

7.2.4.1 Link Between Two Size-Structured Models

The model (7.1), (7.2), (7.5) of a population with natural reproduction can possess the same stationary solutions as the model (7.1)–(7.3) of a fully managed population only if the *reproduction number* of the population

$$R(E) = \int_{l_0}^{l_m} \frac{m(t,l)}{g(t,l)} e^{-\int_{l_0}^{l} \frac{\mu(t,\xi)}{g(t,\xi)} \mathrm{d}\xi} \mathrm{d}l = 1. \tag{7.21}$$

The reproduction number plays an important role in population ecology and biology. Condition (7.21) holds for stable populations. Such a connection between the two models allows us to focus on the analysis of a fully managed population that involves less equations and then extend some obtained results to the model (7.1)–(7.2), (7.5).

7.3 Population Models with Control and Optimization

Control functions are introduced into population models for solving some applied problems, such as rational management of resources, selection of species, harvesting, modification of crops, protection of species, and others. The controls are implemented in different ways and can be age- or size-structured, time-dependent, one- or two-dimensional. In this section, we consider population models with controls that appear in harvesting problems.

7.3.1 *Age-Structured Population Models with Control*

Harvesting decreases the size of a population and can be described as an artificial increase of the mortality. Let us consider the harvesting control that depends on the individual age τ and current time t. The linear age-structured Lotka–McKendrick model (7.4) produces two well-known models of controlled harvesting:

$$\frac{\partial x(t,\tau)}{\partial t} + \frac{\partial x(t,\tau)}{\partial \tau} = -(\mu(t,\tau) + w(t,\tau))x(t,\tau), \tag{7.22}$$

and

$$\frac{\partial x(t,\tau)}{\partial t} + \frac{\partial x(t,\tau)}{\partial \tau} = -\mu(t,\tau)x(t,\tau) - u(t,\tau), \tag{7.23}$$

where τ, t, $x(t,\tau)$, $\mu\,(t,\tau)$ are as in Sect. 7.2.1 and
$w(t,\tau)$—the *harvesting effort*,
$u(t,\tau)$—the *harvesting rate*.

In spite of their similarity, the harvesting models (7.22) and (7.23) have different applied interpretations and investigation techniques. The model (7.23) considers the *harvesting rate* u, while the model (7.22) involves the *harvesting effort* w. The model (7.22) with harvesting effort reflects a so-called *catch-per-unit-effort hypothesis* that the harvesting yield is proportional to the size of a population. Indeed, the probability of catching fish is lower when there is less fish in the lake. Economic harvesting models with controlled effort also assume the cost of harvesting to be proportional to the effort (see (7.48) below) rather than to the harvesting rate. In particular, two well-known applied models of open-access commercial fishing, the Gordon–Schaefer and Beverton–Holt models, use the fishing effort as an endogenous control. However, it is known that in the case of a sole owner of fish resource, the optimal harvesting policy is to utilize the harvesting rate. The control of the harvesting effort is not applicable to *forestry*, where the catch-per-unit-effort hypothesis does not hold. Although optimization of a *harvesting rate* is relevant to practice, it is often avoided in mathematical research because such problems involve active state constraints. We will choose the model (7.23) with harvesting rate for our further analysis in this section.

In both models (7.22) and (7.23), the harvesting control $w(t,\tau)$ or $u(t,\tau)$ is the unknown *decision variable*, the population density $x(t,\tau)$ is the unknown *state variable*, and the mortality $\mu\ (t,\tau)$ is given. Many applied control problems, e.g., in farming, use a fully managed population without natural reproduction and we will consider such populations. The related models have fewer equations due to absence of the fertility equation. Then, the density $x(t,0)$ of zero-age individuals introduced into the population is an endogenous control $p(t)$ with the initial condition

$$x(t,0) = p(t), \quad t \in [0,T), \quad T \leq \infty. \tag{7.24}$$

We consider the *planning horizon* $[0, T)$ to be an infinite time interval $[0,\infty)$, which simplifies the model investigation keeping away undesirable end-of-horizon effects with all disturbances in solutions they cause. The infinite interval also corresponds to sustainable development.

The population density $x(t,\tau)$ is always nonnegative and the amount of introduced species $p(t)$ and harvesting rate $u(t,\tau)$ have their boundaries, which is reflected by

$$0 \leq u(t,\tau) \leq u_{\max}, \quad 0 \leq p(t) \leq p_{\max}, \quad x(t,\tau) \geq 0. \tag{7.25}$$

In order to find the optimal harvesting regime, we maximize the *discount profit of harvesting* over the infinite horizon $[0,\infty)$

$$\max_{u,p,x} I = \max_{u,p,x} \int_0^\infty e^{-rt} \left(\int_0^A c(t,\tau)u(t,\tau)d\tau - k(t)p(t) \right) \mathrm{d}t, \tag{7.26}$$

where

$c(t,\tau)$—the unit market price of the harvesting output,
$k(t)$—the unit cost of introduction of a new population individual,
$r > 0$—the *discount rate*.

The harvesting profit I in (7.26) is the difference between the total discounted price of all harvested individuals and the total discounted cost of all introduced individuals on the infinite planning horizon $[0, \infty)$. If the relations (7.24)–(7.26) are used to describe a population with natural reproduction, then the interpretation of model functions will be slightly different. For instance, $k(t)$ will represent the cost for caring for newborns until they rich a certain age.

7.3.2 Elements of Analysis

The size-structured population model (7.23)–(7.26) with controlled harvesting rate leads to the following linear optimal control problem:

- find the functions $u(t,\tau)$, $p(t,\tau)$, and $x(t,\tau)$, $t \in [0,\infty)$, $\tau \in [0,A)$, that maximize the functional

$$\max_{u,p,x} I = \max_{u,p,x} \int_0^\infty e^{-rt} \left(\int_0^A c(t,\tau)u(t,\tau)d\tau - k(t)p(t) \right) dt,$$

$$\frac{\partial x(t,\tau)}{\partial t} + \frac{\partial x(t,\tau)}{\partial \tau} = -\mu(t,\tau)x(t,\tau) - u(t,\tau), \tag{7.27}$$

$$0 \le u(t,\tau) \le u_{\max}, \quad 0 \le p(t) \le p_{\max}, \quad x(t,\tau) \ge 0,$$

$$x(0,\tau) = x_0(\tau), \quad \tau \in [0,A), \quad x(t,0) = p(t), \quad t \in [0,\infty).$$

Commonly used investigation techniques for optimal control problems include the derivation of extremum conditions, existence and uniqueness of solutions, steady-state analysis and qualitative analysis of optimal trajectories [1, 3, 4, 13]. In this section, we present the necessary condition for an extremum in the form of maximum principle and then provide a steady-state analysis and specify bang–bang regimes. Proofs of some statements presented here are out of the scope of this textbook. They can be found in the references.

7.3.2.1 The Necessary Condition for an Extremum

We assume the unknown functions $x(t,\tau)$, $u(t,\tau)$, and $p(t)$ to be measurable in their domains.

Maximum Principle: If (u^*, p^*) is a solution of the problem (7.27), then

$$\partial I/\partial u \le 0 \text{ at } u^*(t,a) = 0, \quad \partial I/\partial u \ge 0 \text{ at } u^*(t,a) = u_{\max},$$
$$\partial I/\partial u = 0 \text{ at } 0 < u^*(t,a) < u_{\max},$$

$$\begin{aligned}&\partial I/\partial p(t) \leq 0 \text{ at } p^*(t) = 0, \quad \partial I/\partial p \geq 0 \text{ at } p^*(t) = p_{\max}, \\ &\partial I/\partial p = 0 \text{ at } 0 < p^*(t) < p_{\max},\end{aligned} \tag{7.28}$$

where

$$\frac{\partial I}{\partial u} = e^{-rt}(c(t,\tau) - \lambda(t,\tau)), \tag{7.29}$$

$$\frac{\partial I}{\partial p} = e^{-rt}(\lambda(t,0) - k(t)), \tag{7.30}$$

$$\frac{\partial \lambda(t,\tau)}{\partial t} + \frac{\partial \lambda(t,\tau)}{\partial a} = (r + \mu(t,\tau))\lambda(t,\tau) - \eta(t,\tau), \tag{7.31}$$

$$\lim_{t\to\infty} e^{-rt}\lambda(t,\tau) = 0, \quad \tau \in [0,A), \quad \lambda(t,A) = 0, \quad t \in [0,\infty), \tag{7.32}$$

$$\eta(t,\tau) > 0 \text{ at } y^*(t,\tau) = 0, \quad \eta(t,\tau) = 0 \text{ at } y^*(t,\tau) > 0. \tag{7.33}$$

The functions λ and η are called the *dual, co-state,* or *adjoint variables* associated with the constraints (7.23) and (7.25). They represent the marginal cost of violating these constraints. All formulas (7.28)–(7.33) have applied interpretation. The dual variable λ is the possible future revenue from the individual of age τ at time t (the shadow price). The gradient (7.29) is positive when the current market price c is larger than the revenue λ and is negative otherwise. The economic meaning of the dual variable λ requires its positivity. The conditions (7.28) and (7.30) mean that the discounted future value of a young introduced individual should be not less than their initial price k, otherwise, it is not profitable to introduce new individuals at all.

The partial differential equation (7.31) is known as the *dual* or *co-state equation*. It is linear and has a unique solution λ at natural conditions. While the state equation (7.23) is subject to the initial condition (7.24), the dual equation (7.31) is subject to the terminal condition $\lambda(t,A) = 0$, $t \in [0,\infty)$, in (7.32) and is solved backward. The condition $\lambda(t, A) = 0$ shows that population individuals that reach their maximum age have no commercial value: they just die out.

The *transversality condition* (7.32) is common in bioeconomics and reflects the behavior of the process in the long run. It implies that an increase in the value of species has to be smaller than a decrease in a discounting factor. The condition (7.33) is the *condition of complementary slackness* and is an essential part of optimality conditions for the optimal control with state constraints.

An analysis of the extremum conditions (7.28)–(7.31) shows that any nontrivial solution of the optimal control problem (7.27) involves an interval on which the state constraint $x \geq 0$ in (7.25) is active [9]. To obtain more insight into the structure of optimal trajectories, we provide steady-state analysis of the problem.

7.3.2.2 Steady-State Analysis

The steady-state analysis is an important step in the analytic study of population models. The basic idea is the same as in models with homogeneous factors such as the economic models of Chaps. 3 and 4. Namely, we assume that the optimization problem (7.27) is autonomous (all its parameters do not depend directly on the time t):

$$\mu(t,\tau) = \mu(\tau), \quad c(t,\tau) = c(\tau), \quad k(t) = k, \tag{7.34}$$

and look for possible *stationary* (time–independent) *solutions*

$$x(t,\tau) = x(\tau), \quad u(t,\tau) = u(t), \quad p(t) = p \tag{7.35}$$

of the optimization problem (7.27). Under the assumptions (7.34) and (7.35), the state equation in (7.27) becomes the linear ordinary differential equation

$$x'(\tau) = -\mu(\tau)x(\tau) - u(\tau), \tag{7.36}$$

and its exact solution is

$$x(\tau) = pe^{-\int_0^\tau \mu(\xi)d\xi} - \int_0^\tau e^{-\int_a^\tau \mu(\xi)\,d\xi} u(a)\,da. \tag{7.37}$$

Under the assumptions (7.34) and (7.35), the initial value problem (7.31)–(7.32) for the dual equation (7.31) is also autonomous:

$$\lambda'(\tau) = (r + \mu(\tau))\lambda(\tau) - \eta(\tau), \quad \lambda(A) = 0, \tag{7.38}$$

and has the stationary solution

$$\lambda(\tau) = \int_\tau^A e^{-\int_\tau^\nu [r+\mu(\xi)]d\xi} \eta(\nu)\,d\nu, \quad 0 \le \tau \le A. \tag{7.39}$$

From the bioeconomic point of view, optimal controls in harvesting problems can combine *boundary solutions* and exceptional *singular* controls, along which the functional derivatives are zero. The optimal control problem (7.27) is a *linear optimal control problem* and, as such, normally possesses only boundary solutions.

The necessary condition for an extremum (7.28)–(7.31) indicates that the optimal control problem has no interior regime such that $\partial I/\partial p = 0$ or $\partial I/\partial u = 0$. Indeed, expressions (7.29) and (7.30) for the functional derivatives $\partial I/\partial u$ and

$\partial I/\partial p$ do not depend on the unknown controls u and p. So $\partial I/\partial u$ or $\partial I/\partial p$ does not equal zero on some intervals in the stationary case (7.34) as well. To describe possible stationary regime, we use the terminology of bang–bang solutions.

7.3.2.3 Bang–Bang Regimes

Bang–bang regimes play an important role in the structure of solutions of linear optimization problems, such as (7.27). They reflect a situation when a solution of the optimization problem takes mostly boundary values [11]. Bang–bang regimes are also known in other scientific areas, for example, in economics (see Chaps. 3–5). The related mathematical statements are called *bang–bang theorems in a weak form*. The statement that a solution takes *only* boundary values is called a *bang–bang theorem in the strongest form*. It appears that a strong bang–bang principle holds for the steady-state harvesting rate $u(\tau)$ in the model (7.27) under realistic assumptions.

Property 7.1 *(on steady-state bang–bang regime). Let* $u_{\max} \gg 1$,

$$c'(\tau)/c(\tau) > r + \mu(\tau) \text{ at } 0 \le \tau \le \hat{a} \le A, \tag{7.40}$$

and

$$\int_0^A c(\tau) e^{-\int_0^\tau \mu(\xi)\,\mathrm{d}\xi} \mathrm{d}\tau > b. \tag{7.41}$$

Then, the optimization problem (7.27) has the following steady-state regime

$$p^* = p_{\max}, \quad u^*(\tau) = \begin{cases} 0, & 0 < \tau \le a^*, \\ u_{\max}, & a^* < \tau \le a_e, \\ 0, & a_e < \tau \le A, \end{cases} \tag{7.42}$$

$$x^*(\tau) = \begin{cases} > 0, & 0 < \tau \le a_e, \\ = 0, & a_e < \tau \le A, \end{cases} \tag{7.43}$$

where the endogenous harvesting age a^*, $0 < a^* < A$, is determined from

$$a^* = \underset{0 \le \tau \le A}{\operatorname{argmax}} \left[c(\tau) e^{-\int_0^\tau \mu(\xi)\,\mathrm{d}\xi} \right], \tag{7.44}$$

and the endogenous age a_e, $a^* < a_e < A$, is found from the condition $y(a_e) = 0$.

Let us interpret the conditions of Statement 7.1. The condition (7.40) requires the given harvesting price $c(\tau)$ to increase faster than the sum of the discount and mortality rates, at least, for young individuals. This condition is quite natural and holds in applied situations. It guarantees that the harvesting age a^* is larger than the age $\tau = 0$ of introducing new individuals, otherwise, there is no sense to raise the population. The condition (7.41) requests the unit cost of introducing new population members to be smaller than the discounted harvesting price, which is also natural. The endogenous age a^* is the age when the intensive harvesting starts. The condition (7.44) tells us that the price $c(\tau)$ of individuals times the probability of individual survival $e^{-\int_0^\tau \mu(\xi)d\xi}$ is maximal at the age $\tau = a^*$. Under these conditions, the population individuals are kept until the age a^* and then are harvested with maximal possible rate $u_{\max}$ until the age a_e. There are no individuals older than a_e left in the population.

The *strong bang–bang principle* does not allow for singular regimes, along which the derivative (7.29) is zero. Such results possess essential implications for corresponding management policies. For instance, in forestry, they indicate the advantages of *selective logging* regime over *clear cutting*, which is still an open issue in official harvesting policies of many European countries.

7.3.2.4 Relation to Economic Models with Heterogeneous Capital

This textbook emphasizes the versatility of mathematical models. A good example is the age-structured models in ecology and economics. In everyday life, firms buy, use, and sell capital (equipment, machines) that has been installed at different times. The parameters of the capital depend on the current time and the time of installation or the "age" of capital. Usually, newer machines are more productive, environmentally friendlier, and can be even cheaper because of technological innovations and physical deterioration of old machines. Models that consider such effects have recently become popular in economic research. Then, the controlled dynamics of the age-structured (vintage) capital can be described by the age-structured model (7.23) but with different interpretation of parameters. In particular, $x(t,\tau)$ represents the density of capital stock, $p(t)$ turns to be the investment into new capital, the rate $u(t,\tau)$ is the investment at time t into the capital of age τ, $\mu(t,\tau)$ is the depreciation rate of the capital. The functions $x(t,\tau)$, $u(t,\tau)$, and $p(t)$ are endogenous as they used to be, but the constraint (7.24) can be changed to $-u_{\min}(t,\tau) \leq u(t,\tau) \leq u_{\max}(t,\tau)$, because the investment $u(t,\tau)$ can be positive when machines are bought or negative when they are sold. Chapters 4 and 5 discuss such models in detail.

7.3.3 Nonlinear Age-Structured Models of Controlled Harvesting

A bang–bang principle can also be proven for the Lotka–McKendrick model (7.22) with controlled effort and its nonlinear modifications discussed in Sect. 6.3.2. Let us consider the controlled version of the *Gurtin–MacCamy nonlinear population model* (6.26):

$$\frac{\partial x(t,\tau)}{\partial t}+\frac{\partial x(t,\tau)}{\partial \tau}=-\mu(E(t),\tau)x(t,\tau)-w(t,\tau)x(t,\tau), \tag{7.45}$$

$$E(t)=\int_0^A x(t,\tau)\,\mathrm{d}\tau, \tag{7.46}$$

$$x(0,\tau)=x_0(\tau),\quad \tau\in[0,A),\quad x(t,0)=p(t),\quad t\in[0,\infty). \tag{7.47}$$

The dependence of the *mortality rate* $\mu(E,\tau)$ on the *total population size* $E(t)$ reflects the intraspecies competition for limited resources. This dependence is a common example of a so-called *nonlocal nonlinearity* which represents nonlocal effects in the population [7, 8, 14]. In the Gurtin–MacCamy model, the control effort $w(t,\tau)$ raises the mortality factor $\mu(E,\tau)$. In the case of controlled effort, the harvesting yield is wx, so the objective functional (the discounted profit of harvesting) becomes

$$\max_{u,p,x} I=\max_{u,p,x}\int_0^\infty e^{-rt}\left(\int_0^A c(t,\tau)w(t,\tau)x(t,\tau)d\tau-k(t)p(t)\right)\mathrm{d}t, \tag{7.48}$$

subject to constraints $0\le w(t,\tau)\le u_{\max},\ 0\le p(t)\le p_{\max},\ x(t,\tau)\ge 0,\ t\in[0,\infty)$.

The necessary extremum condition for the optimal control problem (7.45)–(7.48) is obtained analogously to Sect. 7.3.2 and looks similar to (7.28)–(7.32). The problem remains linear with respect to the controls w and p, and the corresponding expressions for the functional derivatives $\partial I/\partial w$ and $\partial I/\partial p$ include neither w nor p. So, as in the optimization model (7.27), we have the same reasons to expect a bang–bang structure to hold for the optimal control w^*. The steady-state analysis indicates that in the stationary environment the time-independent optimal control $w^*(\tau)$ is of the form

$$w*(\tau)=\begin{cases}0, & 0\le\tau<a*(t)\\ w_{\max}, & a*(t)\le\tau\le A\end{cases},\quad t\in[0,\infty), \tag{7.49}$$

under $\partial c/\partial\tau>0$ and some other conditions. The key condition $\partial c/\partial\tau>0$ means that the price of individuals increases with their age.

If the condition $\partial c/\partial\tau > 0$ fails, then the bang–bang structure may be of more sophisticated form than (7.49). In particular, if price $c(a)$ is maximal at a certain age $a_{\max} < A$ and is small for young and old individuals: $\lim_{\tau\to 0} c(\tau) = \lim_{\tau\to A} c(\tau) = 0$, then the corresponding optimal harvesting control w^* is

$$w*(\tau) = \begin{cases} 0, & 0 \le \tau < a_1(t), \\ w_{\max}, & a_1(t) \le \tau \le a_2(t), \quad t\in[0,\infty), \\ 0, & a_2(t) < \tau < A, \end{cases} \tag{7.50}$$

with two switching ages $a_1(t)$ and $a_2(t)$, such as $a_1(t) < a_{\max} < a_2(t)$.

The bang–bang regime (7.49) has a simpler structure with just one switching age a^*, compared to the bang–bang control (7.42) in the model (7.27) with harvesting rate u. In particular, the optimal harvesting effort $w^*(t,\tau)$ is maximal after the age a^* for all remaining ages, which was not the case in (7.42). It happens because a harvested population with the controlled effort never dies out: the density $x^*(t,\tau)$ remains positive all the time. Indeed, when the population density decreases to zero, it is harder and harder to provide harvesting (catch an individual), so the effectiveness of harvesting wx in (7.45) decreases to zero as well.

As a result, the state constraint $x \ge 0$ never becomes active in the model (7.45) with controlled harvesting effort w, as opposed to the model (7.27) with harvesting rate u. It occurs because the optimization problem (7.22) is not linear: it includes the quadratic term wx in the model (7.46) and in the objective functional (7.48). Avoiding state constraints makes mathematical analysis simpler. That is why harvesting models with controlled effort are preferred in bioeconomic theory.

According to the general optimization theory, the bang–bang structure of optimal controls is not mandatory in nonlinear controlled dynamic systems. The nonlinearity can be powerful enough to override the bang–bang behavior. However, nonlinear optimal control problem (7.45)–(7.48) remains linear in the controls u and p. So the nonlocal nonlinearity in the Gurtin–MacCamy model caused by resource limitations does not change the bang–bang behavior. In some cases, the nonlinearity can even strengthen the bang–bang structure as in the controlled size-structured models considered in Sect. 7.3.4.

7.3.4 Size-Structured Models with Controls

An optimization version of the size-structured model (7.1)–(7.3) with intraspecies competition is to determine the functions $x(t,l)$, $w(t,l)$, $E(t)$, $p(t)$, $t \in [0,\infty)$, $l \in [l_0,l_m]$, that maximize

$$\max_{w,p,x,E} J = \int_0^\infty e^{-rt}\left\{\int_{l_0}^{l_m} c(t,l)w(t,l)x(t,l)dl - k(t)p(t)\right\} dt, \tag{7.51}$$

subject to:

$$\frac{\partial x(t,l)}{\partial t} + \frac{\partial(g(E(t),l)x(t,l))}{\partial l} = -\mu(E(t),l)x(t,l) - u(t,l), \tag{7.52}$$

$$E(t) = \int_{l_0}^{l_m} \chi(l)x(t,l)\,\mathrm{d}l, \tag{7.53}$$

$$g(E(t),l_0)x(t,l_0) = p(t), \tag{7.54}$$

$$0 \le w(t,l) \le w_{\max}, \quad 0 \le p(t) \le p_{\max}, \tag{7.55}$$

$$x(0,l) = x_0(l), \quad l \in [l_0, l_m], \quad t \in [0,\infty), \tag{7.56}$$

under the given functions $c(t,l)$, $k(t)$, and $\chi(l)$. In forestry, $\chi(l) = \chi l^2$ is common.

Optimization in size-structured models uses the objective functional (7.51) analogous to (7.48), and thus [12], the maximum principle appears to be similar to the one presented in Sect. 7.3.1 for age-structured populations. Because of this similarity, we can expect a bang–bang behavior of the steady-state optimal control $u^*(l)$ similar to (7.32)

$$u*(l) = \begin{cases} 0, & 0 < l \le l*, \\ u_{\max}, & l* < l \le l_e, \\ 0, & l_e < l \le l_{\max}, \end{cases} \tag{7.57}$$

with the harvesting size $l^*(t)$ under certain assumptions about the model functions $g(t,l)$, $\mu(t,l)$, and $c(t,l)$, such as $\partial c(t,l)/\partial l \ge 0$, $\partial\mu(E,l)/\partial E \ge 0$, $\partial\mu(E,l)/\partial l \ge 0$. These assumptions are quite natural and require the price of harvested population to increase in size and the mortality to increase with the size and competition, which holds for many biological populations, e.g., populations of trees and fish. The bang–bang regime (7.57) suggests harvesting only the population individuals starting the size $l^*(t)$.

7.3.4.1 Applications: Forest

Forests present a renewable resource, which provides timber and energy, maintains biological diversity and offers recreation facilities, mitigates climate change and improves air quality. Forests play an essential role in catching and storing greenhouse gases, in particular, in carbon sequastration. Human intervention, natural disturbances, and climate change may cause irreversible and unfavorable changes in the forest dynamics. How to protect the forest and at the same time use its valuable resources? The model (7.51)–(7.56) can be employed to find a rational forest management policy.

If applied to forestry, the objective function (7.51) describes the net benefits of harvesting, which are equal to the revenue from the timber production minus the expenses to plant young trees. By (7.51), the timber revenue is linearly proportional to the amount u of logged timber. It is assumed that the management costs to maintain the forest are included in the revenue.

The steady-state analysis and bang–bang theorems assist forest management in making some qualitative predictions about sustainable harvesting and clarify qualitative properties of the optimal harvesting policies [6, 15]. In particular, there are two major harvesting regimes: clear cutting logging and selective logging. Foresters and economists vigorously advocated clear cutting in the past. However, it seems that there has been a change of mind in the last decades and both harvesting regimes often coexist. Even Finland, where selective logging is prohibited by law, is questioning its previous evaluation. The bang–bang structure (7.57) reflects selective logging when all trees with a certain diameter are to be harvested.

Forest scientists suggest that the climate change primarily affects the growth rate of forests, whereas its effect on the tree mortality cannot be determined unambiguously. The growth rate determines other vital parameters of the forest. Thus, the comparison of the forest dynamics and optimal harvesting regime for different growth rates, keeping in mind that they represent diverse climate scenarios, leads to understanding how the climate change impacts the optimal harvesting rate, the harvesting size (the diameter of a cut tree), the total number of logged trees, and the net benefits of the forest.

Natural disturbances, such as fire, greatly affect the dynamics of a forest population. Forest scientists state that fires influence the mortality rate the most. Thus, the dependence of all model parameters on a mortality rate should be considered to analyze the consequences of natural disturbances.

The size-structured population model (7.51)–(7.56) can be modified in various ways, for instance, by adding benefits from carbon sequestration in timber, soil, and water, or considering other (e.g., recreational) benefits of forest.

Exercises

1. Justify that that the nonlinear equation (7.9) has one trivial solution if $\lambda < \lambda^*$ and two solutions if $\lambda \geq \lambda^*$, where λ^* is defined by (7.10).
2. Prove the maximum principle (7.28)–(7.33) for the optimal control problem (7.27).
3. Derive necessary conditions for the optimality of the control problem (7.26), (7.22), (7.24), (7.25) and show that the dual equation takes the form

$$\frac{\partial\lambda(t,a)}{\partial t} + \frac{\partial\lambda(t,a)}{\partial a} = (r + \mu(E(t),a) + u(t,a))\lambda(t,a) - c(t,a)u(t,a) + \\ + \int_0^{\min(A,T-t)} \frac{\partial}{\partial E}\mu(E(t),s)y(t,s)\lambda(t,s)\,\mathrm{d}s \quad .$$

4. Compare the objective functions (7.27) and (7.48) in age-structured models with controlled rate and effort and explain the similarities and differences.
5. Verify that in the notations (7.34), (7.35), the state equation (7.22) is written as (7.36), which has the steady-state solution (7.37).
6. Prove that in notations (7.34), (7.35), the state equation (7.23) is written as

$$x'(\tau) = -\mu(\tau)x(\tau) - u(\tau)$$

that has the steady-state solution

$$x(\tau) = pe^{-\int_0^\tau \mu(\xi)d\xi} - \int_0^\tau e^{-\int_\nu^\tau \mu(\xi)\,\mathrm{d}\xi} u(\nu)\,\mathrm{d}\nu.$$

7. Check that (7.19) is the solution to the initial problem (7.16), (7.18).
8. Justify that the steady-state solution of the size-structured model (7.11)–(7.13) is possible under the condition (7.15).
9. Provide a detailed proof that (7.20) has a unique solution.
10. Show that (7.39) is the solution to the initial–value problem (7.38).

References[1]

1. Ahmed, N.U., Teo, K.L.: Optimal control of distributed parameter systems. Elsevier/North Holland, New York (1981)
2. Anita, S.: Optimal harvesting for a nonlinear age dependent population dynamics. J. Math. Anal. Appl. **226**, 6–22 (1998)
3. 📖 Anita, S., Arnautu, V., Capasso, V.: An introduction to optimal control problems in life sciences and economics: from mathematical models to numerical simulation with MATLAB. Birkhäuser, Berlin (2011)
4. Barbu, V., Iannelli, M.: Optimal control of population dynamics. J. Optim. Theory Appl. **102**, 1–14 (1999)
5. de Roos, A.: A gentle introduction to physiologically structured population models. In: Tuljapurkar, S., Caswell, H. (eds.), Structured-population models in marine, terrestrial, and freshwater systems, pp. 119–204. Springer-Science+Business, Dordrecht (1997)

[1] The book symbol 📖 means that the reference is a textbook recommended for further student reading.

6. Goetz, R., Hritonenko, N., Mur, R., Xabadia, A., Yatsenko, Y.: Forest management and carbon sequestration in size-structured forests: the case of Pinus Sylvestris in Spain. For. Sci. **56**, 242–256 (2010)
7. Greenhalgh, D.: Some results on optimal control applied to epidemics. Math. Biosci. **88**, 125–158 (1988)
8. Gurtin, M., Murphy, L.: On the optimal harvesting of age-structured populations: some simple models. Math. Biosci. **55**, 115–136 (1981)
9. Hritonenko, N., Yatsenko, Y.: The structure of optimal time- and age-dependent harvesting in the Lotka-McKendrick population model. Math. Biosci. **208**, 48–62 (2007)
10. Hritonenko, N., Yatsenko, Y.: Age-structured PDEs in economics, ecology, and demography: optimal control and sustainability. Math. Popul. Stud. **17**, 191–214 (2010)
11. Hritonenko, N., Yatsenko, Y., Goetz, R., Xabadia, A.: A bang-bang regime in optimal harvesting of size-structured populations. Nonlinear Anal. **71**, 2331–2336 (2009)
12. Kato, N.: Maximal principle for optimal harvesting in linear size-structured population. Math. Popul. Stud. **15**, 123–136 (2008)
13. 📖 Lenhart, S., Workman, J.: Optimal control applied to biological models. Chapman and Hall/CRC (2007)
14. Murphy, L.F., Smith, S.J.: Optimal harvesting of an age structured population. J. Math. Biol. **29**, 77–90 (1990)
15. Tahvonen, O.: Optimal choice between even- and uneven-aged forestry. Nat. Resource. Model. **22**, 289–321 (2009)

Chapter 8
Models of Air Pollution Propagation

The modeling of environmental contamination is a complex subject that requires considering heterogeneous natural and human factors distributed in space and time. Corresponding models involve partial differential equations or their discrete analogues. This chapter presents and analyzes models of pollution propagation in the atmosphere. Section 8.1 reviews basic definitions and properties of pollution propagation in air and water. Starting with simple models, Sect. 8.2 derives partial differential equations for air pollution transport and diffusion with pollution sources. The next two sections are devoted to two air pollution control problems: the location of new plants and control of pollution intensity of existing plants. The last section discusses the structure and features of more complicated atmospheric pollution models. Models of water pollution are considered in Chap. 9.

8.1 Fundamentals of Environmental Pollutions

This section introduces basic definitions used in the modeling of pollution propagation processes in air and water environments [2, 4, 5, 7].

- *Pollution* is the introduction of contaminants into the environment, which causes damage to the environment.
- The *contaminant* (*pollutant*) is a chemical, physical, biological, or other substance (ingredient, agent), which is unusual and harmful for the environment. There are more than two thousands of known pollution agents with negative effect on the environment. The most common ones are the so-called *greenhouse gases* (carbon dioxide CO_2, methane CH_4, nitrous oxide N_2O, chlorofluorocarbons CFCs, water vapor H_2O, and ozone O_3), *soot* (a black substance produced by incomplete combustion of coal, oil, wood, or other fuels), carbon monoxide CO, sulfur dioxide SO_2, and ammonia NH_3.
- *Diffusion* [Lat. *diffusio*—diffusion] is the fundamental process of the penetration of molecules of one substance into another during their contact caused by the

N. Hritonenko and Y. Yatsenko, *Mathematical Modeling in Economics, Ecology and the Environment*, Springer Optimization and Its Applications 88,
DOI 10.1007/978-1-4614-9311-2_8,

heat motion of molecules. The diffusion leads to the spontaneous equilibration of substance concentration in the space.

- *Advection* [Lat. *advectio*—delivering] is a horizontal transfer of liquid or gas together with their properties, such as humidity, heat, pollution, and others.
- *Stratification* [Lat. *stratum*—layer + *facere*—make] is the vertical distribution of the air temperature in accordance with altitude, which determines the intensity of vertical air transfer in the atmosphere. The stratification also occurs in water as the vertical distribution of water layers with different density and affects heat exchange and other physical processes in reservoirs.
- *Temperature inversion* occurs in the atmosphere when its higher layers have a higher temperature compared with the temperature of lower layers (*inversion layers*). It affects the vertical diffusion of pollutants: a warm stratum squeezes the polluted air (a low stratum), which raises a risk of smog.
- *Sedimentation* [Lat. *sedimentum*—settling] is the accumulation of solid particles suspended in a liquid or gas on the bottom surface under the influence of gravity.
- *Dispersion system* [Lat. *dispersus*—dispersed, scattered] is a set of small particles of some substance (the dispersible phase) distributed in a homogeneous medium (dispersing medium), for example mist, smoke, suspension, emulsion, soil, or organic tissues.
- *Dispersibility* is the degree of breaking up a substance into particles. It is higher for smaller particles.
- *Aerosol* [Greek. *aer*—air + germ. *sol*—sault] is a dispersion system that consists of solid or liquid particles (fractions) suspended in a gaseous environment. Examples include smoke, fog, and mist.
- *Smog* [*smoke* + *fog*] is a toxic fog that combines particles of different pollutants, dust, wood smoke, and fog drops.

8.2 Models of Air Pollution Transport and Diffusion

The level of air contamination depends on the presence of pollutants in the atmosphere. The amount of a pollution agent is determined by the number and intensity of pollution sources and meteorological conditions that affect the processes of formation, transport, diffusion, and dispersion of the pollution agent.

In order to describe the process of pollution propagation in space, let us introduce the following notations:

$\mathbf{r} = (x_1, x_2, x_3)$—a point in the three-dimensional space $\mathbf{R}^3$ with coordinates x_1, x_2, x_3,
t—the continuous time,
$\mathbf{v}(\mathbf{r}, t) = \mathbf{v}(x_1, x_2, x_3, t) = (v_1, v_2, v_3)$—the air velocity,
$g(\mathbf{r}, t) = g(x_1, x_2, x_3, t)$—the specific concentration (per a volume unit) of a pollutant at the point $\mathbf{r}$ at time t.

8.2.1 Model of Pollution Transport

We first assume that there is no diffusion in the atmosphere and consider a small space domain of a constant volume that moves with the air. Then, the concentration of the pollution agent in this domain is constant in time t:

$$dg/dt = 0. \tag{8.1}$$

The total derivative of the composite function $g(x_1, x_2, x_3, t)$ with respect to the variable t is equal to:

$$\frac{dg}{dt} = \frac{\partial g}{\partial t} + \frac{\partial g}{\partial x_1}\frac{dx_1}{dt} + \frac{\partial g}{\partial x_2}\frac{dx_2}{dt} + \frac{\partial g}{\partial x_3}\frac{dx_3}{dt}.$$

Then, considering that $v_1 = dx_1/dt$, $v_2 = dx_2/dt$, $v_3 = dx_3/dt$ by the definition of the velocity, we obtain from (8.1) that

$$\partial g/\partial t + v_1 \partial g/\partial x_1 + v_2 \partial g/\partial x_2 + v_3 \partial g/\partial x_3 = 0. \tag{8.2}$$

The *divergence* of the velocity **v** of moving incompressible fluid at the point **r** is

$$\text{div } \mathbf{v} = \partial v_1/\partial x_1 + \partial v_2/\partial x_2 + \partial v_3/\partial x_3.$$

The inequality div $\mathbf{v} > 0$ holds in the presence of a substance *source* at the point **r** (the substance output from the neighborhood of the point **r** is more that its input). Conversely, the point **r** is a *sink* if div $\mathbf{v} < 0$. The equality div $\mathbf{v} = 0$ means the absence of sources and sinks and is commonly considered to hold in lower layers of the atmosphere. The condition div $\mathbf{v} = 0$ is known as the *continuity equation*:

$$\partial v_1/\partial x_1 + \partial v_2/\partial x_2 + \partial v_3/\partial x_3 = 0. \tag{8.3}$$

Applying (8.3) to (8.2), we obtain the following *linear advection equation*:

$$\partial g/\partial t + \text{div}(\mathbf{v}g) = 0, \tag{8.4}$$

as the *model of pollution transport* [3]. If the pollution agent is partially precipitated or decomposed, then $\partial g/\partial t = -\sigma g$ instead of (8.1) and the model (8.4) becomes

$$\partial g/\partial t + \text{div}(\mathbf{v}g) + \sigma g = 0, \tag{8.5}$$

where

σ = const > 0 is a specific *rate of pollution deterioration (decay)*.

If pollution sources exist in the modeling domain, then the model becomes

$$\partial g/\partial t + \operatorname{div} \mathbf{v} g + \sigma g = f, \tag{8.6}$$

where

$f(x_1, x_2, x_3, t)$ is the *intensity function of the pollution sources.*

In particular, the intensify function of the *point source* with a constant intensity Q at the point $\mathbf{r}_0$ is

$$f = Q\delta(\mathbf{r} - \mathbf{r}_0), \tag{8.7}$$

where

$\delta\ (\mathbf{r}-\mathbf{r}_0)$ is the *delta-function* defined by the following statement:

$$\int_C \phi(r)\delta(\mathrm{r} - \mathrm{r}_0)\,\mathrm{dr} = \begin{cases} \phi(r_0), & r_0 \in C, \\ 0, & r_0 \notin C, \end{cases} \tag{8.8}$$

where C is some space domain. Roughly speaking, the delta-function is equal to ∞ at the point $\mathbf{r}_0$ and to 0 at any other point, while the integral of this function is equal to 1. The delta-function is a simple example of the *generalized functions* whose treatment requires some special techniques (see Sect. 8.2.3 below).

In the case of n point pollution sources at $\mathbf{r}_i$, $i = 1,..,n$, with the intensity levels $Q_i(t)$ the intensity function is

$$f = \sum_{i=1}^{n} Q_i(t)\delta(\mathbf{r} - \mathbf{r}_i). \tag{8.9}$$

8.2.2 Model of Pollution Transport and Diffusion

In reality, pollutants also disseminate in the air because of diffusion process. The advection–diffusion equation

$$\frac{\partial g}{\partial t} + \operatorname{div}(\mathbf{v}g) + \sigma g = \frac{\partial}{\partial x_3}\left(\eta \frac{\partial g}{\partial x_3}\right) + \mu \Delta g + f, \tag{8.10}$$

models the transport and diffusion of pollutants. In (8.10),

$\Delta = \partial^2/\partial {x_1}^2 + \partial^2/\partial {x_2}^2$—the two-dimensional *Laplace operator*,
$\mu > 0$ and $\eta > 0$—the *horizontal* and *vertical diffusion coefficients.*

Note that the diffusion coefficients μ and η are different. The vertical diffusion coefficient always depends on the altitude x_3. Estimation of the relation $\eta(x_3)$ is a complicated problem by itself. The horizontal diffusion coefficient μ is assumed to be constant in many modeling cases.

The models (8.5)–(8.10) allow finding the distribution of a pollutant concentration in a given space domain. To solve the partial differential equation (8.3), we need to know the *initial conditions*:

$$g(\mathbf{r}, t_0) = g_0(\mathbf{r}) \quad \text{at} \quad t = t_0, \tag{8.11}$$

and boundary conditions that determine the character of the air interaction with the earth surface. Simple realistic *boundary conditions* are of the form:

$$g_0 = g_S \text{ on } S,\ \partial g/\partial x_3 = ag \text{ at } x_3 = 0,\ \partial g/\partial x_3 = 0 \text{ at } x_3 = H, \tag{8.12}$$

where C is a cylinder and S is its lateral area. These conditions suggest the sedimentation of the pollutant on the bottom surface with the *sedimentation rate* $a \geq 0$, no pollutant transfer through the upper horizontal cylinder boundary, and a given pollution concentration g_S on the vertical cylinder boundary.

A theoretical analysis shows that the boundary problem (8.10)–(8.12) has a unique solution, which is usually found using numerical methods and computer simulation. An analytic solution of the problem (8.10)–(8.12) is possible in special cases only, in particular, in the one illustrated in the next section.

8.2.3 Steady-State Analysis: One-Dimensional Stationary Distribution of Pollutant

To illustrate the qualitative behavior of solutions to the problem (8.10)–(8.12), we consider its special case:

- the stationary (time-independent) distribution of a pollutant in *a one-dimensional infinite medium* (with only one coordinate $x_1 = x$) with one point source of a constant pollution intensity.

Then, $\partial g/\partial t \equiv 0$ and the partial differential equation (8.10) is reduced to the following *ordinary differential equation* with respect to the unknown function $g(x)$:

$$vdg/\mathrm{d}x + \sigma g = \mu d^2 g/\mathrm{d}x^2 + Q\delta(x - x_0), \quad -\infty < x < \infty, \tag{8.13}$$

where

$v = \text{const} > 0$ is the wind velocity (along the coordinate x),
$Q = \text{const} > 0$ is the intensity of the pollution source at the point $x = x_0$.

Despite the presence of the delta-function, (8.13) can be investigated using relatively simple mathematical techniques. Namely, in order to eliminate the delta-function, we analyze (8.13) separately to the right and to the left of the point $x = x_0$. Then (8.13) can be written as the system of the following three equations:

$$\begin{aligned} &\mu d^2 g_-/dx^2 - v dg_-/dx - \sigma g_- = 0 && \text{at } -\infty < x < x_0, \\ &\mu d^2 g_+/dx^2 - v dg_+/dx - \sigma g_+ = 0 && \text{at } x_0 < x < \infty, \\ &\mu dg_+/dx = \mu dg_-/dx + Q && \text{at } x = x_0. \end{aligned} \tag{8.14}$$

The first two linear differential equations (8.14) have exact exponential solutions on the intervals $(-\infty, x_0]$ and $[x_0, \infty)$ respectively. Combining these solutions for $-\infty < x < \infty$ and using the third condition (8.14) and $g_-(x_0) = g_+(x_0)$, we obtain the analytic stationary solution of the model (8.10) over $(-\infty, \infty)$

$$g(x) = \frac{Q}{\sqrt{4\sigma\mu + v^2}} \begin{cases} e^{\left(\sqrt{v^2+4\sigma\mu}+v\right)(x-x_0)/2\mu} & \text{at } x < x_0 \\ e^{\left(\sqrt{v^2+4\sigma\mu}-v\right)(x-x_0)/2\mu} & \text{at } x > x_0 \end{cases}. \tag{8.15}$$

It is illustrated in Fig. 8.1.

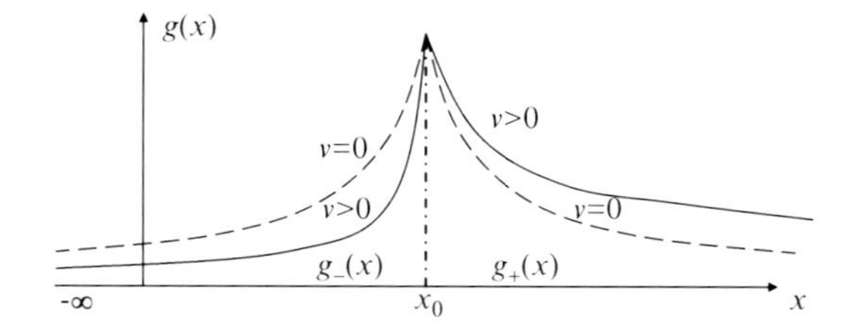

Fig. 8.1 The stationary pollution distribution in the one-dimensional infinite medium with a source at the point x_0 and the wind velocity v

8.2.4 Models of Pollution Transport, Diffusion, and Chemical Reaction

In reality, several pollutants (e.g., carbon dioxide CO_2, carbon monoxide CO, and soot) simultaneously affect the air quality in a region under study. In such cases, the model should consider possible chemical reactions of various contaminants, which are described by so-called *photochemical models*. Let us assume that N relevant pollutants are identified for a specific region. Then, the *model of the transport, diffusion, and chemical transformation of pollutants* can be written as:

$$\begin{aligned} &\partial g_j/\partial t + \operatorname{div}\left(\mathbf{v} g_j\right) + \sigma_j g_j = \partial g_j/\partial x_3\left(\eta \partial g_j/\partial x_3\right) + \mu \Delta g_j + f_j + R_j, \\ &j = 1, \ldots, N, \end{aligned} \tag{8.16}$$

where

$\mathbf{g}(\mathbf{r}, t) = (g_1, \ldots g_N)$ is a vector of concentrations of specific agents at the point $\mathbf{r}$ at time t,

$\mathbf{f}(\mathbf{r}, t) = (f_1, \ldots f_N)$ is a vector of emission rates for the agents (that depend on the locations and intensities of pollution sources),

$\mathbf{R}(\mathbf{g}(\mathbf{r}, t), \mathbf{r}, t) = (R_1, \ldots R_N)$ is a vector of net rates of chemical production from the reactions among pollutants, resulting from a photochemical model.

All other notations are the same as in the model (8.10)–(8.12). The model (8.16) is known in atmospheric sciences as the *atmospheric diffusion equation* [8] and is subjected to certain initial and boundary conditions similar to (8.11)–(8.12).

8.2.5 Control Problems of Pollution Propagation in Atmosphere

Rational management of regional economic–environmental systems puts forward a broad spectrum of important applied modeling problems, such as the following:

- Identifying the most dangerous types of pollutants and modeling their propagation;
- Estimating the joint combined effect of different pollutant emissions in ecologically significant zones of the region, considering the pollutant transport, diffusion, sedimentation and other factors;
- Determining the *maximal allowable concentrations* of pollutants in the region and their *maximal allowable emissions* for each plant in the region;
- Scrapping or modernizing obsolete ecologically dangerous production processes and determining optimal plans for reallocation of existing and construction of new production facilities; and so on.

In the next two sections, we consider two different types of specific air pollution control problems based on pollution models of this section.

8.3 Modeling of Plant Location

In a quite general case, the environmentally rational choice of the location of a new plant can be described as the following control problem:

- *Problem 1*: Find the area $\omega \subset \mathbf{R}^3$ where a new pollution source (a new plant) may be placed so that the pollution contamination level in the ecologically significant areas G_k, $k = 1, \ldots, m$, does not exceed allowed quotas.

The control influence in this problem involves the decision on a new plant location, which satisfies the accepted level of pollution. Here we demonstrate solution techniques for such problems [1].

For clarity, we restrict ourselves to the *one-dimensional case* of Problem 1 with one new pollution source with constant pollution intensity. Then, the model (8.10)–(8.11) of pollution propagation is simplified to

$$\partial g/\partial t + v\partial g/\partial x + \sigma g - \mu\partial^2 g/\partial x^2 = Q\delta(x - x_0), \quad g(x, 0) = g_0(x), \quad (8.17)$$

with respect to the unknown pollution concentration $g(x, t)$, $-\infty < x < \infty$, $0 \le t \le T$, where the pollution intensity function $f(x, t) = Q\delta(x - x_0)$ depends on the position x_0 of the source. The control variable is the coordinate $x_0 \in R^1$ that determines the location of the new pollution source.

The general control Problem 1 is reduced in the one-dimensional case (8.17) to the following problem:

- *Problem 2*: Find the interval $\omega \subset \mathbf{R}^1$ for the location x_0 of a new pollution source, such that the pollution level at a single "ecologically significant" point ξ_1 at a given instant τ_1, $0 \le \tau_1 \le T$, does not exceed the given maximal allowed quota:

$$g(\xi_1, \tau_1) \le C. \quad (8.18)$$

Such problems can be solved numerically using approximate search techniques (based on a repetitive solution of equation (8.17) for different values x_0). However, there exist more efficient mathematical techniques for finding an exact solution of the problem.

One of relevant techniques is the *adjoint method* that has a variety of applications, including location problems, inverse pollution modeling based on measurements, optimization problems, sensitivity analysis, and others. The adjoint method is based on a more general method of Lagrange multipliers, which is also used in Chaps. 5 and 7 for the optimal control in economic and population models.

8.3.1 Adjoint Method

Applied to the one-dimensional diffusion equation (8.17), the adjoint method consists of the following steps:

Step 1. Define a functional that depends on the unknown function g

$$I = \int_0^T \int_{-\infty}^{\infty} p(x, t) g(x, t)\, \mathrm{d}x\, \mathrm{d}t, \quad (8.19)$$

where the function $p(x,t)$ is arbitrary at a specific moment but will be chosen later on to adequately represent the imposed constraint (8.18).

Step 2. Introduce the *adjoint variable* $\lambda(x, t)$, $-\infty < x < \infty$, $0 \le t \le T$, and construct the *Lagrangian* of the problem (8.17)–(8.19) as

$$L = \int_0^T \int_{-\infty}^{\infty} \left\{ p(x,t)g(x,t) - \lambda(x,t)\left[\frac{\partial g(x,t)}{\partial t} + v\frac{\partial g(x,t)}{\partial x} + \sigma g(x,t) - \mu\frac{\partial^2 g(x,t)}{\partial x^2} - f\right]\right\} dx\, dt. \tag{8.20}$$

The adjoint variable λ is also called the *dual variable* or the *Lagrange multiplier* in other applications (Chaps. 3–5 and 7). The key idea of the method of Lagrange multipliers is that we can choose any λ later, because the expression in brackets in (8.20) is zero by (8.17), therefore, $L(p, g, \lambda) = I$ for any λ.

Assuming the pollution concentration $g(x, t)$ to be zero at infinitely remote points x, using the integration by parts, and interchanging integration limits, the Lagrangian is transformed to

$$L = \int_0^T \int_{-\infty}^{\infty} \left\{ p(x,t)g(x,t) + \frac{\partial\lambda(x,t)}{\partial t}g(x,t) + v\frac{\partial\lambda(x,t)}{\partial x}g(x,t) - \sigma\lambda(x,t)g(x,t) + \mu\frac{\partial^2\lambda(x,t)}{\partial x^2}g(x,t) + f(x,t)\lambda(x,t)\right\} dx\, dt - \int_{-\infty}^{\infty} [g(x,T)\lambda(x,T) - g(x,0)\lambda(x,0)]dx,$$

or

$$L = \int_0^T \int_{-\infty}^{\infty} \left\{ \left[p(x,t) + \frac{\partial\lambda(x,t)}{\partial t} + v\frac{\partial\lambda(x,t)}{\partial x} - \sigma\lambda(x,t) + \mu\frac{\partial^2\lambda(x,t)}{\partial x^2}\right] g(x,t). + f(x,t)\lambda(x,t)\right\} dxdt - \int_{-\infty}^{\infty} [g(x,T)\lambda(x,T) - g_0(x)\lambda(x,0)]\, dx. \tag{8.21}$$

Step 3. Now, let us choose the function λ to satisfy the *adjoint equation*

$$-\partial\lambda/\partial t - v\partial\lambda/\partial x + \sigma\lambda - \mu\partial^2\lambda/\partial x^2 = p, \quad -\infty < x < \infty, \quad 0 \le t \le T, \tag{8.22}$$

with the given condition at the right end of the interval $[0, T]$:

$$\lambda(x,T) = 0, \quad -\infty < x < \infty. \tag{8.23}$$

Then, the *Lagrangian* (8.21) becomes

$$L(p,g,\lambda)=\int_0^T\int_{-\infty}^{\infty}f(x,t)\lambda(x,t)\,\mathrm{d}x\,\mathrm{d}t+\int_{-\infty}^{\infty}g_0(x)\lambda(x,0)\,\mathrm{d}x \tag{8.24}$$

and does not depend on the unknown function g. Thus, we find the dependence of the functional $I = L(p, g, \lambda)$ directly on control variables. Indeed, in Problem 2, $f(x, t) = Q\delta(x - x_0)$ and x_0 is the control variable.

For Problem 2, we choose $p(x,t) = \delta(x - \xi_1)\delta(t - \tau_1)$ in (8.19), then the functional (8.19) becomes

$$I=\int_0^T\int_{-\infty}^{\infty}p(x,t)g(x,t)\,\mathrm{d}x\,\mathrm{d}t=g(\xi_1,\tau_1), \tag{8.25}$$

and describes the condition (8.18) as $I \leq C$. On the other side, the representation (8.24) of the functional I at $f(x, t) = Q\delta(x - x_0)$ leads to the following dependence of I on the unknown control variable x_0:

$$I=I(x_0)=Q\int_0^T\lambda(x_0,t)\,\mathrm{d}x\,\mathrm{d}t. \tag{8.26}$$

The analytic solution $\lambda(x, t)$, $-\infty < x < \infty$, $0 \leq t \leq T$, of the linear partial differential equation (8.22) can be found using the Fourier transform as:

$$\lambda(x,t)=\begin{cases}\dfrac{p}{2\sqrt{\pi\mu(\tau_1-t)}}e^{-\frac{\sigma(\tau_1-t)+(x-\xi_1+v(\tau_1-t))^2}{4\mu(\tau_1-t)}} & \text{at}\quad t\in[0,\tau_1],\\ 0 & \text{at}\quad t\in(\tau_1,T].\end{cases} \tag{8.27}$$

In more complicated cases, the solution of adjoint equations can be determined numerically. Knowing the adjoint variable λ, we can construct and evaluate the function $I(x_0)$. The graph of the function $I(x)$ for Problem 2 is given in Fig. 8.2. It is easy to see that the condition $I(x) \leq C$ holds if $x \leq x_1$ or $x \geq x_2$. Hence, the acceptable domain of the new plant locations is $\omega = \{x_0\colon x_0 \leq x_1, x_0 \geq x_2\}$.

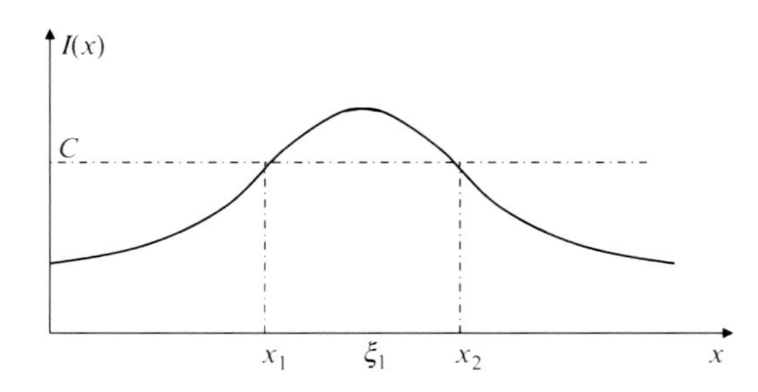

Fig. 8.2 The distribution of pollutions from the source located in the point x around the "ecologically significant" point ξ_1

8.4 Control of Plant Pollution Intensity

Let us consider a region with the area $G \in R^3$ that has n existing plants located at points $\mathbf{r}_i$ with the pollution intensities $u_i(t)$, $i = 1,..,n$. Then, a relevant environmental problem is to determine the *maximal allowable emissions for existing plants*, such that the resulting pollution contamination in ecologically significant areas does not exceed a given maximal allowable level. It can be described as the following control problem:

- *Problem 3*: Determine the maximal emissions u_i, $i = 1,..,n$, for all plants such that the average pollution concentration in a certain ecologically significant area $G_0 \subset G$ does not exceed the given level C.

Let the model of air pollution propagation be given by (8.10) or

$$\partial g/\partial t + \mathrm{div}(\mathbf{v}g) + \sigma g = \partial g/\partial x_3(\eta \partial g/\partial x_3) + \mu \Delta g + f, \tag{8.28}$$

with the initial and boundary conditions

$$g(\mathbf{r}, 0) = 0, \tag{8.29}$$

$$g = g_S \text{ on } S, \quad \partial g/\partial x_{3\mathrm{v}} = ag \text{ at } x_3 = 0, \quad \partial g/\partial x_3 = ag \text{ at } x_3 = H, \tag{8.30}$$

where S is the boundary of the domain G. In accordance with the formula (8.9), the emission intensity function in (8.28) is

$$f(\mathbf{r}, t) = \sum_{i=1}^{n} u_i(t)\delta(\mathbf{r} - \mathbf{r}_i). \tag{8.31}$$

This problem can also be solved using *the adjoint method*. The stationary and dynamic cases of the problem lead to different mathematical problems and are considered separately in the next two sections.

8.4.1 Stationary Control of Air Pollution Intensity

Let us first assume that the maximal plant emissions do not depend on the time t, i.e., the unknown controls are n real numbers $u_i \in R^1$, $i = 1,..,n$.

As applied to the problem (8.28)–(8.30), the ***adjoint method*** is:

1. The *averaged over the time period* $[0,T]$ concentration of the pollution agent in the area G_0 is described by the following functional:

$$I = \int_0^T \int_{G_0} pgdG\,\mathrm{d}t, \quad \text{where } p = p(\mathrm{r}) = \begin{cases} 1/T, & r \in G_0, \\ 0 & r \notin G_0. \end{cases} \tag{8.32}$$

With the chosen function p, the functional I equals the average total pollution concentration that should not exceed the given level C, therefore, $I \leq C$.

2. The adjoint equation for the problem (8.28)–(8.30) is obtained using the method of Lagrange multipliers similarly to Step 2 of Sect. 8.3.1:

$$-\partial\lambda/\partial t - \operatorname{div}(\mathbf{v}\lambda) + \sigma\lambda = \partial g/\partial x_3(\eta\partial\lambda/\partial x_3) + \mu\Delta\lambda + p, \tag{8.33}$$

with the initial condition $\lambda(T, \mathbf{r}) = 0$ at the right end of $[0, T]$ and the certain boundary conditions related to (8.30). This linear equation has a unique solution $\lambda(t, \mathbf{r})$.

3. Similarly to Step 3 of Sect. 8.3.1, using the initial condition (8.29) and assuming a large enough domain G such that $g_S = 0$ in (8.30), we obtain the following relation between g and λ

$$I = \int_0^T \int_{G_0} p(\mathbf{r})g(\mathbf{r}, t)\,\mathrm{d}\mathbf{r}\,\mathrm{d}t = \int_0^T \int_{G_0} f(\mathbf{r}, t)\lambda(\mathbf{r}, t)\,\mathrm{d}\mathbf{r}\,\mathrm{d}t. \tag{8.34}$$

Applying (8.31) and the constraint $I \leq C$ to (8.34), we obtain the condition:

$$I = \int_0^T dt \int_{G_0} \lambda \sum_{i=1}^{n} u_i\delta(\mathbf{r} - \mathbf{r}_i)\,\mathrm{d}\mathbf{r} = \sum_{i=1}^{n} u_i \int_0^T \lambda(\mathbf{r}_i, t)\,\mathrm{d}t = \sum_{i=1}^{n} u_i A_i \leq C, \tag{8.35}$$

where the values $A_i = \int_0^T \lambda(\mathbf{r}_i, t)\mathrm{d}t$ do not depend on the unknown variables u_i, $i = 1,..,n$.

4. Any combination of u_i, $i = 1,..,n$, that satisfies $\sum_{i=1}^{n} u_i A_i = C$ is a solution of the control Problem 3.

The values $\bar{u}_i$, $i = 1,..,n$, such that $\sum_{i=1}^{n} \bar{u}_i A_i = C$, are called the *maximal allowable emissions*. Combinations $\{\bar{u}_1, .., \bar{u}_n\}$ of maximum allowable emissions can be different. Therefore, an additional optimization criterion can be introduced in Problem 3 to find a unique solution, for example, minimization of the total expenses for the pollution cleanup:

$$\min \sum_{i=1}^{n} E_i \bar{u}_i \quad \text{at} \quad \sum_{i=1}^{n} \bar{u}_i A_i = C, \tag{8.36}$$

where E_i is the given expenditure of the i-th plant on decreasing the pollution emission by one unit. The optimization problem (8.36) with n unknowns $\bar{u}_i$, $i = 1,\ldots, n$, belongs to *linear programming problems*.

8.4.2 Dynamic Control of Air Pollution Intensity

In the dynamic case, the statement of Problem 3 and the model (8.28)–(8.30) are the same as at the beginning of Sect. 8.4, but the unknown controls (the maximal allowable emissions for each plant) $u_i = u_i(t)$ depend on the time $t \in [0, T]$. The domain of admissible controls is:

$$0 \le u_i(t) \le u_i^{\max}(t), \quad i = 1, \ldots, n, \quad t \in [0, T]. \tag{8.37}$$

where the functions $u_i^{\max}(t)$, $i = 1,\ldots,n$, are given. This problem can be also analyzed using the adjoint method. First two steps of this method are similar to the previous Sect. 8.4.1. However, the third step leads to a more complex constraint-inequality of the form

$$I(u_1, \ldots, u_n) \le C, \tag{8.38}$$

where I is a nonlinear functional with respect to n unknown functions u_i, $i = 1,\ldots,n$.

As in the static case of Problem 3, there are many possible combinations $(u_1,\ldots, u_n)$ that satisfy the constraint (8.38). So the control problem can be completed with introducing an additional optimization objective. To construct a simple reasonable example of such economic objective, we can define the net profit of the i-th plant during the period $[0, T]$ as the functional

$$P_i(T, u_i, g) = R_i(T) - S_i(T), \quad i = 1, \ldots, n, \tag{8.39}$$

where

$R_i(T) = \int_0^T e^{-rt} h_i(u_i(t))dt$, $h_i \ge 0$, $h_i' > 0$, is the revenue of the i-th plant,
$S_i(T) = \int_0^T e^{-rt} c_i u_i(t)dt$ is the tax for polluting the environment.

The profit $P_i(.)$ considers only the revenue and pollution expenses as the characteristics directly related to the control u_i. It may also include other economic characteristics from Chaps. 2–5. The central planner framework leads to the following *optimization problem*:

$$\max \sum_{i=1}^{n} P_i(T, u_1, \ldots u_n, g), \tag{8.40}$$

subject to the constraints (8.28)–(8.30), (8.38) with respect to n unknown control functions $u_i(t)$, $i = 1,\ldots,n$, $t \in [0, T]$. This problem assumes the *cooperative behavior* of all economic agents (plants).

If all plants are independent economic agents, then the problem of the net profit maximization is solved by each plant independently, which leads to n-person *continuous differential game* with the duration T, which dynamics is described by (8.28) with the unknown functions u_i, $i = 1,..,n$. This problem reflects the competitive behavior of plants.

The qualitative analysis of such problems turns to be quite complicated and is out of the scope of this textbook. The formulated problems can be solved by numerical algorithms, but their computational solution is also challenging.

8.5 Structure of Applied Air Pollution Models

Real processes in the atmosphere are more complex than the models considered in this chapter. Depending on investigation goals, the following aspects can be also taken into account:

- Different interaction of light and heavy pollutants with the surface, gravitational precipitation (sedimentation) of the pollutants, and their wind lift.
- The structural dispersion of pollutants in randomly nonhomogeneous medium of urban buildings (a so-called *surface super-roughness*).
- Structural chemical and physical (optical) properties of pollution agents, composition of their elements, dispersibility, and condensation activity.
- Different structure of pollution sources for various agents, ways and places of pollution entry into the atmosphere.
- Changing technologies of waste processing, the presence and effectiveness of pollution cleanup abatement facilities.
- Meteorological conditions such as atmospheric turbulence, dynamic characteristics of atmospheric layers, wind and temperature changes, humidity, atmospheric stratification, and so on.

The conceptual flow diagram in Fig. 8.3 summarizes the links among the pollution processes in the atmosphere and on the earth surface. Some features of these processes are briefly discussed below.

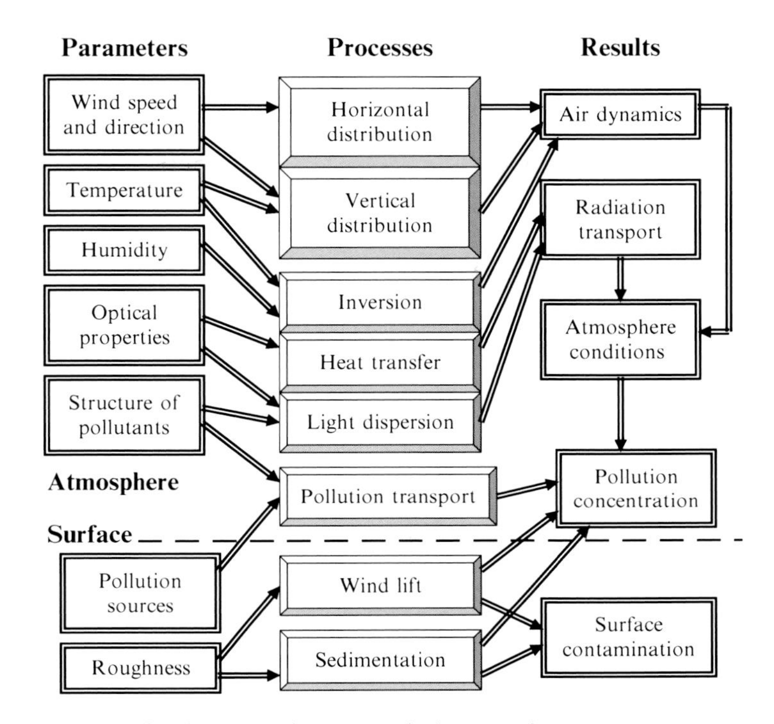

Fig. 8.3 Relations of various dynamic processes in the atmosphere

8.5.1 Interaction with Earth Surface

Applied models for prediction of air contamination in cities and industrial regions should consider the interaction of pollutants with underlying surfaces. Classic solutions of atmospheric diffusion equations similar to given in Sect. 8.2 are valid only for few contamination sources and the air domains distant from the underlying surface. They are not applicable for the description of pollution transport over complex terrains or surfaces covered by urban buildings or trees (*super-rough surfaces*). The complexity of such processes stipulates the development of models based on deterministic and stochastic principles, depending on structural properties of the surface and their correlation with the scale of meteorological processes [4, 6].

The models considered in this chapter describe the pollution propagation with satisfactory accuracy in limited neighborhoods of the pollution source (10–15 km area) and such that a cloud of pollutants (plume) is at some distance from the earth surface. Then, the surface does not affect the pollution propagation process. As the pollution cloud recedes from the source, its lower boundary approaches the earth surface and starts to interact with it. The further propagation of the cloud is influenced by the surface, so more complex models are needed.

From a physical point of view, the air diffusion and transport processes are the most complex for the 100–200 km scale. For such scales, the atmospheric motion depends essentially on surface irregularities, air turbulence characteristics, atmospheric inversions, etc. Diffusion operators in such cases may have a *tensor structure*, i.e., (8.7) may contain terms of the form $\partial^2 g/\partial x_1 \partial x_2$, $\partial^2 g/\partial x_1 \partial x_3$, $\partial^2 g/\partial x_3 \partial x_2$ in addition to the Laplace operator Δ (similar equations will be considered in Sect. 9.3 for the water pollution propagation).

8.5.2 Interaction of Different Air Pollutants

Air contaminants can be divided into the following types:

- Particles of the rigid dispersible phase, such soot or as heavy toxic metals (lead, chrome, and others),
- Aerosol particles,
- Gaseous pollutants,
- Radioisotopes, etc.

Various pollutants from these groups can interact chemically and produce new pollutants that can be more (or less) dangerous for the environment. The corresponding models model should take into account relevant chemical reactions and describe chemical transformation of pollutants (see Sect. 8.2.4). Final products of industrial air contamination are mostly soot and aerosol particles, which play the role of sinks for many gaseous pollutants. The presence of aerosol affects physical, chemical and optical properties of the atmosphere (including its humidity). Main components of urban industrial aerosols are carbon, sulfur, and nitrogen combinations. If a pollutant cloud contains small drops, steam, or small rigid particles, then they are adsorbed on the earth surface. Pollutant particles in the air fly down, deposit on the surface, and do not return into atmosphere. This process is known as the pollution *deposition* or *sedimentation*. Some models of the sedimentation of water pollutants will be considered in Sect. 9.2.

8.5.3 Air Contamination in Cities

Modeling of the vertical transport of pollutants is especially important for cities. The atmospheric temperature normally decreases when the altitude increases. However, a number of cities is characterized by unfavorable meteorological conditions, under which the air pollution (in particular, carbon oxide CO) remains in the urban territory area because of wind absence, while the vertical pollution export is obstructed by *inversion layers* at the altitude of 0.5-2 kilometers. An increase in air temperature with height is known as the *temperature inversion*.

The *surface temperature inversions* cause the maximal concentration of pollutants near the earth surface (smog), while *raised inversions* cause it at a certain altitude (below the inversion layer). The *temperature inversions* are a problem of many large cities like London, New Mexico, Los Angeles, Mumbai, Santiago, and Tehran. The smog caused by surface temperature inversions can be extremely dangerous, for example, *the Great Smog of 1952* in London caused thousands of deaths. Certain cities experience the simultaneous occurrence of near-land and raised inversions. For example, *the CO concentration distribution* in Almaty (Kazakhstan) often includes two layers with maximal concentrations near the surface (because of transport pollution) and at some altitude (caused by hot exhaust of power stations). In addition, many cities are subjected to distinctive daily fluctuations of the CO concentration with a morning maximum because of transport pollution and an evening maximum due to industrial pollution.

Exercises

1. Prove that $\mathrm{div}(\mathbf{v}g) = \mathrm{div}(\mathbf{v})g + \mathbf{v} \cdot \mathbf{grad}g$, where $\mathbf{grad}g = \{\partial g/\partial x_1, \partial g/\partial x_2, \partial g/\partial x_3\}$. Using this formula and (8.3), transform (8.2) to the advection equation (8.4).
2. Propose a modification of the model (8.6) that has one point sink for a pollutant. The sinks can be created using innovative scientific technologies, for example, carbon sequestration.
3. Derive a modification of the model (8.10) in a stationary environment (independent of time).
4. Show that the model (8.10) at $\mu = \eta =$ const can be rewritten using the three-dimensional Laplace operator.
5. Find the exact exponential solution to the first linear differential equation (8.14) on the interval $(-\infty, x_0]$ and to the second equation (8.14) on the interval $[x_0, \infty)$. Combine these solutions with the third condition (8.14) and obtain the formula (8.15).

   HINT: The initial condition $g(x_0)$ is the same for both solutions, while the third formula (8.14) gives the initial condition for the derivative.
6. Take $N = 3$ and present the model (8.16) as a system of equations.
7. Show how the formula (8.21) is obtained from (8.20).
8. Verify that (8.27) is a solution to the problem (8.22)–(8.23).
9. Obtain the adjoint equation (8.33).
10. Explain three steps of the adjoint method for the problem (8.28)–(8.30).

References[1]

1. Drezner, Z. (ed): Facility location: a survey of applications and methods. Springer-Verlag, Heidelberg (1995)
2. 📖 Jacobson, M.Z.: Fundamentals of atmospheric modeling, 2nd edn. Cambridge University Press, New York (2005)
3. 📖 Petrosjan, L.A., Zakharov, V.V.: Mathematical models in environmental policy analysis. Nova Science, New York (1997)
4. Rao, S.T.: Environmental monitoring and modeling needs in the 21st century. EM: Magazine for Environmental Managers **10**, 3–4 (2009)
5. 📖 Sportisse, B.: Fundamentals in air pollution. From processes to modelling. Springer, Dordrecht (2010)
6. 📖 Tiwary, A., Colls, J.: Air pollution: measurement, modelling and mitigation, 3rd edn. Routledge, New York (2010)
7. 📖 Vallero, D.: Fundamentals of air pollution. Academic, London (2007)
8. 📖 Wallace, J.M., Hobbs, P.V.: Atmospheric Science, 2nd edn. Academic, New York (2006)

[1] The book symbol 📖 means that the reference is a textbook recommended for further student reading.

Chapter 9
Models of Water Pollution Propagation

The modeling of pollution dissemination in water is based on a mathematical description of hydrodynamic, hydraulic, and physical–chemical processes that control pollution transfer in water reservoirs. Section 9.1 describes the structure and classification of such models. Section 9.2 discusses a general three-dimensional model of water pollution, which includes transport and diffusion of pollutants in dissolved state, suspension, and ground deposits, adsorption and sedimentation of pollutants, subjected to water dynamics and boundary conditions of water reservoirs, and pollution sources. Section 9.3 describes a two-dimensional horizontal model of water pollution dissemination. Section 9.4 presents a simple one-dimensional model and its analytic solutions for one point source of pollutant. Section 9.5 explores compartmental models of water pollutions and related problems of water pollution control.

9.1 Structure and Classification of Water Pollution Models

Chapter 8 introduced air pollution models starting with the simplest model (a continuity equation) and gradually adding more features such as diffusion, pollution sources, and boundary conditions to obtain realistic models. Modeling of pollution migration in water reservoirs (rivers, oceans, lakes, floods, or storage pools) significantly depends on the geometry and boundary conditions of reservoirs, pollution sources, interaction with air and ground surfaces, and turbulent flows. In this chapter, we start with a general three-dimensional model of water pollution, which has mostly a theoretical value. Next, we discuss its simplified versions, two- and one-dimensional models, often used in practical calculations.

N. Hritonenko and Y. Yatsenko, *Mathematical Modeling in Economics, Ecology and the Environment*, Springer Optimization and Its Applications 88,
DOI 10.1007/978-1-4614-9311-2_9, © Springer Science+Business Media New York 2013

9.1.1 Structure of Models

The distribution of pollutants in water reservoirs involves the following *hydrophysical processes*:

- Wind and water flow currents,
- Dynamics of water stratification,
- Transfer of turbulent characteristics,
- Transport of suspended drifts,
- Transport of involved drifts,
- Distribution and transformation of wind-generated waves,
- Riverside currents generated by waves,
- Sedimentation and lift-off (disturbance) of drifts.

To determine the intensity of transfer mechanisms and pollution accumulation in reservoirs, models of pollution propagation in water reservoirs should contain the *following submodels* (*model blocks*):

- A model of pollution transfer in a dissolved state (solute),
- A model of pollution transfer in suspension,
- A model of pollution accumulation in bottom sediments,
- A model of pollution transfer in erosion–sedimentation processes,
- Models of physical and chemical transformation of pollutants.

Lakes and storage pools differ from surface flows and riverbed flows by a substantially longer stay of pollution in these reservoirs, considerable influence of wind, surface waves, and structure of current field. Turbulence also plays an important role: its intensity in deep reservoirs is considerably different along the vertical coordinate and depends on density stratification parameters. The sedimentation and erosion processes are important for pollution accumulation in water and on the bottom surface. The models require the following input data (as *initial and boundary conditions*):

- Hydrological conditions of the reservoir,
- Morphological data (depth, profile, physical properties of the bottom, etc.),
- Meteorological conditions,
- Locations and intensity of pollution sources.

9.1.2 Classification of Models

Various models of different complexity have been developed and used to describe complex interacting processes of water pollution propagation. These models differ by space dimensions, time interval, level of uncertainty, and mathematical and computer tools [1]. We restrict ourselves to deterministic models in continuous time and provide a classification of known models from the mathematical point of view.

1. *Three-dimensional models* give a detailed description of the processes under study, but are rarely used in practice because of essential challenges in collecting

required detailed data (including an initial state and boundary currents) and high computational complexity.

2. *Two-dimensional horizontal models*. In many practical situations, the vertical transfer of pollutants is low and can be disregarded during modeling process. Specifically, it is true for space scales exceeding several hundred meters. Then, it is enough to compute average by depth characteristics of the water current without obtaining a detailed vertical distribution of pollutions in water.
3. *Two-dimensional vertical models* describe the vertical distribution of pollution concentrations. They are used for investigation of pollutant accumulations in specific areas of fast changing depth, for instance, in river traps.
4. *One-dimensional river-bed models* describe dynamics of pollution concentration along long rivers averaged by a vertical cross section of the river.
5. *One-dimensional vertical models* describe the stratified dynamics of the pollution concentration averaged by a horizontal section.
6. *Zero-dimensional (compartmental) models* split a water reservoir into smaller parts (compartments, cells) and describe the dynamics of pollution flows among the compartments, using ordinary differential or difference equations.

Analytical solutions of considered models are possible only for simple zero-dimensional or simplified versions of one- and two-dimensional models.

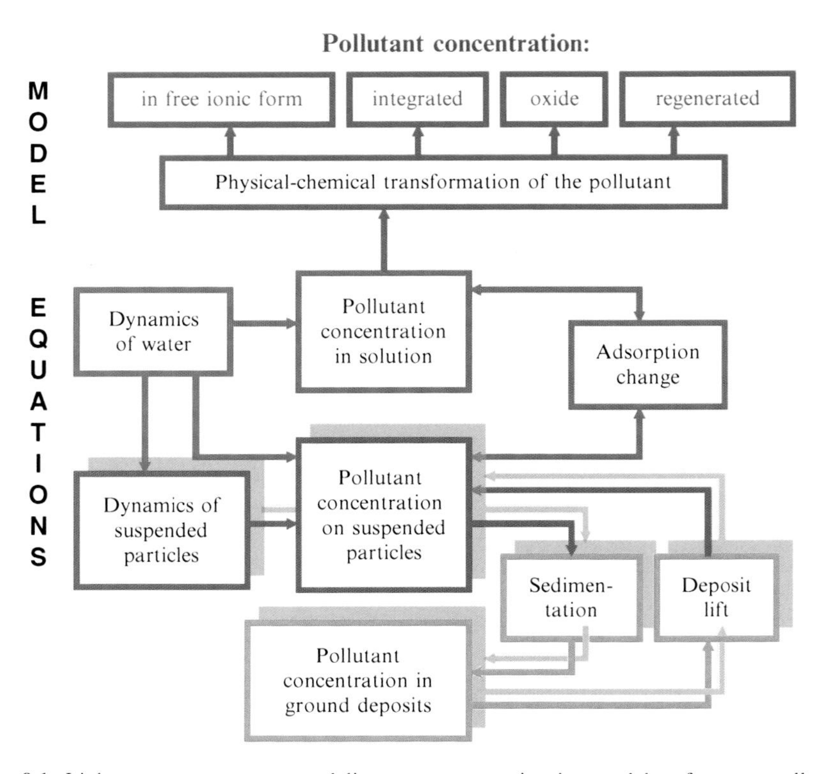

Fig. 9.1 Links among separate modeling components in the models of water pollution propagation

Depending on the dimensionality and complexity, the models of water pollution propagation may include modeling components (blocks) for all or some of the following processes [4, 6]:

- The *equations of water dynamics*,
- The *transport of suspended drifts*,
- The *pollution concentration in dissolute form* C,
- The *pollution concentration in suspension* C^s,
- The *pollution concentration in bottom sediments* C^b,
- *Adsorption interchange, sedimentation and deposit lift processes*,
- *Physical and chemical transformation of pollution agent.*

A flow diagram of these model components is given in the Fig. 9.1. Some of the discussed models of water pollution propagation are explored below.

9.2 Three-Dimensional Model

Here we consider a process of water pollution propagation in the three-dimensional space with the coordinates x_1, x_3, x_3. Corresponding mathematical model is too general for practical use, but it provides an accurate theoretical basis for applied models exposed further.

9.2.1 Models of Adsorption and Sedimentation

All models considered below use common principles of adsorption and sedimentation processes.

Adsorption model: The *adsorption* is the adhesion of atoms, ions, or molecules of a pollutant to the surface of small particles (adsorbents) in the water. The *absorption* means the process of pollutant dissolving in the water, while the *sorption* encompasses both processes. For the purposes of pollution dissemination, the dependence between the concentration of a pollutant in rigid phase and its concentration in solute is well approximated by the following *linear adsorption model*:

$$dC^s/dt = \alpha(C - C^s/K_s), \tag{9.1}$$

where $C(x_1, x_2, x_3, t)$, the concentration of a pollutant in solute (in the dissolved phase); $C^s(x_1, x_2, x_3, t)$, the concentration of a pollutant in suspension (in the rigid phase); α, the coefficient of mass-response in fluid; K_s, the *equilibrium coefficient* for the system "suspension-water."

The coefficient K_s describes the ratio C^s/C in *equilibrium*, when this ratio is constant. This coefficient depends on physical and chemical characteristics of water and suspension, especially on the temperature. In a more general case, K_s can also depend on the concentration C^s, which leads to nonlinear *adsorption models*.

Sedimentation Model: Hydraulic processes of the sedimentation and lift of ground deposits determine the accumulation of the pollutant on the bottom of water reservoirs. The accumulation of pollutant deposits from suspended particles on the bottom is described as

$$\partial M^b / \partial t = g^s - g^b, \tag{9.2}$$

where $M^b(x_1, x_2, x_3, t)$, the amount of the ground deposits on the bottom; $g^s(x_1, x_2, x_3, t)$, the sedimentation flow; $g^b(x_1, x_2, x_3, t)$, the deposit lift flow.

A simple realistic model of the sedimentation and lift processes is of the form:

$$g^s = \begin{cases} k(S - S*) & \text{at } S > S^*, \\ 0, & \text{at } S < S^*, \end{cases} \qquad g^b = \begin{cases} k(S - S^*) & \text{at } S < S^*, \\ 0, & \text{at } S > S^*, \end{cases} \tag{9.3}$$

where $S(x_1, x_2, x_3, t)$, the concentration of suspended particles; $S^*(x_1, x_2, x_3)$, a given equilibrium concentration S; k, a given coefficient of sedimentation intensity.

9.2.2 *Equation of Transport of Dissolved Pollutants*

The propagation of pollutants in water follows the same physical laws and is described similarly to the advection and diffusion processes in air considered in Sect. 8.2. Namely, the advection–diffusion equation for the concentration $C(x_1, x_2, x_3, t)$ of dissolved pollutants is of the form:

$$\frac{\partial C}{\partial t} + \sum_{i=1}^{3} \frac{\partial (v_i C)}{\partial x_i} = \sum_{i=1}^{3} \frac{\partial}{\partial x_i}\left(A_i \frac{\partial C}{\partial x_i}\right) - \alpha_{12}(K_s C - C^s), \quad i = 1, 2, 3, \tag{9.4}$$

where v_i are components of the water flow velocity; A_i are diffusion coefficients, $i = 1,2,3$; α_{12} is the constant intensity of adsorption in the system "water-suspension."

Compared to the air advection–diffusion equation (8.2) of Sect. 8.2, this equation contains the new term responsible for the adsorption exchange process (9.1) in the system "suspension-water." Commonly used *boundary conditions* for (9.4) are of the form:

- on the free water surface $x_3 = \eta$:

$$A_3 \partial C / \partial x_3 = -v_3 C, \tag{9.5}$$

- on the bottom ground surface $x_3 = z_0$:

$$A_{13}\partial C/\partial x_3 = -(1-\varepsilon)D\alpha_{13}\left(K_b C - C^b\right), \tag{9.6}$$

where $z_0(x_1, x_2)$, the given vertical coordinate x_3 of the bottom surface; $\eta(x_1, x_2, t)$, the unknown vertical coordinate x_3 of the free water surface (defined by the model of water dynamics of Sect. 9.2.4); ε, the coefficient of ground porosity; D, the average size of particles; α_{13}, the adsorption intensity in the system "water-ground deposits"; $C^b(x_1, x_2, z_0, t)$, the pollutant concentration in ground deposits.

9.2.3 *Equation of Transport of Suspended Pollutants*

The transport and diffusion of suspended particles in water is described in the framework of the diffusion theory by the following equation:

$$\frac{\partial S}{\partial t} + \sum_{i=1}^{3} \frac{\partial (v_i S)}{\partial x_i} = \sum_{i=1}^{3} \frac{\partial}{\partial x_i}\left(A_i \frac{\partial S}{\partial x_i}\right) - \omega_0 \frac{\partial S}{\partial x_3}, \quad i = 1, 2, 3, \tag{9.7}$$

where $S(x_1, x_2, x_3, t)$, the concentration of suspended particles in water; ω_0, the so-called *hydraulic size* of particles.

The boundary condition on the free water surface $x_3 = \eta(x_1, x_2, t)$ reflects the condition that the vertical flow of particles of the size ω_0 is zero:

$$\text{at } x_3 = \eta: \quad A_3\partial S/\partial x_3 = (\nu_3 - \omega_0)S. \tag{9.8}$$

The diffusion accumulation of particles on the bottom is equal to the sedimentation equilibrium flow from Sect. 9.1.1:

$$\text{at } x_3 = z_0: \quad A_3\partial S/\partial x_3 = -\omega_0 S^*, \tag{9.9}$$

where $S*$ is the equilibrium concentration of particles determined by the transporting capacity of the flow.

The equation of the pollutant transport on suspended particles is formulated as

$$\frac{\partial (SC^s)}{\partial t} + \sum_{i=1}^{3} \frac{\partial (v_i SC^s)}{\partial x_i} = \sum_{i=1}^{3} \frac{\partial}{\partial x_i}\left(A_i \frac{\partial (SC^s)}{\partial x_i}\right) - \omega_0 \frac{\partial (SC^s)}{\partial x_3} + \alpha_{12} S(K_s C - C^s), \tag{9.10}$$

where $C^s(x_1, x_2, x_3, t)$ is the unknown *concentration of pollutant on suspended particles*. The corresponding boundary conditions are as follows:

$$\text{at } x_3 = \eta: \quad A_3 \partial(SC^s)/\partial x_3 = (\nu_3 - \omega_0)(SC^s), \tag{9.11}$$

$$\text{at } x_3 = z_0: \quad A_3 \partial(SC^s)/\partial x_3 + \omega_0(SC^s) = C^s g^s - C^b g^b, \tag{9.12}$$

where the sedimentation flow g^s and deposit lift flow g^b are determined in Sect. 9.2.1, and $C^b(x_1, x_2, x_3, t)$ is the unknown concentration of the pollutant in ground deposits.

9.2.4 *Equations of Surface Water Dynamics*

The previous model equations include the vector (v_1, v_2, v_3) of *water velocity*. The way of finding the velocity is the major difference between pollution propagation models in the atmosphere and water. In the air pollution models of Chap. 8, the wind velocity can be usually measured or estimated with sufficient accuracy by weather prediction models, and then the pollution model calculates the propagation of pollutants under the given air velocity. It is not always the same in water pollution modeling. The water velocity highly depends on the geometry of reservoirs and is larger in shallow places. Near the bottom, it depends on the physical properties and profile of the bottom. Near the surface, it depends on the wind velocity. So the water currents are often determined from a separate modeling block.

As an example of such a block, let us consider the dynamics of surface water described by the *Reynolds equations for a fluid with free surface*:

$$\frac{\partial v_i}{\partial t} + \sum_{j=1}^{3} v_j \frac{\partial v_i}{\partial x_j} = -\frac{1}{\rho}\frac{\partial P}{\partial t} + G_i + \sum_{j=1}^{3} \frac{\partial \left\langle v_i', v_j' \right\rangle}{\partial x_i}, \quad i = 1, 2, 3, \tag{9.13}$$

$$\partial v_i / \partial x_i = 0, \quad i = 1, 2, 3, \tag{9.14}$$

where P, the water pressure; ρ , the water density; $\mathbf{G} = (0,0,-g)$, the gravity force vector; the 3×3-matrix $<v'_i, v'_j>$ is the *Reynolds turbulent stress tensor*.

The Reynolds tensor is related to the *strain velocity tensor* $<(\partial v_i/\partial x_j)(\partial v_j/\partial x_i)>$ as:

$$\left\langle v_i', v_j' \right\rangle = -A_{ij} \left\langle \left(\partial v_i / \partial x_j\right)\left(\partial v_j / \partial x_i\right) \right\rangle - 2K\delta_{ij}/3, \quad i, j = 1, 2, 3, \tag{9.15}$$

where A_{ij}, the turbulence coefficients; K, the turbulent energy; δ_{ij}, the *Kronecker delta symbol* ($\delta_{ii} = 1$, $\delta_{ij} = 0$ at $i \neq j$).

Equations (9.13)–(9.15) of water dynamics are obviously too general for practical use and require a lot of complex input data. They should be solved with respect to the unknown velocity vector, water temperature, and water pressure with proper initial and boundary conditions, which are not provided here. In practical problems, these equations are reduced to simpler ones. We will consider such equations of water dynamics in Sect. 9.3.

9.2.5 *Modeling of Pollutant Transport in Underground Water*

Mathematical models of *pollution migration in underground water* are even more complicated as compared to the models of pollution propagation in surface waters. The soil consists of a *skeleton* and *pores*, and water (solute) disseminates through pores at a free state. The water can fill the pores completely (a *saturated soil*) or partially (a *non-saturated soil*). In the latter case, the capillary and surface effects create an additional strain and pressure in solute. The elasticity of soil skeleton is taken into account for pressured water layers. Correspondingly, models of pollution propagation in underground water include equations of the underground water flows *in aeration zone*, in *pressured and non-pressured water layers* as well as the equations of migration of a pollutant between solute and soil rigid skeleton.

These models take into account transport of the water supplied by rivers, atmosphere sediments, industrial and agricultural pollutants, mineral fertilizers, depletion and destruction of land, and other related processes. Some pollutants (such as radionuclide, heavy metals, etc.) propagate in soil in a chemically connected (with oxide) form. Propagation models for such pollutants should contain additional models of oxide transport and chemical ionic-oxide equilibrium in the soil.

9.3 Two-Dimensional Horizontal Model

The two-dimensional pollution propagation model of this section describes the *averaged in depth concentrations* $C(x_1, x_2, t)$, $C^s(x_1, x_2, t)$, $C^b(x_1, x_2, t)$ of pollutants in solute, suspension, and ground deposits. It assumes that there is no vertical flow of water and pollutants and the concentrations C, C^s, C^b are the same at any depth. This model can be formally obtained by averaging the three-dimensional model (9.3)–(9.14) with related boundary conditions over the flow depth. However, this process requires tedious transformations. In this section, we illustrate and interpret the equations of a two-dimensional model and their links to the three-dimensional model of the previous section.

9.3.1 *Equation of Ground Deposit Accumulation*

The dynamics of the amount (thickness) $M(x_1, x_2, t)$ of the ground deposit layer is described by (9.2):

$$\partial M/\partial t = g^s - g^b, \tag{9.16}$$

where

$$g^s(x_1, x_2, t) = \begin{cases} B(S - S^*) & \text{at } S > S^*, \\ 0, & \text{at } S < S^*, \end{cases} \quad \text{—the sedimentation flow,}$$

$$g^b(x_1, x_2, t) = \begin{cases} B(S - S^*) & \text{at } S < S^*, \\ 0, & \text{at } S > S^*, \end{cases} \quad \text{—the deposit lift flow.}$$

Then, the pollutant concentration $C^b(x_1, x_2, t)$ in the ground deposit layer on the bottom is described by the following equation:

$$\partial\left(MC^b\right)/\partial t = -\hat{\alpha}_{13}\left(K_d C - C^b\right) - C^b g^b + C^s g^s, \tag{9.17}$$

which combines the near-bottom adsorption processes in the system "water-ground deposits" (described by the boundary condition (9.6) in the three-dimensional model) with the sedimentation and lift flows of suspended particles.

9.3.2 *Equation of Transport of Dissolved Pollutants*

Integrating (9.4) over the vertical space coordinate x_3 (the flow depth) and taking into account the boundary conditions (9.5)–(9.6), we obtain the following *two-dimensional equation of the pollutant transport in a dissolved phase*:

$$\frac{\partial(hC)}{\partial t} + \sum_{i=1}^{2} \frac{\partial(v_i hC)}{\partial x_i} = \sum_{i=1}^{2} \frac{\partial}{\partial x_i}\left(A_i \frac{\partial(hC)}{\partial x_i}\right) - \hat{\alpha}_{12}(K_s C - C^s) + \hat{\alpha}_{13}\left(K_b C - C^b\right), \qquad i = 1, 2, 3, \tag{9.18}$$

where $v_1(x_1, x_2, t)$ and $v_2(x_1, x_2, t)$, the horizontal water velocities averaged over x_3; $h(x_1, x_2, t) = \eta - z_0$, the unknown depth of water flow; α_{12}, the adsorption intensity in the system "water-suspension"; α_{13}, the adsorption intensity in the system "water-ground deposits"; $C^b(x_1, x_2, z_0, t)$, the pollutant concentration in ground deposits.

Other notations are the same as in Sect. 9.2. Equation (9.18) is written in the terms of the average pollution flow hC through a vertical cross section of the reservoir. It includes the variable water depth $h(x_1, x_2, t)$ and means that the pollution concentration $C(x_1, x_2, t)$ will be higher in the areas with lower depth.

9.3.3 Equation of Transport of Suspended Pollutants

The transport equation for suspended particles is obtained by averaging (9.7) over the coordinate x_3:

$$\frac{\partial(hS)}{\partial t} + \sum_{i=1}^{2} \frac{\partial(v_i hS)}{\partial x_i} = \sum_{i=1}^{2} \frac{\partial}{\partial x_i}\left(A_i \frac{\partial(hS)}{\partial x_i}\right) - B\omega_0\left(\hat{S}^* - S\right). \tag{9.19}$$

The new parameter is the averaged in depth equilibrium concentration $\hat{S}^*$.

Using (9.10)–(9.12), the averaged in depth concentration $C^s(x_1, x_2, t)$ of the pollutant on suspended particles is described by the following equation:

$$\frac{\partial(hSC^s)}{\partial t} + \sum_{i=1}^{2} \frac{\partial(v_i hSC^s)}{\partial x_i} = \sum_{i=1}^{2} \frac{\partial}{\partial x_i}\left(A_i \frac{\partial(hSC^s)}{\partial x_i}\right) + \hat{\alpha}_{12} S(K_s C - C^s) + C^b g^b - C^s g^s. \tag{9.20}$$

9.3.4 Equations of Water Dynamics

In the case of floods or shallow lakes, the unknown water velocities $v_1(x_1, x_2, t)$ and $v_2(x_1, x_2, t)$ and the unknown boundary $\eta(x_1, x_2, t)$ of the free water surface can be determined from the following nonlinear *equations of shallow water theory* [2]:

$$\frac{\partial v_i}{\partial t} + \sum_{j=1}^{2} v_j \frac{\partial v_i}{\partial x_j} + g\frac{\partial \eta}{\partial x_i} = -\lambda v_i |\mathbf{v}| + G_i - \lambda_W W_i |\mathbf{W}|, \quad i = 1, 2, \tag{9.21}$$

$$\frac{\partial \eta}{\partial t} + \sum_{j=1}^{2} v_j \frac{\partial(hv_i)}{\partial x_j} = R, \tag{9.22}$$

where $\mathbf{v} = (v_1, v_2)$, the unknown vector of the water velocity; $\mathbf{W} = (W_1, W_2)$, the given vector of the wind velocity over water surface; λ, the ground friction coefficient; λ_W, the coefficient of the wind friction on the free water surface; R, the intensity function of distributed water sources and sinks (influx, sediments, evaporation).

The equations of water dynamics (9.21)–(9.22) require setting proper initial and boundary conditions. Analysis of such equations is difficult even in stationary time-independent cases under the assumptions of no water sources and sinks and neglected wind influence. Such stationary problems arise in prediction of spring floods and calculation of a spring flood plan. Under the given boundary and initial conditions, the equations can be solved using approximate numeric methods.

9.4 One-Dimensional Pollution Model and Its Analytic Solutions

Analytic solutions are possible for one-dimensional models of pollutant transport and diffusion in special cases. To demonstrate analytic techniques for such models, we disregard pollution adsorption-sedimentation processes and physical-chemical transformation of pollutants and consider the only one advection–diffusion equation for the concentration C of dissolved pollutants. Let us also suppose that the pollutant concentration is constant in the vertical cross section, i.e., $\partial C/\partial x_2 = \partial C/\partial x_3 = 0$. Then, (9.4) of the three-dimensional model and (9.18) of the two-dimensional model lead to the one-dimensional equation for dissolved pollutant concentration $C(x_1,t)$ of the form:

$$\partial C/\partial t = A\partial^2 C/\partial x^2 - V\partial C/\partial x - kC + f(x,t), \tag{9.23}$$

where $x = x_1$, the space coordinate (longitude); $A = \text{const} > 0$, the longitudinal diffusion coefficient; $V = \text{const} > 0$, the water velocity; $k = \text{const} \geq 0$, the pollution deterioration coefficient; $f(x, t)$, the intensity function of pollutant sources.

Compared to two- and three-dimensional equations (9.4) and (9.18), the new coefficient k in (9.23) describes (in an aggregate form) a possible decrease of the pollution concentration C because of adsorption, sedimentation, suspension exchange, and chemical transformation of pollutants.

One-dimensional models similar to (9.23) satisfactorily describe processes of pollution propagation in rivers depending on the scale of involved hydrological processes. Equation (9.23) is known as the *convective diffusion equation with the source*. It can be solved analytically under certain initial and boundary conditions that describe the water flow formation and pollution sources (see also Sect. 8.2). Below we consider several analytic solutions of the one-dimensional pollution model (9.23) in special meaningful cases, such as pollution exhaust or a permanent point source of pollutant.

The partial differential equation (9.23) at $V = 0$ and $k = 0$ is known as the one-dimensional *heat equation* because it also describes the propagation of heat in a long thin one-dimensional bar. There exists an important connection between solutions of (9.23) and the heat equation:

9.4.1 *Link Between Convective Diffusion Equation and Heat Equation*

The solution of (9.23) is of the form

$$C(x,t) = u(x,t)\exp(\mu x - \beta t), \quad \mu = V/2A, \quad \beta = k + V^2/4A, \tag{9.24}$$

where $u(x, t)$ is the solution of the *nonhomogeneous heat equation*

$$\partial u/\partial t = A\partial^2 u/\partial x^2 + \exp(-\mu x + \beta t)f(x,t) \tag{9.25}$$

with corresponding boundary and initial conditions.

Equation (9.25) with $f = 0$ is known as the *homogeneous heat equation*.

9.4.2 Mathematical Preliminary: Heat Equation

General analytic solutions of the heat equation (9.25) can be found by the method of separation of variables. However, they usually are complicated and are expressed through the Fourier series or the Fourier transform. As a result, there are dozens of known analytic solutions of the heat equation (9.23) in various special cases [7]. Here we provide several such analytic solutions, which will be used in the next section.

9.4.2.1 Initial Value Problem for the Homogeneous Heat Equation

$$\partial u/\partial t = A\partial^2 u/\partial x^2, \quad t > 0, \tag{9.26}$$

over the space domain $-\infty < x < \infty$ with the initial condition $u(x, 0) = \varphi(x)$ possesses the following solution:

$$u(x,t) = \frac{1}{2\sqrt{A\pi t}}\int_{-\infty}^{\infty}\varphi(\xi)\exp\left[\frac{-(x-\xi)^2}{4At}\right]d\xi, \quad t > 0, \quad -\infty < x < \infty. \tag{9.27}$$

9.4.2.2 Boundary Problem on (0,∞) for the Homogeneous Heat Equation with Nonhomogeneous Dirichlet Boundary Condition

$$u(0,t) = h(t), \quad t > 0, \tag{9.28}$$

over the semi-infinite space domain $(0,\infty)$ with the zero initial condition $u(x,0) = 0$ possesses the following solution for $t > 0$, $0 \le x < \infty$:

$$\begin{aligned} u(x,t) = h(t) &- \frac{h(0)}{2\sqrt{A\pi t}}\int_{-\infty}^{\infty}\varphi(\xi)\exp\left[\frac{-(x-\xi)^2}{4At}\right]d\xi \\ &+ \int_0^t \frac{h'(\tau)}{2\sqrt{A\pi(t-\tau)}}\int_{-\infty}^{\infty}\exp\left[\frac{-(x-\xi)^2}{4A(t-\tau)}\right]d\xi d\tau. \end{aligned} \tag{9.29}$$

9.4.2.3 Initial Value Problem for the Nonhomogeneous Heat Equation

$$\partial u/\partial t = A\partial^2 u/\partial x^2 + f(x,t), \quad t > 0, \tag{9.30}$$

over the space domain $-\infty < x < \infty$ with the initial condition $u(x,0) = \varphi(x)$ possesses the following solution:

$$u(x,t) = \frac{1}{2\sqrt{A\pi t}} \int_{-\infty}^{\infty} \varphi(\xi) \exp\left[\frac{-(x-\xi)^2}{4At}\right] d\xi + \int_0^t \frac{1}{2\sqrt{A\pi(t-\tau)}} \int_{-\infty}^{\infty} f(\xi,\tau) \exp\left[\frac{-(x-\xi)^2}{4A(t-\tau)}\right] d\xi d\tau. \tag{9.31}$$

Similar formulas can be obtained for other possible combinations of initial and boundary conditions.

9.4.3 *Instantaneous Source of Pollutant*

The case of an instantaneous point pollution source corresponds to a one-time sudden release of a finite amount of pollutant at a certain location, such as exhaust, explosion, or an emergency incident at a manufacturing plant or power station. Without loss of generality, we can assume that the location of the pollution is $x = 0$ and the time is $t = 0$.The instantaneous point source of a pollutant in the origin $x = 0$ at time $t = 0$ can be described by (9.23) with the following initial condition:

$$C(x,0) = I\delta(x), \tag{9.32}$$

where I, the total cumulative intensity of the pollutant exhaust; $\delta(.)$, the delta-function (see its definition in Sect. 8.2).

Let us assume the infinite space region $(-\infty, \infty)$. Then, the solution of the one-dimensional pollution (9.23) is given by the formula (9.24), where $u(x, t)$ is the solution to the homogeneous heat equation (9.26) with the corresponding initial condition $u(x,0) = \exp(-\mu x)I\delta(x)$, or

$$C(x,t) = \exp\left[\frac{Vx}{2A} - \left(kt + \frac{V^2 t}{4A}\right)\right] \frac{I}{2\sqrt{A\pi t}} \int_{-\infty}^{\infty} \delta(\xi) \exp(-\mu\xi) \exp\left[\frac{-(x-\xi)^2}{4At}\right] d\xi$$
$$= \frac{I}{2\sqrt{A\pi t}} \exp\left[-kt + \frac{1}{4At}\left(2Vxt - V^2t^2 - x^2\right)\right],$$

or

$$C(x,t) = \frac{I}{2\sqrt{\pi A t}} \exp\left(-kt - \frac{(x - Vt)^2}{4At}\right), \quad x \in (-\infty, \infty), \quad t \in [0, \infty), \tag{9.33}$$

which is a so-called *Gaussian function.*

The solution (9.33) demonstrates that the maximal pollutant concentration $C(x,t)$ over the entire domain $(-\infty, \infty)$ is reached at the point $x = Vt$ and decreases exponentially in time t with the rate k. Snapshots of the distribution of the pollutant concentration $C(x,t_i)$ in the space coordinate x for several consecutive instants $t_i > 0$, $i = 1,2,3,\ldots$ is illustrated in Fig. 9.2.

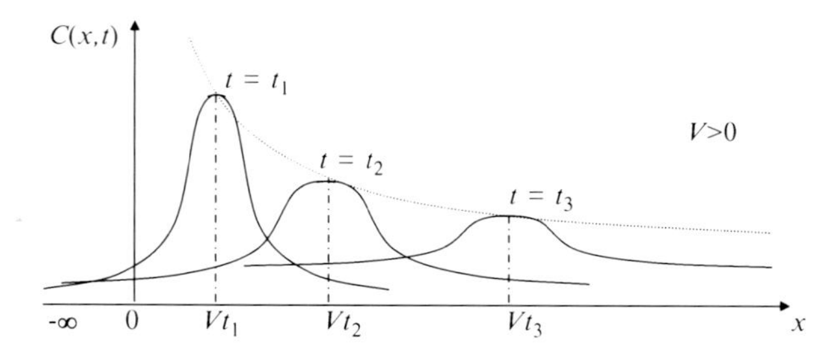

Fig. 9.2 The pollution distribution in the one-dimensional space x from an instantaneous pollutant source ("pollutant exhaust") at the point $x = 0$ at time $t = 0$

Alternatively, the instantaneous point pollution source can be described by the a pollution source with the pollution intensity function

$$f(x,t) = I\delta(x)\delta(t) \tag{9.34}$$

in (9.23) and zero initial condition $C(x,0) = 0$. This problem leads to the same result (9.33) but requires more complicated transformations.

9.4.4 Pollutant Source with Constant Intensity

The space dynamics of the pollutant propagation from a pollutant source with the constant intensity $I > 0$ at the origin $x = 0$ in the one-dimensional infinite medium $(-\infty,\infty)$ can be described by (9.23) with the source function

$$f(x,t) = I\delta(x). \tag{9.35}$$

The solution of (9.23) with a non-zero intensity $f(x,t)$ is given by the formula (9.24), where $u(x,t)$ is now the solution (9.31) of the *nonhomogeneous heat equation*

(9.30) with the pollution intensity $\exp(-\mu x + \beta t)f(x, t) = I\exp(-\mu x + \beta t)\delta(x)$ and the zero initial condition $u(x,0) = 0$. Combining formulas (9.24) and (9.31), we obtain that

$$\begin{aligned} C(x,t) &= \exp[\mu x - \beta t]\int_0^t \frac{I}{2\sqrt{A\pi(t-\tau)}}\int_{-\infty}^{\infty}\exp[\mu\xi - \beta\tau]\delta(\xi)\exp\left[\frac{-(x-\xi)^2}{4A(t-\tau)}\right]d\xi d\tau \\ &= \int_0^t \frac{I}{2\sqrt{A\pi(t-\tau)}}\exp\left[\frac{Vx}{2A} - \left(k(t-\tau) + \frac{V^2(t-\tau)}{4A}\right)\right]\exp\left[\frac{-x^2}{4A(t-\tau)}\right]d\tau \\ &= I\int_0^t \frac{\exp(-k(t-\tau))}{2\sqrt{A\pi(t-\tau)}}\exp\left[-\frac{1}{4A(t-\tau)}(x - V(t-\tau))^2\right]d\tau, \end{aligned}$$

or

$$C(x,t) = I\int_0^t \frac{\exp(-k\tau)}{2\sqrt{A\pi\tau}}\exp\left[-\frac{1}{4A\tau}(x - V\tau)^2\right]d\tau, \quad t > 0, \quad -\infty < x < \infty. \tag{9.36}$$

The pollutant concentration (9.36) is always maximal at the origin $x = 0$. It is zero at the initial moment $t = 0$ and converges to a finite positive value for each fixed point x when the time t increases indefinitely. Then, the pollutant concentration (9.34) approaches the *stationary* (time-independent) pollution distribution in the one-dimensional infinite medium with a source of constant intensity, analyzed in Sect. 8.2 and illustrated in Fig. 8.1.

Alternatively, we can consider the half-infinite water medium $(0,\infty)$ with a rigid boundary at $x = 0$ and no boundary to the right. Then, the pollutant source of constant intensity can be modeled by the pollution (9.23) with the boundary condition $C(0, t) = I$. The corresponding pollution distribution $C(x, t)$ is found from the formula (9.24), where $u(x, t)$ is the solution (9.29) of the homogeneous heat equation (9.26) with the nonhomogeneous Dirichlet boundary condition $u(0, t) = I\exp(\beta t)$ over the space domain $0 < x < \infty$ and the zero initial condition $u(x, 0) = 0$. The final analytic formula for pollution distribution is more complicated than (9.36) and is not provided here.

Analytical solutions of one-dimensional diffusion equations are possible in some other special cases, but the corresponding formulas are more complicated. In general, numeric algorithms and computer simulation are major techniques for applied modeling of water pollution propagation. The analytical models provide their input as tools for getting new insight into the pollution dynamics at the stage of preliminary assessment and design of pollution abatement policies (see Chap. 11).

9.5 Compartmental Models and Control Problems

The compartmental models split a water reservoir into smaller parts (compartments, cells) and assume that the pollution concentration is constant inside separate compartments [3]. Such models describe the change of pollution concentration among the compartments and use ordinary differential or difference equations [1].

The compartmental models are especially convenient when there is a real physical partition of the water reservoir into relatively isolated parts. Then, the compartments correspond to those parts. For example, a cascade of artificial reservoirs (water storage pools) can be conveniently described as a sequence of compartments, each of which corresponds to one reservoir (or its fragment). The construction of such models is based on averaging of the two-dimensional model (9.16)–(9.22) over separate compartments. In other words, we assume the full intermixing, when the average pollution concentration inside a compartment is equal to the concentration in the output flow from the compartment.

9.5.1 *Equations of Water Balance*

Let us divide a river into the series of compartments, assuming that the water flows from the upper $(i-1)$-th compartment to the lower i-th compartment, $i = 1,\ldots, M$. The river has a number of inflows j, $j = m(i),\ldots,n(i)$, in each compartment i. The continuity equation (9.22) under the assumption of full intermixing in separate compartments leads to the following equation of the water balance in the i-th compartment:

$$\frac{dV_i}{dt} = Q_{i-1} - Q_i + R_i + \sum_{j=m(i)}^{n(i)} Q_i^t - Q_i^w, \qquad (9.37)$$

where $V_i(t)$, the water volume in the i-th compartment; $Q_i(t)$, the water flow into the lower compartment; $Q_{i-1}(t)$, the water flow from the upper compartment; $R_i(t)$, the vector of external flows (rains, evaporation, water consumption); $Q_j^t(t)$, the water flow from each inflow j, $j = mi),\ldots,ni)$, entered compartment i; $Q_i^w(t)$, the total water consumption from the compartment i.

9.5.2 *Equations of Suspension Balance*

After averaging the concentration of suspended particles over each compartment, the two-dimensional equation (9.19) of the transport of suspended particles leads to the following equation:

$$\frac{d(V_iS_i)}{dt} = Q_{i-1}S_{i-1} - Q_iS_i + q_i^{\,b} - q_i^{\,s} + R_i^{\,h} + \sum_{j=m(i)}^{n(i)} Q_i^{\,t}S_i^{\,t} - S_iQ_i^{\,w}, \tag{9.38}$$

where S_i, the concentration of suspended particles (averaged over the compartment volume); S_{i-1}, the average flux of suspended particles entered from the upper compartment; $R^h{}_i$, the influx of particles because of bank erosion; $S^t{}_j$, the concentration of particles in the river inflows; $q^b{}_i$ and $q^s{}_i$, the average sedimentation and deposit lift flows in compartment i.

9.5.3 *Equations of Pollution Propagation*

The change of the amount G_i^b of the dynamic layer of ground deposits in the i-th compartment under the influence of suspended particles is described by (9.20):

$$\frac{dG_i^{\,b}}{dt} = q_i^{\,s} - q_i^{\,b}, \tag{9.39}$$

The averaging of the two-dimensional equations (9.17), (9.18), and (9.19) over each reservoir area with additional pollutant sources on reservoir boundaries leads to the following ordinary differential equations with respect to the concentrations $C_i(t)$, $C_i^{\,s}(t)$, $C_i^{\,b}(t)$ of the pollutant in the i-th compartment in solute, in suspension, and on ground deposits correspondingly:

$$\begin{aligned}\frac{d(V_iC_i)}{dt} &= Q_{i-1}C_{i-1} - Q_iC_i + \sum_{j=m(i)}^{n(i)} Q_i^{\,t}C_i^{\,t} \\ &\quad - \alpha_{12}S_i(K_dC_i - C_i^{\,s}) + \alpha_{13}(K_dC_i - C_i^{\,b}),\end{aligned} \tag{9.40}$$

$$\begin{aligned}\frac{d(V_iS_iC_i^{\,s})}{dt} &= Q_{i-1}S_{i-1}C_{i-1}{}^{s} - Q_iS_iC_i^{\,s} + \sum_{j=m(i)}^{n(i)} Q_i^{\,t}S_i^{\,t}C_i^{\,s} \\ &\quad + \alpha_{12}S_i(K_dC_i - C_i^{\,s}) + C_i^{\,b}q_i^{\,b} - C_i^{\,s}q_i^{\,s} - C_i^{\,s}S_iQ_i^{\,w},\end{aligned} \tag{9.41}$$

$$\frac{d(G_i^{\,b}C_i^{\,b})}{dt} = C_i^{\,s}q_i^{\,s} - C_i^{\,b}q_i^{\,b} - \alpha_{13}S_i(K_dC_i - C_i^{\,b}). \tag{9.42}$$

The parameters are the same as in the two-dimensional model (9.18)–(9.22).

9.5.4 Control Problems of Water Pollution Propagation

The control problems of water pollution propagation and water contamination are usually solved together with other water management problems and use simple types of the water pollution propagation models, such as one-dimensional or compartmental models. Formulation of such problems depends on control goals and possibilities.

We illustrate the specifics of space-distributed problems of pollution propagation control on an example of the optimal management regime for a cascade of M connected water reservoirs. Then, it is convenient to describe the water dynamics and the process of pollution propagation by the compartment model (9.37)–(9.42), which reflects the structure of connected reservoirs $i = 1,\ldots,M$, i.e., each reservoir corresponds to a separate compartment in the model.

A practical approach to the rational water management is to consider the process dynamics in discrete time $k = 1,2,3\ldots$. Let a planning horizon consist of N discrete time intervals. Then, the optimal water management regime can be described by a *finite-dimensional optimization* problem. After the time discretization, the differential equations of model (9.37)–(9.42) are reduced to their difference analogues in the discrete time. The essential dynamic parameters are the water consumptions Q_{ik}, water levels H_{ik}, and the pollution concentrations C_{ik}, $k = 1,\ldots,N$, for each reservoir $i = 1,\ldots,M$, under the given initial conditions H_{i0} and C_{i0}.

The control problem under study is to determine the most rational management regime for all reservoirs subjected to given exploitation conditions of the reservoirs. It involves a sequential (or simultaneous) solution of the following optimization problem for each reservoir $i = 1,\ldots,M$.

The optimal management regime for each reservoir $i = 1,\ldots,M$, is determined by the vectors $\mathbf{Q}_i = (Q_{i1},\ldots, Q_{iN})$, $\mathbf{H}_i = (H_{i1},\ldots, H_{iN})$, $\mathbf{C}_i = (C_{i1},\ldots, C_{iN})$ that minimize a certain objective function

$$\min F_i(\mathbf{H}_i, \mathbf{Q}_i, \mathbf{C}_i), \tag{9.43}$$

subject to some restrictions:

$$G_l(\mathbf{H}_i, \mathbf{Q}_i, \mathbf{C}_i) < 0, \quad l = 1, \ldots, L. \tag{9.44}$$

The constraints (9.44) describe relevant water balance conditions, minimal water consumption, economic and hydrological and meteorological conditions (including influxes, sediments, and evaporation), hydro safety specifications, ecological norms, sufficient water depth for watercraft security, and others.

The objective function (9.43) should combine the priorities of a decision-maker, who usually has several objectives. The objective function of the i-th reservoir, $i = 1,\ldots,M$, can be represented as

$$F_i(\mathbf{H}_i, \mathbf{Q}_i, \mathbf{C}_i) = \sum_{k=1}^{L} p_l f_l, \tag{9.45}$$

where f_k are partial criteria that reflect particular objectives of water management; $p_l > 0$ are the weight coefficients for each objective, $l = 1,\ldots,L$, $p_1 + \ldots + p_L = 1$.

The following functions f_l describe possible particular objectives for the i-th reservoir:

- $f_1 = \max(C_{ik})$—reduction of the peak concentration of pollutants;
- $f_2 = \sum_{k=1}^{N} \left(H_{ki} - \widetilde{H}_{ki}\right)^2$—obtaining a water regime maximally close to the given recommended regime $\widetilde{H}_{ki}$, $k = 1,\ldots,N$;
- $f_3 = \sum_{k=1}^{N} (Q_{ki} - Q_{k-1i})^2$—preventing sharp fluctuations in the water level and obtaining the most uniform water regime;
- $f_4 = -\sum_{k=1}^{N} {Q_{ki}}^2$—minimization of water consumption and the retention of water in the reservoir;
- $f_5 = \max(Q_{ik})$—water peak cutoff, and so on.

The convolution (9.45) converts a multi-criteria *optimization* problem into a standard *nonlinear programming problem* for each reservoir. Finally, if the cascade of reservoirs is governed by a central management body, then the cooperative management of reservoirs is described by the following *finite-dimensional* optimization problem:

$$\min \sum_{i=1}^{n} F_i(\mathbf{H}_i, \mathbf{Q}_i, \mathbf{C}_i) \tag{9.46}$$

Zsubject to the constraints (9.37)–(9.42), (9.44) with respect to NM unknown discrete controls Q_{ik}, C_{ik}, H_{ik}, $i = 1,\ldots,M$, $k = 1,\ldots,N$.

If the reservoir owners are independent economic agents, then the minimization problem (9.43) is solved by each owner independently, which leads to an M-person discrete-time *game* with duration N. This model reflects the competitive behavior of reservoir owners. Such problems are usually solved using computer simulation and numerical algorithms of mathematical programming.

Exercises

1. Prove that the linear adsorption model (9.1) has a time-independent steady-state (equilibrium) solution $C^s(x_1, x_2, x_3) = K_s C(x_1, x_2, x_3)$, if the concentration $C(x_1, x_2, x_3)$ is time-independent.

2. Substitute formulas (9.3) into (9.2) and obtain a relation between the deposit amount $M^b(x_1, x_2, x_3, t)$ and suspension concentration $S(x_1, x_2, x_3, t)$. Prove that the M^b does not depend on time in the case of the equilibrium concentration $S = S^*$.
3. Assuming $A_1 = A_2 = A_3 = 0$ (no diffusion) and $v_1 = v_2 = v_3 = 0$ (motionless water) in the partial differential equation (9.7), obtain a linear ordinary differential equation for the concentration $C(x_1, x_2, x_3, t)$ of dissolved pollutants.
4. Solve the linear ordinary differential equation from previous exercise and show that the concentration $C(x_1, x_2, x_3, t)$ approaches $C^s(x_1, x_2, x_3)/K_s$ when t approaches ∞.
   HINT: rewrite the differential equation for the new variable $y = K_s C - C^s$ and use the formula (1.25) to solve it.
5. Note the similarity of the models (9.7)–(9.9) and (9.10)–(9.12). Explain the meaning of the additional terms in (9.10) and (9.12).
   HINT: Consider the presence of the pollutant in dissolved form and on ground deposits.
6. The horizontal two-dimensional model (9.17)–(9.22) of pollution propagation involves the vertical position $z_0(x_1, x_2)$ of the bottom. Noticing that $z_0(x_1, x_2)$ does not depend on time t, explain why the unknown depth $h(x_1, x_2, t)$ of water flow can depend on the time.
7. Substitute (9.24) into the pollution model (9.23) and obtain the nonhomogeneous heat equation (9.25).
8. Verify that (9.27) satisfies the initial value problem for the homogeneous heat equation (9.26).
9. Verify that the solution of the one-dimensional pollution (9.23) in the case $f(x, t) = I\delta(x)$ of a pollutant source at the origin $x = 0$ is given by the formula (9.34).
10. Consider the formula (9.33) for the one-dimensional pollutant concentration $C(x,t)$ and demonstrate that the maximal pollutant concentration $C(x,t)$ over the entire domain $(-\infty, \infty)$ decreases exponentially in time t.
    HINT: Use Fig. 9.2.

References[1]

1. 🕮 Fulford, G., Forrester, P., Jones, A.: Modelling with differential and difference equations. Cambridge University Press, New York (1997)
2. 🕮 Pedlosky, J.: Geophysical fluid dynamics. Springer, New York (1990)
3. Godfrey, K.: Compartmental models and their application. Academic Press, New York (1983)
4. James, A. (ed.): Mathematical models in water pollution control. Wiley, New York (1980)
5. James, I.D.: Modelling pollution dispersion, the ecosystem and water quality in coastal waters: a review. Environ. Model. Software **17**, 363–385 (2002)

[1] The book symbol 🕮 means that the reference is a textbook recommended for further student reading.

6. Kachiashvili, K.J., Gordeziani, D.G., Lazarov, R.G., Melikdzhanian, D.I.: Modeling and simulation of pollutants transport in rivers. Appl. Math. Model. **31**, 1371–1396 (2007)
7. Polyanin, A.D., Zaitsev, V.F.: Handbook of linear partial differential equations for engineers and scientists. Chapman & Hall/CRC Press, Boca Raton, London (2002)

Part III
Models of Economic-Environmental Systems

Chapter 10
Modeling of Nonrenewable Resources

This chapter is devoted to an important economic–environmental problem: the modeling of the optimal extraction and utilization of nonrenewable (exhaustible) resources. Nonrenewable resources are natural resources that cannot be replaced as quickly as they are being consumed. Examples of such resources include fossil fuels (petroleum, coal, and natural gas) and mineral resources (iron, gold, and other). The models of Section 10.1 consider the resource extraction process in isolation from other economic activities. Section 10.2 investigates the economic growth model with a two-factor Cobb–Douglas technology that uses physical capital and exhaustible resource to produce aggregate product.

10.1 Aggregate Models of Nonrenewable Resources

Nonrenewable (exhaustible) resources are natural resources that have been formed in the environment throughout thousands, or even millions, of years and cannot be regenerated as quickly as they are being consumed. The quantities of such resources are fixed and, thus, the more is used today, the less is available for use in the future. Common examples of exhaustible resources are fossil fuels (petroleum, coal, and natural gas) and mineral resources (iron, gold, and so on). In contrast, *renewable resources*, such as soil, water, forests, plants and animals, wind and solar energy, can be replaced by natural processes in the environment. Some economic models with renewable resources are discussed in Chap. 7.

In this chapter, we consider the extraction and consumption of the *nonrenewable resources*. Let us introduce the following characteristics:

$R(t)$—the total stock (storage, deposit) of a certain nonrenewable resource,
R_0—the amount of the resource held at the initial time $t = 0$,
$E(t)$—the quantity of the resource extracted per time period (the intensity of resource extraction, the rate of resource extraction, the flow of resource),

N. Hritonenko and Y. Yatsenko, *Mathematical Modeling in Economics, Ecology and the Environment*, Springer Optimization and Its Applications 88,
DOI 10.1007/978-1-4614-9311-2_10,

$p(t)$—the *market price* of the unit of the resource extracted,
$C(t)$—the extraction cost (the cost of the extraction of one resource unit).

10.1.1 Models of Optimal Resource Extraction

The extraction process of an exhaustible resource is described by deterministic models in cases when discovery and exploration of new resource deposits is not feasible. A simple deterministic model of the extraction of the exhaustible resource R is described by the linear ordinary differential equation (ODE):

$$R'(t) = -E(t), \tag{10.1}$$

$$R(0) = R_0 > 0. \tag{10.2}$$

The intensity of resource extraction E is usually considered as an endogenous control.

The majority of optimization problems in the model (10.1)–(10.2) are based on the following framework. Let us consider a firm (industry, corporation) that exploits a deposit of a nonrenewable resource R and sells the extracted resource E as its only product. The firm chooses the extraction path of $E(t)$ in order to maximize its total discounted profit over the future planning horizon $[0,T]$, $T \leq \infty$. Then, the objective functional has the following form:

$$\max_E \int_0^T e^{-rt}[p(t)E(t) - C(E(t), R(t))]\, dt, \tag{10.3}$$

where the total *extraction cost* $C(E, R)$ depends on the resource extraction intensity and resource stock, $\partial C/\partial E > 0$, $\partial C/\partial R \leq 0$, and the resource price $p(t)$ is given. The function $E(t)$, $t \in [0, T]$, is the unknown control function that satisfies the restriction

$$0 \leq E(t) \leq E_{\max}, \tag{10.4}$$

where the maximal value $E_{\max}$ always exists in real problems because of technical and financial restrictions. Despite the relative simplicity of the model (10.1)–(10.4), it produces qualitatively different results depending on additional assumptions about the given and unknown model functions and parameters, as it is shown in the sections below.

10.1.2 Linear Model with No Resource Extraction Cost

Let us first consider the optimization problem (10.1)–(10.4) under the assumption of costless extraction, i.e., when the *extraction cost* $C(E,R)$ is zero. Then, the objective functional (10.3) is

$$\max_{E} \int_{0}^{T} e^{-rt} p(t) E(t) \, \mathrm{d}t, \tag{10.5}$$

subject to the state equation (10.1), the initial condition (10.2) and the restriction (10.4). Thus, the optimization problem is to find the unknown control $E(t)$, $t \in [0, T]$, that maximizes (10.5) and satisfies

$$R'(t) = -E(t), \quad R(0) = R_0 > 0, \quad 0 \leq E(t) \leq E_{\max}. \tag{10.6}$$

It is clear that the entire resource can be completely extracted during a finite interval $[0,T]$ depending of the extraction process capacity. If it happens, then $R(T) = 0$. So, finding the optimal T is a part of the problem (10.5), (10.6).

10.1.2.1 Model Analysis

The analysis of optimization problems in this chapter employs maximum principles of Sect. 2.4. The Hamiltonian (2.53) of the optimal control problem (10.5), (10.6) is constructed as

$$H(E(t), R(t), \lambda(t)) = e^{-rt} p(t) E(t) - \lambda(t) E(t). \tag{10.7}$$

The function (10.7) does not depend on R. Then, the dual equation (2.42) is $\lambda' = -\ \partial H / \partial R = 0$, therefore, the present-value dual variable $\lambda(t)$ is constant over time. In accordance with Sect. 2.4, the optimality condition $R' = \partial H / \partial \lambda$ coincides with the state equation (10.1). By the maximum principle for ODEs from Sect. 2.4, the control $E(t)$ shall maximize $H(E, R, \lambda)$. Because (10.7) is linear in E, the Hamiltonian (10.7) is maximal when

$$E(t) = \begin{cases} 0 & \text{if } \ p(t) e^{-rt} < \lambda(t), \\ E_{\max} & \text{if } \ p(t) e^{-rt} \geq \lambda(t). \end{cases} \tag{10.8}$$

The optimal solution (10.8) is known as the *bang–bang solution* because the control $E(t)$ takes only maximal or minimal possible values. Bang–bang controls appear in some other optimization problems (see Chaps. 5 and 7).

The problem (10.5)–(10.6) is an optimization problem with the free terminal time T (see Sect. 2.4) and its transversality condition has the form (2.62) or

$$H(E(T), R(T), \lambda) = \left[e^{-rT} p(T) - \lambda \right] E(T) = 0. \tag{10.9}$$

Because $E^*(T) = E_{\max}$ (otherwise, T is not the end of the extraction period), (10.9) implies

$$\lambda = e^{-rT} p(T). \tag{10.10}$$

Combining (10.8) and (10.10), the optimal resource extraction intensity E has to follow the rule

$$E^*(t) = \begin{cases} 0 & \text{if } \; p(t) < p(T)e^{r(t-T)}, \\ E_{\max} & \text{if } \; p(t) \geq p(T)e^{r(t-T)}, \end{cases} \tag{10.11}$$

which means that the optimal extraction intensity $E^*(t)$ is zero on the intervals where $p(t) < p(T)e^{-r(t-T)}$ and is maximal $E_{\max}$ otherwise.

Let us interpret the optimal rule (10.11) for the exponential price $p(t) = p_0 e^{ct}$. Then, the firm does not extract and sell resource ($E^*(t) = 0$) if the price increases faster than e^{rt} (the discount rate r is smaller than c). The firm sells as much resource as possible ($E^*(t) = E_{\max}$) if the price increases slower than e^{rt} or does not increase at all.

In the case of a non-exponential resource price $p(t)$, the future values of the resource price play a critical role. In particular, by (10.11), the optimal policy is no extraction at all if $p(t) < p(T)e^{r(t-T)}$ for all t. If the resource price $p(t)$ is not exponential, then the solution (10.11) can possess several alternate intervals with maximal and zero extraction. To switch from zero extraction to maximal extraction at time t, the relative increase rate $p'(t)/p(t)$ of price $p(t)$ should be larger than the discount rate r. The case with one such switching point is illustrated in Fig. 10.1. Namely, if the resource price $p(t)$ follows the path presented in Fig. 10.1, then the optimal extraction intensity has two switching points T_0 and T. It is shown in the top Fig. 10.1. The end T of the extraction period is uniquely determined from the condition

$$\int_0^T E^*(t)\,\mathrm{d}t = R_0, \tag{10.12}$$

where $E^*(t)$ depends on the unknown value T itself.

10.1.3 Models with Resource Extraction Cost

We restrict ourselves to the optimization problem (10.1)–(10.4) with a positive *extraction cost* $C(E)$ that nonlinearly depends on the resource extraction intensity E, $dC/dE > 0$. Then, the maximizing functional is

$$\int_0^T e^{-rt}[p(t)E(t) - C(E(t),t)]\,\mathrm{d}t \tag{10.13}$$

subject to (10.1), (10.2), and (10.4). The optimization problem is to find the unknown control $E(t)$, $t \in [0,T]$, that maximizes (10.13) and satisfies

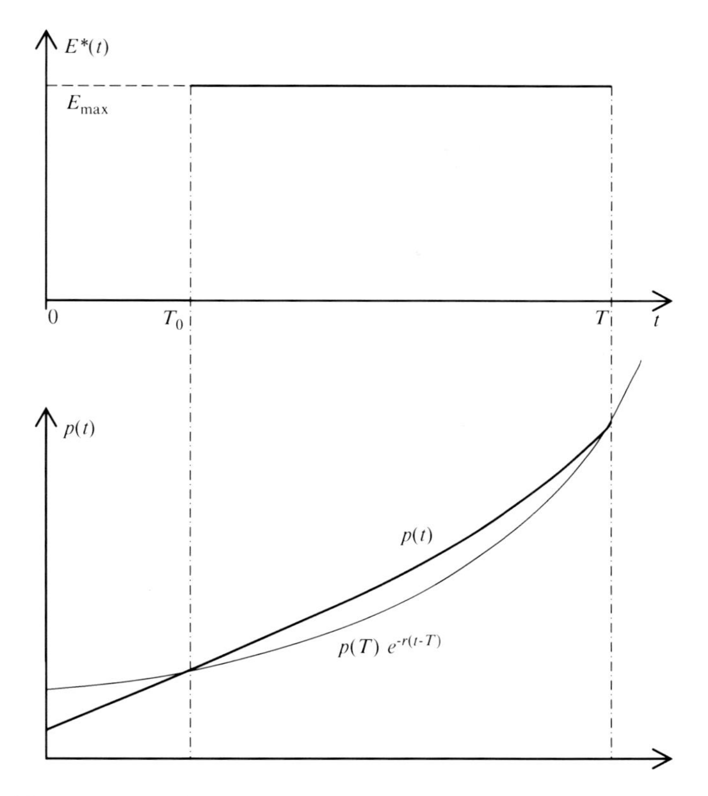

Fig. 10.1 The dependence of the optimal resource extraction E^* on the dynamics of the given resource price p

$$R'(t) = -E(t), \quad R(0) = R_0 > 0, \quad 0 \leq E(t) \leq E_{\max}. \tag{10.14}$$

Following Sect. 2.4, the Hamiltonian (2.53) of the optimal control problem (10.13)–(10.14) is constructed as

$$H(E(t), R(t), \lambda(t)) = e^{-rt}[p(t)E(t) - C(E(t), t)] - \lambda(t)E(t). \tag{10.15}$$

The dual equation (2.42) is again $\lambda' = -\partial H/\partial R = 0$, so the dual variable $\lambda(t)$ is constant. So the Hamiltonian (10.15) is

$$H(E(t), \lambda) = e^{-rt}[(p(t) - \lambda e^{rt})E(t) - C(E(t), t)]. \tag{10.16}$$

By the maximum principle of Sect. 2.4, the control $E(t)$ shall maximize $H(E(t), \lambda)$ for each time $t > 0$. We consider two cases with a different structure of solutions.

10.1.3.1 Model with Linear Extraction Cost

Let the total extraction $C(E)$ linearly depend on the resource extraction intensity E:

$$C(E) = c(t)E(t),$$

where $c(t) > 0$ is the given *unit extraction cost* (also known as the *marginal extracting cost*). Then, the Hamiltonian (10.16) becomes

$$H(E(t), \lambda) = e^{-rt}[p(t) - c(t) - \lambda(t)e^{rt}]E(t). \tag{10.17}$$

If the extraction cost $c(t)$ is larger than the resource price $p(t)$, then the optimal strategy is no extraction at all. Indeed, because $\lambda \geq 0$ by the maximum principle, then the maximum of the Hamiltonian (10.17) is reached at $E(t) = 0$.

Let $c(t) < p(t)$. Then, analogously to (10.10) we determine the constant $\lambda = e^{-rT}[p(T) - c(T)] \geq 0$ and analogously to (10.11) the optimal extraction intensity is

$$E * (t) = \begin{cases} 0 & \text{if} \quad p(t) - c(t) < [p(T) - c(T)]e^{r(t-T)}, \\ E_{\max} & \text{if} \quad p(t) - c(t) \geq [p(T) - c(T)]e^{r(t-T)}. \end{cases} \tag{10.18}$$

The bang–bang solution (10.18) is similar to (10.11) if we replace $p(t)$ with the difference $p(t)-c(t)$. The firm does not sell resource ($E^*(t) = 0$) if the increase of the function $p(t)-c(t)$ over any interval $[0,T]$ is larger than e^{rT}. Depending on the given dynamics of $p(t)-c(t)$, the solution (10.18) can alternate between maximal and zero extraction as shown in Fig. 10.1 (where p is replaced with $p-c$).

10.1.3.2 A Model with Nonlinear Extraction Cost *C*(*E*)

In the case of nonlinear extraction cost $C(E)$, the solution is not necessarily bang–bang. Let us assume that $\partial C(E, t)/\partial E > 0$ (the cost is larger for larger intensity) and $C(0, t) = 0$ (no cost occurs when there is no extraction). If $p(t) < \lambda e^{rt}$, then the Hamiltonian (10.16) is negative and, therefore, the optimal $E(t) = 0$. However, if $p(t) > \lambda e^{rt}$, then the Hamiltonian can be maximal at a certain positive value $E(t)$ for some or all $t \in [0,\infty]$. To be interior in the interval $[0,E_{\max}]$, this value $E(t)$ should satisfy the extremum condition $\partial H(E, t)/\partial E = 0$ or

$$\frac{\partial C(E(t), t)}{\partial E} = p(t) - \lambda e^{rt}. \tag{10.19}$$

As before, we use the transversality condition (2.62) to determine the constant λ. By (2.62), $H(E(T), \lambda) = 0$, therefore $E(T) = 0$ by (10.16) and $C(0,t) = 0$. Substituting $E(T) = 0$ into (10.19) at $t = T$, we obtain

$$\lambda = e^{-rT}\left[p(T) - \frac{\partial C(0,T)}{\partial E}\right]. \tag{10.20}$$

Finally, substituting λ from (10.20) into (10.19), we obtain the nonlinear equation for the interior value $E(t)$. Because now the function $\partial C(E,t)/\partial E$ depends on E, (10.19) can have a solution $E(t)$ such that $0 < E(t) < E_{max}$. Below, we consider a special case of the model, when the optimal solution $E(t)$ is positive and interior for all $t \in [0,\infty]$.

Depending on the given dynamics of price and cost, the solution in a general case can be entirely interior or alternate between zero and non-zero less-than-maximal extraction. For example, if the resource price $p(t)$ follows the path presented in Fig. 10.2b, then the optimal extraction intensity is shown in Fig. 10.2b. If $E(t) < E_{max}$ (the solid curve), then the constraint $E(t) \le E_{max}$ is called non-binding (inactive). If $E(t)$ eventually

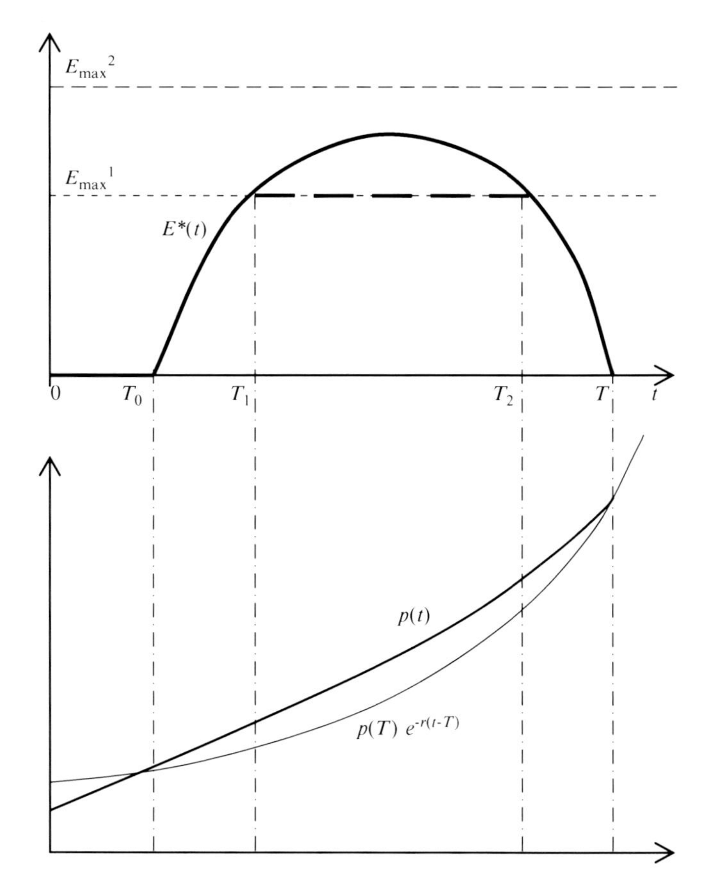

Fig. 10.2 The structure of the resource extraction problem (10.13)–(10.14) with nonlinear extraction cost $C(E,t)$. The solid curve demonstrates the optimal resource extraction intensity $E^*(t)$ in the case of a non-binding constraint $E(t) \le E_{max}^2$. The dashed boundary part occurs in the $E^*(t)$ in the case of a stricter constant E_{max}^1

reaches E_{max}, then the constraint $E(t) \leq E_{max}$ is binding. Such optimal control E is sometimes called a *weak bang–bang* because it combines intervals with bang–bang and interior solutions.

The length T of the entire extraction period is determined from the condition (10.12), where E^* depends on T itself.

10.1.3.3 Model with Constant Resource Price and Quadratic Extraction Cost

Let

$$C(E,t) = E^2/2, \quad p(t) = p = \text{const.} \tag{10.21}$$

Then, the extremum condition (10.19) is simply

$$E(t) = p - \lambda e^{rt} \tag{10.22}$$

and (10.20) is $\lambda = pe^{-rT}$. So if $E_{max} > p$, then the optimal extraction intensity is

$$E^*(t) = p\left[1 - e^{r(t-T^*)}\right], \quad 0 \leq t \leq T^*. \tag{10.23}$$

Substituting (10.23) into the integral restriction (10.12), we obtain the nonlinear equation

$$e^{-rT} + rT - 1 = R_0 r/p \tag{10.24}$$

to determine the optimal T^*. Equation (10.24) often occurs in economic applications (see Chap. 5) and has a unique positive solution T^* for any positive parameters R_0, p, and r.

If the price p is much smaller than the initial stock: $p \ll R_0 r$, then the approximate solution of (10.24) is

$$T^* \approx R_0/p. \tag{10.25}$$

To prove (10.25), we notice that the right-hand part of (10.24) $R_0 r/p \gg 1$ and, dividing (10.24) by $R_0 r/p$, we obtain $rT/(R_0 r/p) \approx 1$.

If the discount rate r is small: $r \ll 1$, then

$$T^* \approx \sqrt{2R_0/(rp)}. \tag{10.26}$$

10.1.3.4 Other Nonlinear Resource Extraction Models

The dynamics of optimal trajectories in the last two sections displays a common situation in applied optimal control when the nonlinearity of the problem gradually softens the bang–bang structure of solutions and leads to the appearance of interior trajectories.

In resource extraction problems, another possible cause for the existence of interior solutions is a nonlinearity of the profit-utility function $U(E(t))$. For instance, it can be of the form:

$$\max_{E} \int_{0}^{T} e^{-rt}[U(E(t)) - C(E(t), t)]\, dt. \tag{10.27}$$

There are several economic reasons for the nonlinearity of the function $U(E)$. The first one is the *market power* of a firm. When such a firm extracts and sells more resources E, the unit resource price p decreases following a decreasing *demand curve*: $p = p(E)$. As a result, the profit in the objective function (10.27) becomes the nonlinear function $U(E) = p(E)E$, $\partial U/\partial E > 0$, $\partial^2 U/\partial E^2 < 0$ [1, 6].

Another economic reason for having a nonlinear function $U(E)$ in (10.27) occurs when we consider a social objective of maximizing the consumer utility instead of maximizing the profit. The nonlinear utility $U(E)$ is discussed in more detail in Sect. 2.4 and possesses the same properties $\partial U/\partial E > 0$ and $\partial^2 U/\partial E^2 < 0$.

Concluding Sect. 10.1.3, we shall notice the special role of an increasing resource price in determining the optimal firm extraction policy. Specifically, in all cases considered above, it is profitable for firms to sell resource when the rate p'/p of the resource price $p(t)$ is sustainably smaller than the discount rate r. Otherwise, firms stockpile their resource stock for the future. In general equilibrium economic models, market competitive forces tend to adjust the dynamics of resource price to this threshold level. We consider one such model in the next section.

10.1.4 Hotelling's Rule of Resource Extraction

Hotelling's model of natural resource exploitation [5] was the first and probably the most influential optimization model in the economics of exhaustible resources. It is a *general equilibrium model* with perfectly competitive product and capital market. Here we provide its simplified version that focuses on the production side. The model is based on the following assumptions:

1. The resource extraction sector consists of a number of identical representative firms. It is modeled in isolation and is subjected to a given economy-wide interest rate $r > 0$.
2. The stock of the natural resource in situ is a capital asset to its owner. It must earn the same rate of return as any other capital asset in the competitive capital market equilibrium.
3. The extraction cost is negligible.

4. The representative firm chooses the extraction path $E(t) \geq 0$ in order to maximize its total discounted profit over *a given finite* interval $[0,T]$.
5. The extracted resource is a product whose supply must equal the given isoelastic demand on the competitive product market.

The asset price $\pi(t)$ of the resource (its marginal value in the ground) is

$$\pi(t) = p(t) - c(t), \tag{10.28}$$

where

$p(t)$ is the market price of the unit of resource extracted,
$c(t)$ is the cost of extracting this unit.

Then, Assumption 2 affirms that

$$\pi'(t)/\pi(t) = r. \tag{10.29}$$

Since $c(t) = 0$ by Assumption 3, (10.28) and (10.29) lead to the famous *Hotelling's no-arbitrage rule* for the market price of resource

$$p'(t)/p(t) = r, \tag{10.30}$$

or $p(t) = p_0 e^{rt}$, where the initial constant p_0 is determined later. The exponential price $p_0 e^{rt}$ is known as the *Hotelling price.*

Assumption 4 means that the equilibrium condition

$$\int_0^T e^{-rt} p(t) E(t)\, dt = p_0 R_0 \tag{10.31}$$

holds for the industry at whole, where R_0 is the given initial industry-wide resource reserve.

Finally, Assumption 5 suggests that the extracted resource quantity $E(t)$ is equal to the given demand for the resource:

$$E(t) = d_0 p^{-\alpha}(t), \tag{10.32}$$

where $\alpha > 0$ is the elasticity of demand. Substituting (10.32) and (10.30) into (10.31), we obtain

$$d_0 \int_0^T \left(p_0 e^{rt}\right)^{-\alpha} dt = R_0 \ \text{ or } \ p_0 = \left(d_0 \frac{1 - e^{-r\alpha T}}{R_0 r \alpha}\right)^{1/\alpha}.$$

Thus, the exact Hotelling resource price is

$$p(t) = \left(d_0 \frac{1 - e^{-r\alpha T}}{R_0 r \alpha}\right)^{1/\alpha} e^{rt}. \tag{10.33}$$

Combining formulas (10.33) and (10.32), we obtain that the optimal quantity of the resource extracted and traded by industry at whole

$$E(t) = \frac{r\alpha R_0}{1 - e^{-r\alpha T}} e^{-r\alpha t} \tag{10.34}$$

exponentially decreases and is interior in the domain $E \geq 0$.

The formulas (10.30), (10.33), and (10.34) constitute the *Hotelling's model*. The theoretical attractiveness of the Hotelling's model is that it delivers the closed-form solution to the optimal resource extraction problem.

In contrast to extraction models of Sect. 10.1.3, the extraction period length T is given in the Hotelling's model. If we assume the infinite extraction period, then the growth rates of the optimal resource price $p(t)$ and the extracted resource $E(t)$ remain the same, but their factors are even simpler than in (10.33) and (10.34):

$$p(t) = (r\alpha R_0/d_0)^{-1/\alpha} e^{rt}, \quad E(t) = r\alpha R_0 e^{-r\alpha t}. \tag{10.35}$$

10.1.5 Modifications of Hotelling's Model

The most famous and controversial part of the Hotelling's model is the *Hotelling's no-arbitrage rule* $p'(t)/p(t) = r$ for the market price of resource. In particular, estimated dynamics of the US prices for ten major nonrenewable minerals and fossil fuels in 1870–2004 did not display a positive trend suggested by the Hotelling's no-arbitrage rule (10.30) [2]. In reality, those prices were volatile with the average rate around zero. This fact has caused several modifications of the original Hotelling's model (10.28)–(10.34), which include extraction costs, technological change, market uncertainty, and other factors. Below we describe two such modifications.

10.1.5.1 Extraction Costs

Let the total *extraction cost* $C(E)$ in (10.27) be linearly proportional to the extraction quantity:

$$C(E) = c(t)E(t) + c_f,$$

where $c(t)$ is the unit cost of resource extraction; $c_f > 0$ is a fixed cost.

If the unit extraction cost is constant: $c(t) = c > 0$, then, by (10.28) and (10.29), the market price of resource

$$p'(t)/p(t) = r(1 - c/p(t)) \tag{10.36}$$

and, therefore, the relative rate of $p(t)$ is lower than the discounting rate r as in (10.33) (but asymptotically tends to it).

10.1.5.2 Cost-Saving Technological Change

The economic evidence demonstrates that the average cost of extracting resources has possessed a declining trend for the last one hundred years because of technological change.

For simplicity, let us describe this progress as an exponential decrease of the unit extraction cost: $c(t) = ce^{-at}$, $a > 0$. Then, by (10.28) and (10.29), the market price of resource is even lower than (10.36):

$$p'(t)/p(t) = r(1 - ce^{-at}/p(t)) - ace^{-at}/p(t). \tag{10.37}$$

If the extraction cost is essential as compared to the market price, then the relative rate of $p(t)$ is significantly lower than the discounting rate r on an initial part of the planning horizon, and it can be even negative when $a > r$. However, the rate will increase later and still asymptotically tends to the rate r, which gives a gloomy prediction for the continuing increase of fossil prices since 2000.

10.1.6 Stochastic Models of Resource Extraction

Both deterministic and stochastic models (see Sect. 1.2.1) are used for mathematical description of natural resource extraction. Stochastic models are helpful, in particular, to describe the possibility of discovery of new resources.

A simple stochastic model with possibility of the discovery of new resources can be constructed as follows. Let us introduce two new dynamic characteristics:

$D(t)$—the intensity of the discovery of new resources (the quantity of new resource discovered per time unit),
$W(t)$—the investment intensity into new resource discovery (the investment into the discovery of one unit of new resource per time unit).

The variable $D(t)$ is of stochastic nature and obviously depends on the investment $W(t)$. For simplicity, we assume the random variable $D(t)$ at a fixed t to be discrete and the probability of new resource discovery to be very small. Then, the stochastic process of new resource discovery can be described by the *Poisson distribution*:

$$p[D = d] = \mathrm{e}^{-\varepsilon W}(\varepsilon W)^d/d!, \tag{10.38}$$

where

$p[D = d]$ is the probability of $D = d$,
$\varepsilon > 0$ is a given efficiency parameter of the investment into resource discovery.

The probabilistic relation (10.38) means that D stochastically increases in W and the mean of the random variable $D(t)$ at time t is $\mu[D(t)] = \varepsilon W(t)$. This description is obviously simplified and does not take into account some relevant features, such as time delays in the process of discovery of new resources. Then, in the terms of expected values, the resource dynamics with possibility of new resources discovery can be described as $R'(t) = -E(t) + \mu[D(t)]$ or

$$R'(t) = -E(t) + \varepsilon W(t), \quad R(0) = R_0 > 0, \tag{10.39}$$

where R_0 is the initial stock of resource.

The firm chooses the extraction rate $E(t)$ and the investment rate $W(t)$ in order to maximize its discounted profit over the planning horizon $[0, T)$, $T \leq \infty$:

$$\max_{E,W} \int_0^T e^{-rt}[p(t)E(t) - W(t)]\,dt \tag{10.40}$$

under the restrictions $0 \leq E(t) \leq E_{max}$ and $0 \leq W(t) \leq W_{max}$.

10.1.6.1 Elements of Model Analysis

The problem (10.39)–(10.40) is a linear optimization problem and, as such, can possess only corner solutions, i.e., bang–bang solutions. Similarly to resource extraction problems from Sect. 10.1.1, its optimal dynamics depends on the dynamics of the given resource price p. Indeed, replacing E with the new unknown variable $x = E - \lambda A$, the optimization problem (10.39)–(10.40) can be transformed to the problem

$$\max_{x,A} \int_0^T e^{-rt}[p(t)x(t) + (\lambda p(t) - 1)A(t)]\,dt, \quad R'(t) = -x(t). \tag{10.41}$$

By (10.41), the optimal control $A^*(t) = A_{max}$ when $p(t) \geq 1/\lambda$ and $A^*(t) = 0$ otherwise (then, the search for a new resource is not effective). Under the known $A^*(t)$, the optimal extraction intensity $E^*(t)$ follows the same bang–bang rule (10.11) as in the optimization problem (10.5)–(10.6) with costless resource extraction. However, the length T of the extraction period is now determined from the modified condition

$$\int_0^T [E^*(t) - \lambda A^*(t)]\,dt = R_0. \tag{10.42}$$

Simple models of resource extraction are often used as components in more complex models of economic–environmental interaction. Such models are considered in the next section and Chap. 12.

10.2 Dasgupta–Heal Model of Economic Growth with Exhaustible Resource

In Sect. 10.1, the economics of resource extraction is modeled in isolation and does not include production. Now, we investigate the economic growth model with a Cobb–Douglas technology that uses physical capital K and exhaustible resource R to produce aggregate product Q (known as the *Dasgupta–Heal* model).

Following the modeling framework of Chaps. 2 and 3, we construct a social planner problem in continuous time t. Namely, a benevolent social planner maximizes the discounted utility over the infinite horizon $[0,\infty)$

$$\max_{I,E,C} \int_0^\infty e^{-rt} \frac{C(t)^{1-\gamma}}{1-\gamma} \, dt \tag{10.43}$$

in the following model:

product output:

$$Q(t) = AK^{\alpha}(t)E^{1-\alpha}(t), \tag{10.44}$$

product distribution:

$$Q(t) = I(t) + C(t), \tag{10.45}$$

capital accumulation:

$$K'(t) = -\mu K(t) + I(t), \tag{10.46}$$

resource depletion:

$$R'(t) = -E(t). \tag{10.47}$$

The initial conditions

$$R(0) = R_0 > 0, \quad K(0) = K_0 > 0, \tag{10.48}$$

are given. The inequality-constraints are

$$I(t) \geq 0, \quad E(t) \geq 0, \quad C(t) \geq 0, \quad R(t) \geq 0. \tag{10.49}$$

The restriction $R(t) > 0$ holds automatically for all t, because $K > 0$ for all t and the production function (10.44) satisfies the Inada condition on E. The flow diagram of the model (10.43)–(10.49) is illustrated in Fig. 10.3.

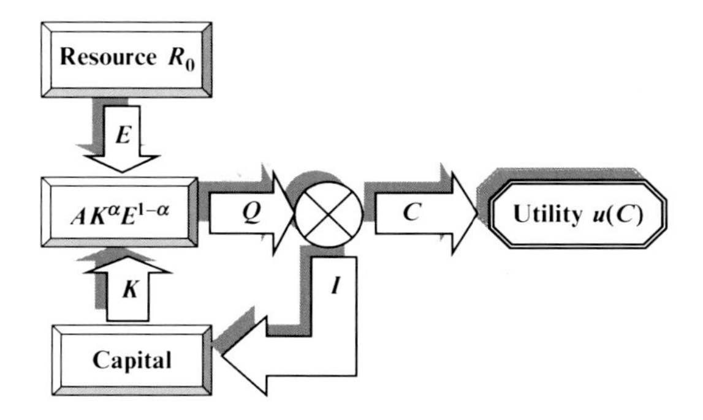

Fig. 10.3 The flow diagram of the Dasgupta–Heal model (10.43)–(10.49) with exhaustible resource

After eliminating Q, the nonlinear optimal control problem (10.43)–(10.49) has three controls (C, E, I) and two state variables (K, R). This problem allows for a complete closed-form solution, which provides an essential insight into the economics of exhaustible resource extraction.

Historic comments. The model (10.43)–(10.49) was analyzed in [4] for a general production function $F(K, E)$ in (10.43) and a general nonlinear utility $U(C)$ in (10.44). Later [4] switches to the Cobb–Douglas function and isoelastic utility $U(C) = C^{1-\gamma}/(1-\gamma)$ and points out that "only the Cobb–Douglas form has reasonable properties" and that "the Cobb–Douglas case is particularly interesting since the analysis can relatively easily be taken further."

10.2.1 Optimality Conditions

To simplify the derivation of the optimality condition, we temporarily eliminate both Q and $C = AK^{\alpha}E^{1-\alpha} - I$, then the nonlinear optimal control problem (10.43)–(10.49) becomes

$$\max_{I,E} \int_0^{\infty} e^{-rt} \frac{\left(AK(t)^{\alpha}E(t)^{1-\alpha} - I(t)\right)^{1-\gamma}}{1-\gamma} \mathrm{d}t \tag{10.50}$$

and involves two controls I, E and two state variables K, R related by two equalities (10.46) and (10.47). By the maximum principle of Sect. 2.4, the Hamiltonian for the this problem is

$$H(I, E, K, R, \lambda_1, \lambda_2) = \left(AK^{\alpha}E^{1-\alpha} - I\right)^{1-\gamma}/(1-\gamma) + \lambda_1(I - \mu K) + \lambda_2 E, \tag{10.51}$$

where $\lambda_1(t)$ is the dual variable for the equality (10.46) and $\lambda_2(t)$ is the dual variable for the equality (10.47).

Following Corollary 2.2 from Sect. 2.4, the first-order optimality conditions for the interior solutions I, E, K are:

$$\frac{\partial H}{\partial I} = 0, \quad \frac{\partial H}{\partial E} = 0, \quad \lambda_1{}' = r\lambda_1 - \frac{\partial H}{\partial K}, \quad \lambda_2{}' = r\lambda_2 - \frac{\partial H}{\partial R}, \tag{10.52}$$

or

$$-C^{-\gamma} + \lambda_1 = 0, \tag{10.53}$$

$$(1-\alpha)A\left(\frac{E}{K}\right)^{-\alpha} C^{-\gamma} = \lambda_2, \tag{10.54}$$

$$\lambda_1{}' = (r+\mu)\lambda_1 - \alpha A\left(\frac{E}{K}\right)^{1-\alpha} C^{-\gamma}, \tag{10.55}$$

$$\lambda_2{}' = r\lambda_2. \tag{10.56}$$

We use the variable C in (10.53)–(10.56) for simplicity only. The transversality conditions are

$$\lim_{t\to\infty} e^{-rt}\lambda_1(t)K(t) = 0, \quad \lim_{t\to\infty} e^{-rt}\lambda_2(t)E(t) = 0. \tag{10.57}$$

It is possible to formally prove a complete maximum principle for the nonlinear optimal control problem (10.43)–(10.49), however, simpler conditions (10.52)–(10.57) for interior solutions are sufficient for the purposes of our analysis.

10.2.2 Analysis of Model

The economic interpretation of the dual variable $\lambda_1(t)$ is the shadow price (future rental value) of the capital K. By (10.53),

$$\lambda_1(t) = C^{-\gamma}(t) \tag{10.58}$$

The economic meaning of the dual variable $\lambda_2(t)$ is the shadow price of the resource R. By (10.56),

$$\lambda_2(t) = \lambda_{20}e^{rt} \tag{10.59}$$

therefore, the resource price $\lambda_2(t)$ follows the *Hotelling no-arbitrage rule* (10.30) (see Sect. 10.1.3).

The major idea of the subsequent mathematical treatment is to exclude the dual variables λ_1 and λ_2 and solve the remaining nonlinear differential equations analytically. First, by (10.55) and (10.53),

$$r + \mu - \alpha A_2 \left(\frac{E(t)}{K(t)}\right)^{1-\alpha} = \lambda'(t)/\lambda(t) \tag{10.60}$$

or

$$\gamma \frac{C'(t)}{C(t)} = \alpha A \left(K \frac{E(t)}{k(t)}\right)^{1-\alpha} - r.$$

Introducing the new unknown variable $x(t) = \frac{K(t)}{E(t)}$, we have

$$\gamma \frac{C'(t)}{C(t)} = \alpha A x^{\alpha-1}(t) - r. \tag{10.61}$$

Next, by (10.53), $(1 - \alpha)Ax^{\alpha}(t)\lambda_1(t) = \lambda_{20}e^{rt}$ and $\lambda_{20} = (1 - \alpha)A{x_0}^{\alpha}{C_0}^{-\gamma}$ at $t = 0$. Therefore, $x^{\alpha}(t)\lambda_1(t) = {x_0}^{\alpha}{C_0}^{-\gamma}e^{rt}$.

Taking the logarithm of the last equation and differentiating it, we have

$$\alpha x'(t)/x(t) = -{\lambda_1}'(t)/\lambda_1(t) + r$$

or, by (10.60), the following equation

$$x'(t)/x(t) = Ax^{\alpha-1}(t) \tag{10.62}$$

for the unknown x only. The nonlinear ordinary differential equation (10.62) can be solved analytically, its solution is

$$x(t) = \left[(1 - \alpha)At + {x_0}^{1-\alpha}\right]^{1/(1-\alpha)} \tag{10.63}$$

and increases up to ∞. Substituting (10.63) into (10.61), we have

$$\frac{C'(t)}{C(t)} = -\frac{r}{\gamma} + \frac{\alpha A}{\gamma[(1 - \alpha)At + {x_0}^{1-\alpha}]}.$$

The obtained nonlinear ordinary differential equation with respect to C can also be solved analytically. Solving it over the interval $[0,t]$, we obtain

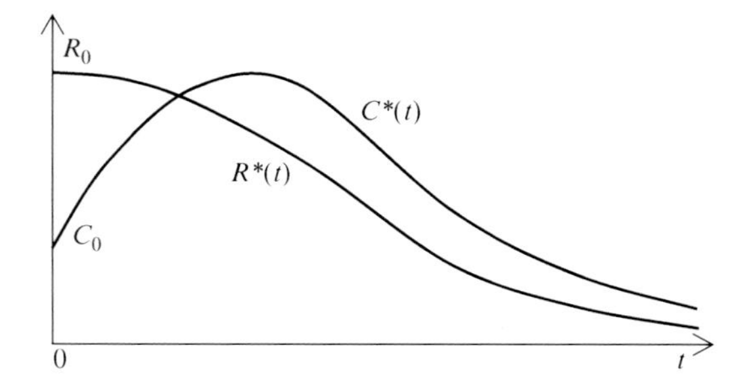

Fig. 10.4 The optimal dynamics of consumption C and remaining resource stock R in the model (10.43)–(10.49) with exhaustible resource

$$\ln C(t) - \ln C_0 = -\frac{r}{\gamma}t + \frac{\alpha}{\gamma(1-\alpha)}\ln\left(1 + (1-\alpha)A{x_0}^{\alpha-1}t\right)$$

or

$$C(t) = C_0 e^{-\frac{r}{\gamma}t}\left(1 + (1-\alpha)A{x_0}^{\alpha-1}t\right)^{\frac{\alpha}{\gamma(1-\alpha)}}. \tag{10.64}$$

By (10.64), if the initial values K_0 and R_0 are large enough, then the optimal consumption C can increase during a certain initial period $[0,T_0]$ as shown in Fig. 10.4. However, the consumption C ultimately decreases and asymptotically approaches zero with the rate $-r/\gamma$ when $t \to \infty$.

The optimal ratio $E(t)/K(t) = x^{-1}(t)$ also asymptotically approaches zero at $t \to \infty$ but remains positive, $x(t) > 0$, on $[0,\infty)$. Finally, we obtain from (10.44)–(10.46) that

$$K'(t) = E'(t)x(t) + E(t)x'(t) = Ax^{\alpha}(t)E(t) - C(t),$$

which leads in combination with (10.62) to

$$E'(t) = -C(t)/x(t). \tag{10.65}$$

So the optimal resource extraction $E(t)$ also decreases and asymptotically approaches zero when $t \to \infty$.

Formulas (10.63)–(10.65) determine the complete optimal dynamics of the problem (10.43)–(10.49). To obtain a unique optimal trajectory (I, K, E, C), we need to know the optimal values E_0 and C_0. Because of the nonlinear nature of the problem (10.43)–(10.49), it is not possible to obtain analytic formulas for the optimal values E_0 and C_0. However, as shown in [4], such *unique* positive values E_0 and C_0 exist and satisfy the transversality conditions (10.57) over the infinite horizon $[0,\infty)$.

10.2.3 *Interpretation of Results*

The most important conclusions about optimal growth in the economy (10.43)–(10.49) with an exhaustible resource are the following:

- The endogenous price of resource follows the *Hotelling non-arbitrage rule* (10.30), which means that it exponentially increases with the discount rate r. So the increasing scarcity of the resource increases its market price.
- The optimal policy is to retain some portion of the exhaustible resource over the infinite horizon $[0,\infty]$ rather than to consume it over a finite interval $[0,T]$.
- The optimal resource extraction decreases and asymptotically approaches zero when the time indefinitely increases.
- The optimal consumption ultimately decreases and asymptotically approaches zero when the time indefinitely increases.

The last prediction is quite pessimistic, that is why the model is often considered in a combination with another alternative technology.

Exercises

1. Explain why the total resource extraction cost $C(E,R)$ should depend on the resource extraction intensity E and resource stock R. Do you think that the assumptions $\partial C/\partial E > 0$ and $\partial C/\partial R \leq 0$ are realistic?
2. Following Sect. 2.4, construct the Hamiltonian (10.7) and derive optimality conditions for the optimal control problem (10.5), (10.6).
3. Solve the initial value problem (10.1), (10.2) and estimate its solution using restriction (10.4).
4. Assume that the resource price $p(t)$ is constant in the problem (10.5), (10.6) with costless resource extraction. Show that the optimal extraction intensity $E^*(t) = E_{\max}$ over the entire interval $[0,T^*]$ and the length of the entire extraction period is $T^* = R/E_{\max}$.
   HINT: use formulas (10.11) and (10.12).
5. In the problem (10.5), (10.6), (10.21) with constant resource price and quadratic resource extraction cost, show that the nonlinear equation (10.24) for the length of extraction period has a unique solution $0 < T^* < \infty$ for any positive parameters R_0, p, and r.
   HINT: use plot graphs of the left- and right-hand sides of the nonlinear equation (10.24) and show that they have a unique interception point.
6. In the optimization problem (10.5), (10.6), (10.21) with constant resource price and quadratic resource extraction cost, show that the optimal extraction length T^* is determined by the approximate formula (10.26) if the discount rate r is small: $r \ll 1$.
   HINT: expand the function e^{-rT} in (10.24) into the Taylor series and neglect terms higher than $(rT)^2$.

7. In Dasgupta–Heal model (10.43)–(10.49), express the variables (C, E, I) in terms of the state variables (K, R).
8. In Dasgupta–Heal model (10.43)–(10.49), verify the optimality conditions (10.53)–(10.57) for the interior solutions I, E, K.
9. Solve the nonlinear ODE (10.62) using the separation of variables. Show that the solution of (10.62) has the form (10.63).
10. Demonstrate (analytically and/or numerically) that the function (10.64) can increase during a certain initial period, but it decreases later and asymptotically approaches zero with the rate $-r/\gamma$ when $t \to \infty$.

References[1]

1. 🕮 Clark, C.W.: Mathematical bioeconomics: The optimal management of renewable resources. Wiley, New York (1976)
2. Gaudet, G.: Natural resource economics under the rule of Hotelling. Can. J. Econ. **40**, 1033–1059 (2007)
3. 🕮 Dasgupta, P., Heal, G.: The economic theory and exhaustible resources. Cambridge University Press, Cambridge, UK (1979)
4. Dasgupta, P., Heal, G.: The optimal depletion of exhaustible resources. Rev. Econ. Stud. **41**, 3–28 (1974)
5. Hotelling, H.: The economics of exhaustible resources. J. Polit. Econ. **39**, 137–175 (1931)
6. 🕮 Hoy, M., Livernois, J., McKenna, C., Rees, R., Stengos, T.: Mathematics for economists, 2nd edn. The MIT Press, Massachusetts (2001)
7. Simpson, D., Toman, M.A., Ayres, R.U. (eds.): Scarcity and growth revisited: Natural resources and the environment in the new millennium. Resources for the Future, Washington, DC (2005)

[1] The book symbol 🕮 means that the reference is a textbook recommended for further student reading.

Chapter 11
Modeling of Environmental Protection

This chapter is devoted to analytic modeling of the relations between the economy, society, and the environment, with a focus on combating climate change. Section 11.1 describes simple models of these relationships and discusses mitigation and adaptation as two major human responses to environmental damage. Sections 11.2 and 11.3 investigate economic models with optimal investments into mitigation of environmental pollutions. Section 11.4 analyzes optimal investments into mitigation and adaptation against environmental damage. These economic–environmental models are formulated as social planner problems with mitigation and adaptation investments as separate variables. The steady-state analysis of optimal investments leads to essential implications for associated long-term environmental policies.

11.1 Mutual Influence of Economy and Environment

Modeling of relationships between human society and the environment should be based on a clear understanding of modeling objectives and priorities, to distinguish primary processes from numerous details. For clarity, this chapter focuses on climate change, which has been a subject of systematic modeling efforts during the last two decades.

11.1.1 Climate Change and Environmental Strategies

We start with discussing major environmental threats and ways of protecting the society and economy against them.

N. Hritonenko and Y. Yatsenko, *Mathematical Modeling in Economics, Ecology and the Environment*, Springer Optimization and Its Applications 88,
DOI 10.1007/978-1-4614-9311-2_11,

11.1.1.1 Global Warming

The increase of average global temperature on the Earth is the major factor of climate change that causes economic and welfare losses. The average temperature has risen nearly 1 °C during the past century and about 0.6 °C in the past 30 years. Global warming causes changes in the environment and affects different aspects of human life. In 2005, sea ice in the northern hemisphere reached its lowest levels, Greenland's glaciers lost 220 km^3 of ice, and the Gobi Desert expanded by 26,000 km^2. The global warming has forced millions of Chinese farmers to move to more fertile land and natives of the Arctic region to go further north to follow prey. Tourism industry has lost millions of dollars because of the extreme heat and thousands of residents in different parts of the world were asked to leave their homes due to wild fires. Further increase of the global temperature will lead to more severe storms, hurricanes, biodiversity loss, reduced agricultural yields, and other threats to human well-being. Climate science experts claim that the global warming is a result of emissions of greenhouse gases. However, the average global temperature during 2005–2013 has been unchanged, while carbon emissions into atmosphere have substantially increased. It means that the connection between emissions of greenhouse gases and global warming is not quite consistent, at least, in the short run.

11.1.1.2 Greenhouse Effect

The Earth receives solar radiation from the Sun. Approximately 70 % of incoming solar radiation is absorbed by the Earth's surface and atmosphere. The Earth's surface emits some of long-wave thermal radiation back into space. However, an essential portion of this radiation is absorbed by the so-called *greenhouse gases* (water vapor H_2O, carbon dioxide CO_2, methane CH_4, nitrous oxide N_2O, chlorofluorocarbons, and ozone O_3), which keeps the Earth warmer (the "*natural greenhouse effect*").

During the last centuries, atmospheric concentrations of the greenhouse gases have increased due to different anthropogenic activities. This increase causes an additional increase of the global average temperature of the Earth (the "*enhanced greenhouse effect*"). It is still unknown how fast such a temperature increase will occur and how it will impact the biosphere and human society. In general, scientific knowledge of the global Earth system is insufficient to predict results of human impact. Certain human activities can create multiple interacting effects in the environment in a complex and unpredictable way.

11.1.1.3 Pollution Mitigation and Environmental Adaptation

Observed changes in the atmosphere demonstrate increasing concentrations of greenhouse gases, which may change the global climate. The society has two major long-term potential strategies to deal with climate change: to mitigate greenhouse gas emissions and/or to adapt to the global warming.

Mitigation is an anthropogenic intervention to reduce sources or enhance sinks of the greenhouse gases [8, p.750]. Mitigation is closely related to the concept of *abatement*. Specifically, the *mitigation* refers to a reduction in the *net emissions* of greenhouse gases, whereas the *abatement* means a reduction in the *gross* emissions. The majority of theoretical models of environmental sciences does not consider sinks and, therefore, describe only emission abatement opportunities.

The environmental *adaptation* is an adjustment in natural or human systems in response to actual or expected environmental changes, which moderates the harm from those changes or exploits beneficial opportunities. In practice, the *adaptation* covers a large range of various actions such as investment in a coastal protection infrastructure, diversification of crops, implementation of warning systems, improvement in water resource management, development of new insurance instruments, air-cooling devices, and many others [2].

11.1.2 Modeling of Economic Impact on Environment

The anthropogenic activities effect the environment in many ways. This section describes simple aggregate models of this impact. Let us introduce the following dynamic characteristics:

$T(t)$—the global average temperature,
$P(t)$—the concentration of greenhouse gases in the atmosphere,
$B(t)$—the *investment* into *mitigation (abatement, cleanup)* measures that decrease the industrial emissions of greenhouse gases,
$D(t)$—the *investment* into *adaptation* measures that reduce the negative impact of climate change.

The increase of the average temperature $T(t)$ depends on the atmospheric pollution $P(t)$ with some lags [8]. *Simulation-based economic–environmental models* with exogenous (given) dynamics of climate change usually consider $T(t)$ as an aggregate dynamic parameter of climate damage, for example, see the integrated assessment models of Sect. 12.2. *The analytic economic–environmental models* prefer to use $P(t)$ as the aggregate dynamic parameter of climate damage, because controlled dynamics of greenhouse gases pollution is better understood than the one of the temperature increase.

The choice of a law of motion for the environmental pollution P is critical. In the case of greenhouse gases, the pollution is accumulated as a *stock*:

$$P'(t) = -\delta_P P(t) + \Phi(Q(t), K(t), B(t)), \ P(0) = P_0, \tag{11.1}$$

where $\delta_P > 0$ is a constant deterioration (decay) rate of pollutions; $\Phi(Y, K, B)$ is the *pollution emission function*; $Q(t)$ is the economic production output; $K(t)$ is the capital amount.

The emission function $\Phi(Y, K, B)$ in (11.1) describes the *net flow of pollution* resulting from productive activities and reflects the environmental dirtiness of the economy. The pollution emission Φ grows with the output Y and capital K, and declines when the abatement investment B increases. It can also depend on other economic parameters.

By (11.1), the pollution stock P increases with the pollution emission flow Φ and decays at a fixed natural rate $\delta_P > 0$, so the environmental damage is assumed to slowly decrease due to the regenerative capacity of the environment.

The industrial pollution emission and abatement in (11.1) can take different forms [4–6, 11]. Their common and quite general specifications are

$$\Phi(K, B) = \gamma K^{\bar{\omega}} B^{-\lambda}, \quad \Phi(Q, B) = \gamma Q^{\bar{\omega}} B^{-\lambda}, \quad \Phi(Q, B) = \gamma Q^{\bar{\omega}}/(B_0 + B)^{\lambda}, \tag{11.2}$$

where γ, $\lambda \geq \bar{\omega}$, and B_0 are positive constants. The emission factor $\gamma > 0$ describes the environmental dirtiness of the economy.

A simple analytical case happens at $\lambda = \bar{\omega} = 1$, then the pollution net emission is proportional to the output Q and the law of pollution motion is given by:

$$P'(t) = -\delta_P P(t) + \gamma Q(t)/B(t), \quad P(0) = P_0. \tag{11.3}$$

The nonlinear differential equation (11.3) captures major qualitative features of the air pollution contamination and abatement activities. We analyze the long-term growth in an economic–environmental system with the pollution stock (11.3) and spending on abatement activities, in Sect. 11.3.

For simplicity, some economic–environmental models consider the pollution as the endogenous flow $\Phi(Q,K,B)$ rather than the stock P, for example, see Sect. 11.2. Some other models include an endogenous "technology index," which nonlinearly impacts both output and pollution. Some authors introduce a generic "environmental quality level" that depends on an aggregate "environmental protection expenditure" and do not distinguish between abatement and adaptation investments and their specific impact on the economy, utility, and the environment.

11.1.3 Modeling of the Environmental Impact on Economy and Society

Despite the relatively young age of climate change economics [1, 10, 13], several aggregate analytic approaches have been developed to model a negative impact of undesirable environmental changes on the economy and human welfare.

11.1.3.1 Approach 1: Decreased Productive Value

A simple economically oriented approach emphasizes that the temperature increase causes direct global losses of the economic output. It introduces a *damage function* $G(T), 0 < G(T) \leq 1$, that translates the temperature increase T into global losses of the product output Q. Next, a multiplicative utility function $U(G(T)C)$ is employed instead of the standard utility $U(C)$ in related optimization models (see Sect. 2.3.3 about the utility functions). For instance, the simulation-based models DICE, RICE, and WITCH of Sect. 12.2 use the *gross damage functions* of the form

$$GD(T) = \alpha_0 T + \alpha_1 T^{\alpha}. \tag{11.4}$$

Alternatively, the output Q can be assumed to negatively depend on the pollution P, for instance, as $Q(P) = AK^{\alpha}P^{-\beta}$, $\beta > 0$, with subsequent impact on the consumption C and utility U.

11.1.3.2 Approach 2: Decreased Amenity Value

The other popular approach emphasizes that the global warming causes direct welfare losses. Then, the utility function U depends on the global temperature T or the greenhouse gases concentration P. The majority of such models uses the concentration P, then the utility function is $U = U(C, P)$. To specify the negative impact of the environment, it is convenient to choose the utility $U(C, P)$ to be additively separable:

$$U(C, P) = U_1(C) - U_2(P). \tag{11.5}$$

where $U_1(C)$ is a standard utility function (see Sect. 2.3.3), say, the isoelastic utility $U_1(C) = \frac{C^{1-\gamma}}{1-\gamma}$ or the logarithmic utility $U_1(C) = \ln C$. The function $U_2(P)$ increases in P, so the utility function $U(C,P)$ decreases in P; this property is known as the *disutility of environmental pollution*. If $-U_2''(P) < 0$, then the function (11.5) describes an *increasing marginal disutility* of pollution. For example, the utility function U can be taken in the form

$$U(C, P) = \ln C - \eta \frac{P^{1+\theta}}{1+\theta}, \tag{11.6}$$

where the parameter $\eta > 0$ represents the *environmental vulnerability* of the economy; the parameter $\theta > 0$ reflects the increasing marginal disutility of pollution.

The assumption of increasing marginal disutility of pollution is common in environmental economics [12], although some simpler models, say, in the game theory [9], assume a linear disutility $U_2(P) = \eta P$ of the environment damage.

More accurate models with environmental adaptation should combine both approaches [7], but it requires more accurate modeling data.

11.1.4 Modeling of Mitigation and Adaptation

The concept of environmental adaptation means that the environmental damage can be reduced by investing in adaptation measures. The adaptation control D is usually the amount of *environmental adaptation capital*, which can be modeled as a stock or flow. The specific description of adaptation process depends on the chosen Approach 1 or 2 from the previous section. Namely, the *efficiency of adaptation measures* in protecting people from climate change adverse impacts can be described by introducing a dependence of the gross damage function $G(T)$ in (11.4) or the environment damage disutility $U_2(P)$ in (11.5) on the adaptation control D.

In Approach 1, the *damage function* $0 < G(T) \leq 1$ is decreased by a *adaptation efficiency function* $\eta(D) > 0$, so the final utility becomes $U(\eta(D)G(T)C)$. In particular, the integrated-assessment models with adaptation (Sect. 12.2) empirically estimate and use an adaptation efficiency function of the form $\eta(D) = 1/(1 + D)$.

In Approach 2, the final utility $U(C,P,D)$ depends on the consumption C, pollution P, and adaptation D. In the case of the additively separable utility (11.5), the *adaptation efficiency function* $\eta(D)$ appears in (11.5) as

$$U(C, P, D) = U_1(C) - \eta(D)U_2(P). \tag{11.7}$$

If we take the utility function (11.6), then

$$U(C, P, D) = \ln C - \eta(D)\frac{P^{1+\mu}}{1+\mu}, \quad \mu > 0. \tag{11.8}$$

In both approaches, the functional form for $\eta(D)$ should satisfy some realistic *properties of adaptation investments*:

(i) No damages are reduced without adaptation;
(ii) The infinite adaptation can reduce almost all or all damages;
(iii) The more adaptation is used, the less effective it will be.

The assumption (iii) of *decreasing returns of adaptation* is justified by technological and economic reasons. Indeed, first adaptation measures are usually more efficient. The recent adaptation-related integrated-assessment models (see Sect. 12.2) empirically estimate and use the *adaptation efficiency function*

$$\eta(D) = 1/(1 + D). \tag{11.9}$$

It is easy to see that (11.9) possesses the above properties (i)–(iii).

In Sect. 11.4 below, we will employ the exponential vulnerability function

$$\eta(D) = \underline{\eta} + \left(\overline{\eta} - \underline{\eta}\right)e^{-aD}, \quad \overline{\eta} > \underline{\eta} > 0, \quad a > 0. \tag{11.10}$$

that satisfies properties (i)–(iii). In (11.10), the term $\left(\overline{\eta} - \underline{\eta}\right)$ is the range of physical adaptation opportunities, i.e., the benefits in terms of vulnerability reduction associated with adaptation measures. The parameter a represents the efficiency of adaptation. The function (11.10) monotonically decreases in D from a maximal value $\eta(0) = \overline{\eta} > \underline{\eta}$, when there is no adaptation at all, to a minimal value $\eta(\infty) = \underline{\eta} > 0$, when adaptation efforts approach infinity. The graph of (11.10) is shown in Fig. 11.1 and approaches the horizontal asymptote $\eta = \underline{\eta} > 0$ when D grows indefinitely. The derivative

$$\eta'(D) = -a\left(\overline{\eta} - \underline{\eta}\right)e^{-aD}$$

represents the *marginal efficiency* of adaptation, which is higher for the first adaptation measures, and then decreases gradually in D (see Fig. 11.1).

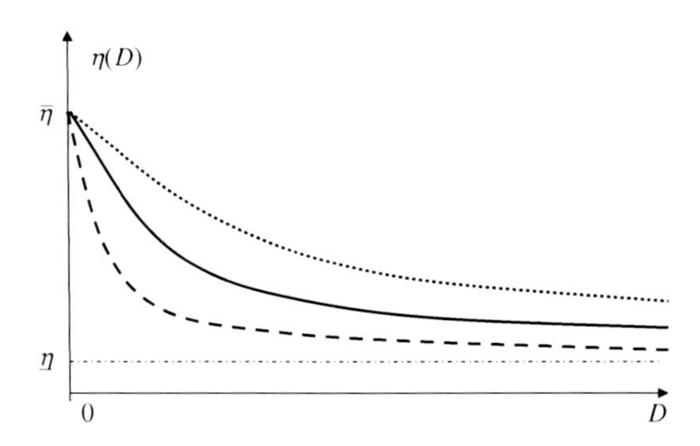

Fig. 11.1 The dependence of the environmental vulnerability η on adaptation D. The *dashed curve* has a larger adaptation efficiency parameter a than the *solid curve*. The *dotted curve* has a smaller parameter a

In the next three sections, we construct and analyze macroeconomic models that consider the environmental quality and investments into environmental protection. Along with capital accumulation, the models involve investments into mitigation and adaptation to combat the climate change. We employ the central planner framework and steady-state analysis from Chap. 2 to explore relationships between a country's per-capita income and environmental quality.

11.2 Model with Pollution Emission and Abatement

In this section, a dynamic model of an economic–environmental system (a country or region) with physical capital, pollution, and abatement expenses is developed. Following Sect. 11.1.2, we assume that the environmental quality is characterized by pollution. We employ the Solow one-sector growth framework of Chap. 2, in which the economy uses a Cobb-Douglas technology with constant returns to produce a single final good Q. The social planner allocates the final product

Q across the consumption C, the investment I in physical capital K, and the emission abatement expenditures B in order to maximize the infinite-horizon discounted utility that depends on the consumption C and pollution P:

$$\max_{I,C} \int_0^\infty e^{-rt}\left(\ln C - \eta\frac{P^{1+\theta}}{1+\theta}\right)dt, \quad I(t) \geq 0, \quad C(t) \geq 0, \tag{11.11}$$

subject to the following constraints:

$$Q(t) = AK^{\alpha}(t) = I(t) + B(t) + C(t), \tag{11.12}$$

$$K'(t) = I(t) - \mu K(t), \quad K(0) = K_0, \tag{11.13}$$

where $r > 0$ is the discount rate; $A > 0$ and $0 < \alpha < 1$ are parameters of the Cobb-Douglas production function; $\mu \geq 0$ is the deterioration rate of the physical capital K.

For simplicity, we assume in this section that the pollution P is a flow proportional to the output Q and can be reduced by the abatement B:

$$P(t) = \gamma\frac{Q(t)}{B(t)} = \gamma A\frac{K(t)^{\alpha}}{B(t)}, \tag{11.14}$$

where the emission factor $\gamma > 0$ describes the environmental dirtiness of the economy. The accumulation of pollution P as a stock is considered in the next section. The abatement activity B in the model is also a flow.

The utility U in the objective functional (11.11) depends on the consumption and pollution and is taken as (11.6). The flow diagram of our model is illustrated in Fig. 11.2. To simplify the analysis of the problem (11.11)–(11.14), we can temporarily exclude the unknown variable P from (11.11) and the unknown Q from (11.12). Then, the optimization problem includes two decision variables I, C and two state variables K, B, determined from the state equations (11.12)–(11.13).

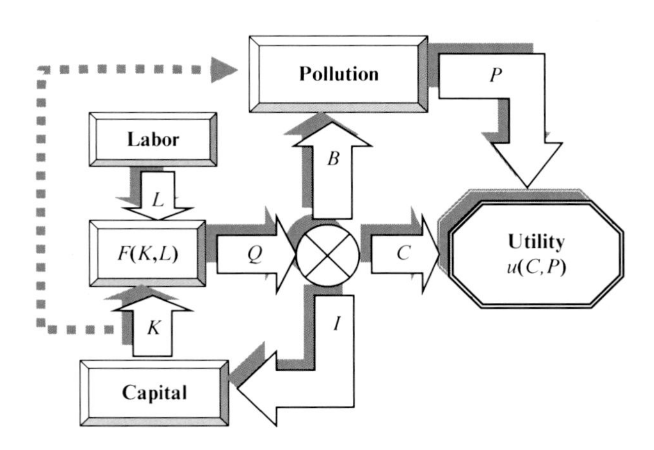

Fig. 11.2 The flow diagram of the economic–environmental model (11.11) with physical capital, pollution, and abatement

11.2.1 Optimality Conditions

For clarity sake, we restrict ourselves with the analysis of interior trajectories of the model. It means that we disregard the constraints $I \geq 0, C \geq 0$ and look for positive controls I and C only. A complete dynamic analysis can be provided similarly to the one for *Solow-Ramsey model* in Sect. 2.3 but is technically more complicated and is out of the scope of this textbook.

Following Sect. 2.4, the present-value Hamiltonian for the problem (11.11)–(11.14) is given by

$$H(C,I,K,B,t) = e^{-rt}\left[\ln C - \eta A^{1+\theta}\frac{\gamma^{1+\theta}K^{\alpha(1+\theta)}}{(1+\theta)B^{1+\theta}}\right] + \lambda_1(AK^\alpha - I - B - C) + \lambda_2\big(I - \mu K\big), \tag{11.15}$$

where the dual variables λ_1 and λ_2 are associated with state equations (11.12) and (11.13). Employing Corollary 2.1 of Sect. 2.4, the first order extremum conditions for the two decision variables I and C are $\partial H/\partial I = 0$, $\partial H/\partial c = 0$, or :

$$-\lambda_1 + \lambda_2 = 0, \tag{11.16}$$

$$C^{-1}e^{-rt} - \lambda_1 = 0, \tag{11.17}$$

or

$$\lambda_1 = \lambda_2 = C^{-1}e^{-rt}. \tag{11.18}$$

The first order conditions for the state variables K and B are $\partial H/\partial K = 0$ and $\partial H/\partial B = 0$, or, respectively:

$$e^{-rt}\alpha\eta A^{1+\theta}\gamma^{1+\theta}\frac{K^{\alpha(1+\theta)-1}}{B^{1+\theta}} + \alpha\lambda_1 AK^{\alpha-1} - \mu\lambda_2 = -\lambda_2', \tag{11.19}$$

$$e^{-rt}\eta A^{1+\theta}\gamma^{1+\theta}\frac{K^{\alpha(1+\theta)}}{B^{2+\theta}} - \lambda_1 = 0, \tag{11.20}$$

and the transversality conditions are $\lim_{t\to\infty}\lambda_1(t) = \lim_{t\to\infty}\lambda_2(t) = 0$.

11.2.2 Analysis of Model

Excluding λ_1 and λ_2 from (11.16)–(11.20) and using (11.12)–(11.13), we obtain the system in K, B, and C:

$$AK^{\alpha} = \mu K + B + C + K', \tag{11.21}$$

$$\alpha AK^{\alpha-1} - \alpha \frac{B}{K} = \mu + r, \tag{11.22}$$

$$B^{2+\theta} = \eta \gamma^{1+\theta} A^{1+\theta} K^{\alpha(1+\theta)} C. \tag{11.23}$$

The system of nonlinear equations (11.21)–(11.23) determines the optimal dynamics of the interior time-dependent trajectories $K(t)$, $B(t)$, and $C(t)$. Finding analytic solution for this system is challenging.

At $\eta = 0$, the problem under study is similar to the Solow-Ramsey model of Sect. 2.4, except for the logarithmic utility function in (11.11). In the case of small values of η, the impact of environmental damage on the utility is small. Therefore, we can expect the dynamics of our problem to be close to the Solow-Ramsey model (2.53), which has a unique *stationary state* (2.42) at $\eta = 0$. Below we prove a similar result for the model (11.11)–(11.14).

11.2.2.1 Steady-State Analysis

Let us analyze whether the model (11.11)–(11.14) has *stationary states* of the form

$$K(t) = \overline{K},\ C(t) = \overline{C},\ B(t) = \overline{B}, \tag{11.24}$$

where $\overline{K}$, $\overline{C}$, $\overline{B}$ are positive constants. The substitution of (11.24) into (11.21)–(11.23) leads to the following system of three nonlinear equations with respect to $\overline{K}, \overline{C}, \overline{B}$:

$$A\overline{K}^{\alpha} = \mu \overline{K} + \overline{B} + \overline{C}, \quad \alpha A\overline{K}^{\alpha-1} - \alpha \frac{\overline{B}}{\overline{K}} = \mu + r, \tag{11.25}$$

$$\eta \gamma^{1+\theta} A^{1+\theta} \overline{C} = \frac{\overline{B}^{2+\theta}}{\overline{K}^{\alpha(1+\theta)}}. \tag{11.26}$$

By (11.25), we express the steady-state components $\overline{B}$ and $\overline{C}$ in terms of $\overline{K}$ as

$$\overline{B} = A\overline{K}^{\alpha} - \overline{K}(\mu + r)/\alpha, \tag{11.27}$$

$$\overline{C} = \mu \overline{K}(1-\alpha)/\alpha + \overline{K} r/\alpha. \tag{11.28}$$

Substituting these formulas into (11.26), we obtain the following equation for $\overline{K}$:

$$\frac{\left[A - \overline{K}^{1-\alpha}(\mu + r)/\alpha\right]^{2+\theta}}{\overline{K}^{1-\alpha}} = \eta \gamma^{1+\theta} A^{1+\theta} [(\mu + r)/\alpha - \mu]. \tag{11.29}$$

To show that (11.29) has a unique solution, let us introduce the new unknown variable $x = \frac{\mu+r}{\alpha A}\overline{K}^{1-\alpha}$. Then, (11.29) takes the following dimensionless form:

$$\frac{(1-x)^{2+\theta}}{x} = \kappa, \quad \text{where } \kappa = \eta\gamma^{\theta+1}\left(1 - \frac{\alpha\mu}{r+\mu}\right). \tag{11.30}$$

The left-hand side $F(x) = \frac{(1-x)^{2+\theta}}{x}$ of (11.30) strictly decreases from ∞ at $x = 0$ to $F(1) = 0$ and intersects the horizontal line $G(x) = \kappa > 0$ at a certain point $0 < x^* < 1$, which is the unique solution of the nonlinear equation (11.30). Therefore, (11.29) has a unique solution $\overline{K}$, $0 < \overline{K} < (\alpha A/(\mu + r))^{\frac{1}{1-\alpha}}$. We summarize the obtained result in the following statement:

Statement 11.1 The problem (11.11)–(11.14) possesses the unique steady state (11.27)–(11.28), where the unique $\overline{K}$, $0 < \overline{K} < (\alpha A/(\mu + r))^{\frac{1}{1-\alpha}}$, is found from the nonlinear equation (11.29).

The upper bound $(\alpha A/(\mu + r))^{\frac{1}{1-\alpha}}$ for $\overline{K}$ is the maximal capital stock level: the size of the economy with parameters A, α, μ, and r in the case with no pollution.

Compared to the Solow-Ramsey model of Sect. 2.4, our problem (11.11)–(11.14) contains one more unknown variable B and the nonlinear equation (11.14). Because of this additional complexity, we cannot obtain exact analytic formulas for the steady state such as the golden rule (2.42) but define the steady state as the solution of the nonlinear equation (11.29). Statement 11.1 establishes the existence of a unique steady state $(\overline{K}, \overline{B}, \overline{C})$, but does not guarantee convergence of optimal trajectories $K(t)$, $B(t)$, $C(t)$, and $P(t)$ to the steady state. Such convergence can be shown similarly to the Solow-Ramsey model, but it is more difficult to prove because of the larger number of variables.

11.2.3 Interpretation of Results

The comparative static analysis of formulas (11.27)–(11.29) displays the following qualitative relations between the optimal long-term abatement policy and model parameters.

- A lower pollution intensity γ increases the size of the economy $\overline{K}$ and leads to smaller abatement efforts $\overline{B}$. Thus, a cleaner technology is always good for the economy. When γ approaches 0 (strong decarbonization), then $\overline{K}$ tends to its maximum level $(\alpha A/(\mu + r))^{\frac{1}{1-\alpha}}$, while $\overline{B}$ and $\overline{P}$ tend to zero.
- The economy size $\overline{K}$ is larger, the abatement expenditure is smaller, and the pollution level is larger for a smaller vulnerability η of the economy to climate change.

- When the productivity A increases, both optimal capital level $\overline{K}$ and abatement level $\overline{B}$ increase even faster, namely, as $(A)^{\frac{1}{1-\alpha}}$, which justifies a common conclusion that an increasing growth leads to a better environmental care.

11.3 Model with Pollution Accumulation and Abatement

In this section, we modify the model with abatement from the previous section adding the process of pollution accumulation to it. As in Sect. 11.2, we assume that the environmental quality is characterized by pollution and use a Cobb-Douglas technology to describe the production output. The social planner maximizes the infinite-horizon discounted utility:

$$\max_{I,C} \int_0^\infty e^{-rt}\left(\ln C - \eta\frac{P^{1+\theta}}{1+\theta}\right) dt, \quad I(t) \geq 0,\ C(t) \geq 0, \tag{11.31}$$

subject to the constraints:

$$Q(t) = AK^{\alpha}(t) = I(t) + B(t) + C(t), \tag{11.32}$$

$$K'(t) = I(t) - \mu K(t), \quad K(0) = K_0, \tag{11.33}$$

where r, A, α, and μ have the same meaning as in the model (11.11)–(11.13).

In contrast to the previous section, we assume that the pollution P is accumulated as a stock. The pollution emission $\gamma Q/B$ is still proportional to the output Q and is reduced by the abatement B. Then, by Sect. 11.1.2, the pollution equation is (11.3) or

$$P'(t) = -\delta_P P(t) + \gamma Q(t)/B(t), \quad P(0) = P_0. \tag{11.34}$$

with a constant natural pollution decay rate $\delta_P > 0$ and the emission factor $\gamma > 0$. The abatement activity B is still modeled as a flow.

The optimization problem (11.31)–(11.34) possesses two decision variables I, C and three state variables K, B, P connected by three state equations (11.32)–(11.34).

11.3.1 Analysis of Model

Compared to the problem (11.11)–(11.14) from Sect. 11.2, the problem (11.31)–(11.34) includes one more additional state equation (11.34) and, correspondingly, will have one more dual variable. It adds some mathematical complexity but produces similar results. The problem is handled in the same way as the previous problem. Namely, the present-value Hamiltonian for the problem (11.31)–(11.34) is

$$H(C,I,K,B,P,t) = e^{-rt}\left(\ln C - \eta\frac{P^{1+\theta}}{1+\theta}\right) + \lambda_1\left(AK^\alpha - I - B - C\right) + \lambda_2(-\delta_P P + \gamma AK^\alpha/B) + \lambda_3\left(I - \mu K\right), \tag{11.35}$$

where the dual variables λ_1, λ_2, λ_3 are associated with the state equations (11.32)–(11.34). Using the first order extremum conditions $\partial H/\partial I = 0$, $\partial H/\partial c = 0$, $\partial H/\partial K = 0$, $\partial H/\partial B = 0$, $\partial H/\partial P = 0$, and excluding λ_1, λ_2, λ_3 from them, we obtain the system of four nonlinear differential equations for K, B, C, and P:

$$AK^\alpha = \mu K + B + C + K', \tag{11.36}$$

$$\alpha AK^{\alpha-1} - \alpha\frac{B}{K} - \frac{C'}{C} = \theta + r, \tag{11.37}$$

$$P' + \delta_P P = \gamma\frac{AK^\alpha}{B}, \tag{11.38}$$

$$\eta P^\theta = \frac{B}{\gamma AK^\alpha C}\left(B(\delta_P + r) + \alpha\frac{B}{K}K' + \frac{B}{C}C' - 2B'\right). \tag{11.39}$$

Compared to (11.21)–(11.23), the system (11.36)–(11.39) contains one more equation (11.38) and one more unknown function P. Nevertheless, qualitative dynamics of the problem (11.31)–(11.34) appears to be close to the problem (11.11)–(11.14) without pollution accumulation from Sect. 11.2.

The steady-state analysis of the optimization problem (11.31)–(11.34) leads to the following expressions for the steady-state components $\overline{B}$, $\overline{C}$, and $\overline{P}$:

$$\overline{B} = A\overline{K}^\alpha - \overline{K}(\mu + r)/\alpha, \tag{11.40}$$

$$\overline{C} = \mu\overline{K}(1-\alpha)/\alpha + \overline{K}r/\alpha, \tag{11.41}$$

$$\overline{P} = \frac{\gamma A}{\delta_P\left[A - \overline{K}^{1-\alpha}(\mu + r)/\alpha\right]}, \tag{11.42}$$

where the steady-state $\overline{K}$ is determined from the single nonlinear equation

$$\frac{\left[A - \overline{K}^{1-\alpha}(\mu + r)/\alpha\right]^{2+\theta}}{\overline{K}^{1-\alpha}} = \frac{\gamma^{1+\theta}A^{1+\theta}\eta[(\mu + r)/\alpha - \mu]}{(\delta_P + r)\delta_P{}^\theta}. \tag{11.43}$$

Using the unknown variable $x = \frac{\mu+r}{\alpha A}\overline{K}^{1-\alpha}$, (11.43) is converted to the same dimensionless equation (11.30) from Sect. 11.2 with

$$\kappa = \eta \left(\frac{\gamma}{\delta_P}\right)^{\sigma+1} \frac{1 - \alpha\mu/(\rho+\mu)}{1+\rho/\delta_P}. \tag{11.44}$$

The only difference is in the parameter κ. As shown in Sect. 11.2, the nonlinear equation (11.30) has a unique solution $\overline{K}$, $0 < \overline{K} < (\alpha A/(\delta+\rho))^{\frac{1}{1-\alpha}}$. Thus, Proposition 1 holds for the optimization problem (11.31)–(11.34) with pollution accumulation.

11.3.2 Interpretation of Results

The comparative statics analysis of formulas (11.40)–(11.43) displays the following relations between the optimal long-term abatement policy and model parameters.

- If the natural pollution decay rate δ_P becomes smaller, then the optimal pollution level $\overline{P}$ increases and the optimal capital $\overline{K}$ shrinks. The capital level $\overline{K}$ characterizes the optimal size of the economy. In reality, greenhouse gases remain in the atmosphere for a very long time and their natural decay rate is very low.
- The natural decay rate plays a critical role in our model. This feature results from our specification of the limited abatement efficiency in the pollution motion law (11.34), where the abatement can keep the pollution emission stable, but cannot decrease it to zero, which is realistic. Similar qualitative dynamics is common in other economic–environmental models with more detailed description of pollution accumulation and assimilation.
- As in the model with no pollution accumulation of Sect. 11.2, the economy size $\overline{K}$ is smaller, abatement expenditure is larger, and the pollution level is smaller for larger pollution intensity γ and/or a larger vulnerability η of the economy to climate change. When the productivity A increases, both the optimal economy size $\overline{K}$ and abatement level $\overline{B}$ increase.

11.4 Model with Pollution Abatement and Environmental Adaptation

The purpose of this section is to develop an aggregate dynamic model to analyze optimal proportions between adaptation and mitigation in long-term climate policies. We modify the one-sector economic–environmental model (11.31)–(11.34) with abatement expenses by adding a new sector of environmental adaptation D.

The model still includes the final good Y, consumption C, physical capital K, investment I_K in K, emission abatement expenditures B, and pollution stock P. The social planner problem maximizes the discounted utility:

$$\max_{I_K, I_D, C} \int_0^\infty e^{-rt} U(C(t), P(t), D(t))\,\mathrm{d}t, \tag{11.45}$$

$$I_K(t) \geq 0,\ I_D(t) \geq 0,\ C(t) \geq 0,$$

subject to the constraints:

$$Q(t) = AK^{\alpha}(t) = I_K(t) + I_D(t) + B(t) + C(t), \tag{11.46}$$

$$K'(t) = I_K(t) - \mu_K K(t), \quad K(0) = K_0, \tag{11.47}$$

$$D'(t) = I_D(t) - \mu_D D(t), \quad D(0) = D_0, \tag{11.48}$$

$$P'(t) = -\delta_P P(t) + \gamma Y(t)/B(t), \quad P(0) = P_0. \tag{11.49}$$

The new state equation (11.48) describes the accumulation of adaptation capital D, where I_D is the investment into the capital D, and $\mu_D \geq 0$ is the deterioration coefficient for the adaptation capital. The utility U in (11.45) depends now on the consumption C, pollution P, and environmental adaptation capital stock D. Following Sect. 11.1.4, we take the utility function as

$$U(C, P, D) = \ln C - \eta(D)\frac{P^{1+\theta}}{1+\theta}. \tag{11.50}$$

This function captures the fact that environmental damage can be reduced by investing in adaptation measures where $\eta(D)$ is the *adaptation efficiency*.

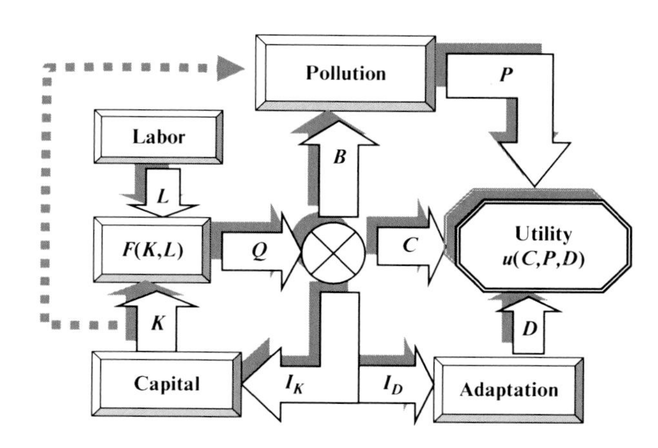

Fig. 11.3 The flow diagram of the economic–environmental model (11.45)–(11.46) with pollution, abatement, and environmental adaptation

The model (11.45)–(11.50) incorporates key ingredients of the mitigation-adaptation problem. It allows us to see how the presence of adaptation changes the qualitative results obtained in Sect. 11.3. The model links are illustrated in Fig. 11.3. The optimization problem (11.45)–(11.50) now includes three decision variables (I_K, I_D, C) and four state variables (K, D, B, P) connected by four state equations (11.46)–(11.49). Because of mathematical complexity of the problem, we cannot obtain such clear analytic results as in previous sections. The present section summarizes the steady-state analysis of the problem (11.45)–(11.50) and offers a look ahead into further directions of the economic–environmental modeling.

11.4.1 Optimality Conditions

As in Sect. 11.2, we restrict ourselves with the case of interior trajectories of the model. For tractability, let $\mu_K = \mu_D = \mu$. Following Sect. 2.4, the present-value Hamiltonian for the optimization problem (11.45)–(11.50) is:

$$H = e^{-rt}\left(\ln C - \eta(D)\frac{P^{1+\theta}}{1+\theta}\right) + \lambda_1\left(AK^{\alpha} - I_K - I_D - B - C\right)$$
$$+ \lambda_2(-\delta_P P + \gamma AK^{\alpha}/B) + \lambda_3\left(I_K - \mu K\right) + \lambda_4\left(I_D - \mu D\right), \tag{11.51}$$

where the dual variables λ_1, λ_2, λ_3, λ_4, are associated with (11.46)–(11.49). The first order conditions for I_K, I_D, and C give

$$\lambda_1 = \lambda_3 = \lambda_4 = C^{-1}e^{-rt}. \tag{11.52}$$

The first order conditions for the state variables K, B, P, and D are, respectively:

$$\alpha\lambda_1 AK^{\alpha-1} + \lambda_2\frac{\gamma}{B}\alpha AK^{\alpha-1} - \mu\lambda_3 = -\lambda_3', \tag{11.53}$$

$$-\lambda_1 - \lambda_2\frac{\gamma}{B^2}AK^{\alpha} = 0, \tag{11.54}$$

$$-\eta(D)P^{\theta}e^{-rt} - \delta_P\lambda_2 = -\lambda_2', \tag{11.55}$$

$$-\eta'(D)P^{\theta+1}e^{-rt}/(\theta+1) - \mu\lambda_4 = -\lambda_4'. \tag{11.56}$$

Excluding λ_1, λ_2, λ_3, and λ_4 from (11.53)–(11.56) and using (11.46)–(11.49), we obtain the system of five nonlinear ordinary differential equations

$$AK^{\alpha} = \delta K + K' + \mu D + D' + B + C, \tag{11.57}$$

$$\alpha AK^{\alpha-1} - \alpha\frac{B}{K} - \frac{C'}{C} = \mu + r, \tag{11.58}$$

$$P' + \delta_P P = \gamma \frac{AK^\alpha}{B}, \tag{11.59}$$

$$\eta(D)P^\theta = \frac{B}{\gamma AK^\alpha C}\left(B(\delta_P + r) + \alpha\frac{B}{K}K' + \frac{B}{C}C' - 2B'\right), \tag{11.60}$$

$$-\eta'(D)\frac{P^{\theta+1}}{\theta+1} = \frac{1}{C}\left(r + \mu + \frac{C'}{C}\right) \tag{11.61}$$

for the optimal time-dependent interior trajectories K, B, C, D, and P.

11.4.2 Steady-State Analysis

Let us analyze a possibility of stationary states of the form

$$K(t) = \overline{K}, \quad C(t) = \overline{C}, \quad B(t) = \overline{B}, \quad P(t) = \overline{P}, \quad D(t) = \overline{D} \tag{11.62}$$

in the optimization model (11.45)–(11.50). Substituting (11.62) into the differential equations (11.57)–(11.61) produces the system of five nonlinear equations:

$$A\overline{K}^\alpha = \overline{B} + \overline{C} + \mu\overline{K} + \mu\overline{D}, \quad \alpha A\overline{K}^{\alpha-1} - \alpha\frac{\overline{B}}{\overline{K}} = \mu + r, \quad \delta_P\overline{P} = \gamma\frac{A\overline{K}^\alpha}{\overline{B}}, \tag{11.63}$$

$$\eta(\overline{D})\overline{P}^\theta = \frac{\overline{B}^2}{\gamma A\overline{K}^\alpha\overline{C}}(\delta_P + r), \quad -\eta'(\overline{D})\frac{\overline{P}^{\theta+1}}{\theta+1} = \mu + r\overline{C} \tag{11.64}$$

with respect to a steady state $(\overline{K}, \overline{C}, \overline{B}, \overline{P}, \overline{D})$. The system (11.63)–(11.64) can be reduced to two nonlinear equations with respect to $\overline{K}$ and $\overline{D}$:

$$\frac{\overline{K}^\alpha\left(A - \overline{K}^{1-\alpha}(\mu+r)/\alpha\right)^{2+\theta}}{\overline{K}(\mu+r)/\alpha - \mu\left(\overline{K} + \overline{D}\right)} = \frac{\gamma^{1+\theta}A^{1+\theta}}{(\delta_P + r)\delta_P{}^\theta}\left[\underline{\eta} + \left(\overline{\eta} - \underline{\eta}\right)e^{-a\overline{D}}\right], \tag{11.65}$$

$$a\left(\overline{\eta} - \underline{\eta}\right)e^{-a\overline{D}}\frac{\gamma^{\theta+1}A^{\theta+1}}{\delta_P{}^{\theta+1}(\theta+1)} = \frac{r\left(A - \overline{K}^{1-\alpha}(\mu+r)/\alpha\right)^{\theta+1}}{\overline{K}(\mu+r)/\alpha - \mu\left(\overline{K} + \overline{D}\right)}. \tag{11.66}$$

The other three steady-state components $\overline{C}, \overline{B}, \overline{P}$ are expressed in the terms of $\overline{K}$ and $\overline{D}$ as

$$\overline{B} = A\overline{K}^\alpha - \overline{K}(\mu+r)/\alpha, \tag{11.67}$$

$$\overline{C} = \overline{K}(\mu + r)/\alpha - \mu(\overline{K} + \overline{D}), \tag{11.68}$$

$$\overline{P} = \frac{\gamma A}{\delta_P\left(A - \overline{K}^{1-\alpha}(\mu + r)/\alpha\right)}. \tag{11.69}$$

We cannot analytically prove the existence of a solution $(\overline{K}, \overline{D})$ to the nonlinear equations (11.65)–(11.66). Even the steady-state analysis appears to be challenging for growth models with several sectors. To keep the analysis tractable, we should make some simplifying assumptions. In particular, the existence of a unique steady state in the model (11.45)–(11.50) without capital depreciation, i.e., at $\mu = 0$, is stated in the following proposition from [3].

Statement 11.2 The optimization problem (11.45)–(11.50) at $\mu = 0$ possesses a unique steady state $(\overline{K}, \overline{C}, \overline{B}, \overline{P}, \overline{D})$ with $0 < \overline{K} < (\alpha A/r)^{\frac{1}{1-\alpha}}$, and $\overline{D} \geq 0$. If $\overline{K}$ lies within an interval $[0, \overline{K}_c]$, then the optimal $\overline{D} = 0$. The critical value $\overline{K}_c$ is the solution of nonlinear equation (11.65) at $\overline{D} = 0$. Otherwise, the unique $\overline{K} > 0$ and $\overline{D} > 0$ are determined from the system of two equations (11.65) and (11.66).

11.4.3 Discussion of Results

An analysis of the steady state in the model (11.45)–(11.50) demonstrates how the presence of the adaptation as a new policy instrument changes the qualitative results obtained in previous sections.

- The relevance of the adaptation depends on the stage of development of a specific country.
- It is not optimal for a country with low global productivity and low capital to engage itself in adaptation. Then, the corresponding optimal capital $\overline{K}$ is determined from (11.29) in the model without adaptation of Sect. 11.3.
- The optimal adaptation is positive starting with a certain positive level of capital. Proposition 2 provides the range for a positive adaptation in terms of the endogenous variable $\overline{K}$ but says nothing about the dynamics of adaptation.

To analyze specific relations between the optimal long-term investments into emission abatement and environmental adaptation, numeric simulation can be used, which is a common technique in economic–environmental modeling.

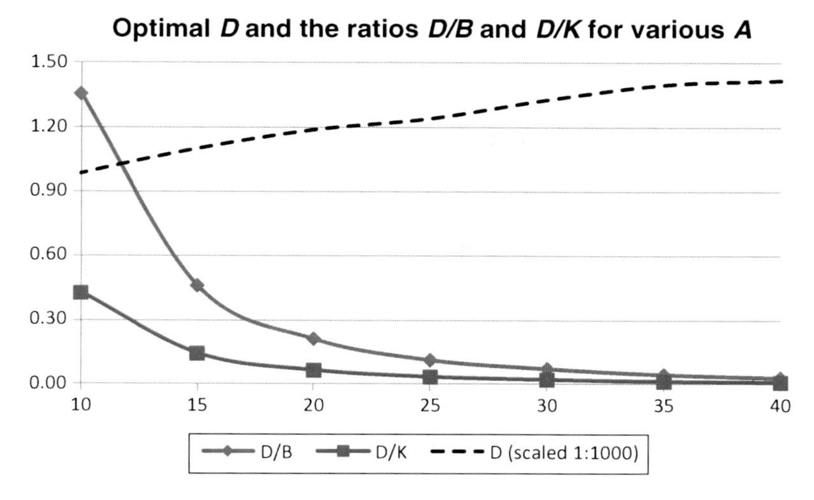

Fig. 11.4 The simulated worldwide optimal adaptation investment D (in billion US dollars), the adaptation–abatement ratio D/B, and the adaptation–capital ratio D/K for various values of global productivity A (in billion US dollars)

Numeric Simulation. Using model calibration parameters from [3], the numeric solution of (11.65)–(11.66) demonstrates a steady decrease of the optimal ratios D/B and D/K between the long-term steady-state adaptation D, abatement B, and capital K (see Fig. 11.4). Specifically, the optimal adaptation–abatement ratio D/B between changes from 1.36 at $A = 10$ to 0.03 at $A = 40$. At the same time, the adaptation D in absolute units increases for larger A (but B and K increase faster). The optimal abatement effort, expressed as the ratio B/K, is 0.60 in the absence of adaptation, and it drops to 0.31 with adaptation. It leads to the following qualitative conclusions:

- The abatement and adaptation are substitutes: a positive adaptation reduces necessary emission abatement efforts.
- The optimal policy mix between adaptation and mitigation (abatement) is lower for countries with higher economic efficiency.
- The optimal adaptation investment is larger in absolute units for a country with larger productivity A (but the capital level K increases faster).

11.4.3.1 A Look Ahead

The model (11.45)–(11.50) with adaptation and abatement may be modified in several directions. Adding a population growth will not change the structure of obtained results. Then, the endogenous variables I_K, I_D, C, K, D, B, P can be expressed *per capita* and the actual variables will grow with the population growth rate. The introduction of exogenous or endogenous technological change (Chap. 3) and learning (Chap. 5) to the production process would alter the results significantly and make them more optimistic.

The models of Sect. 11.2–11.4 describe a closed (one-country) economy. An essential extension is to consider the economy with several countries. The environment is a global public good, which is vital for all countries. It is important to understand how the optimal mitigation/adaptation policy of a country depends on its position in the international area. Extending the model (11.45)–(11.50) to an n-country model with strategic behaviors involves the game theory.

Exercises

1. Show that the utility function (11.6) describes diminishing marginal utility of consumption C.
2. Show that the utility function (11.6) describes increasing marginal disutility of pollution P.
3. Prove that the adaptation efficiency function (11.9) possesses the properties (i)–(iii) of Sect. 11.1.4.
4. Prove that the environmental vulnerability function (11.10) possesses the properties (i)–(iii) of Sect. 11.1.4.
5. Sketch the function (11.9) and compare it with the graph of (11.10) in Fig. 11.1. Which function is more general? Find specific values of the parameters of (11.10) when these two functions have a similar qualitative behavior.
6. Reduce the system of five nonlinear equations (11.63)–(11.64) to two nonlinear equations (11.65) and (11.66) with respect to $\overline{K}$ and $\overline{D}$.
7. Following Sect. 2.4, verify that the present-value Hamiltonian for the optimization problem (11.11)–(11.14) is given by the formula (11.15).
8. Following Sect. 2.4, verify that the first-order extremum conditions for the problem (11.11)–(11.14) are given by (11.19)–(11.20).
9. Using the present-value Hamiltonian (11.35) of the optimization problem (11.31)–(11.34) and corresponding first-order extremum conditions, obtain the nonlinear differential equations (11.36)–(11.39) for the optimal interior trajectories K, B, C, and P.
10. Obtain the formulas (11.40)–(11.43) for the steady state $(\overline{K}, \overline{B}, \overline{C}, \overline{P})$ of the optimization problem (11.31)–(11.34) from the nonlinear differential equations (11.36)–(11.39) for interior trajectories.

References[1]

1. 📖Ackerman, F., Stanton, E.: Climate economics: The state of the art. Routledge Studies in Ecological Economics. Routledge, London, New York (2013)
2. Agrawala, S., Bosello, F., de Carraro, C., Cian, E., Lanzi, E.: Adapting to climate change: Costs, benefits, and modelling approaches. Int. Rev. Environ. Resource Econ. **5**, 245–284 (2011)
3. 📖Bréchet, T., Hritonenko, N., Yatsenko, Y.: Adaptation and mitigation in long-term climate policy. Environ. Resource Econ. **55**, 217–243 (2013)
4. Byrne, M.: Is growth a dirty word? Pollution, abatement and economic growth. J. Dev. Econ. **54**, 261–284 (1997)
5. Chimeli, A.B., Braden, J.B.: Total factor productivity and the environmental Kuznets curve. J. Environ. Econ. Manage. **49**, 366–380 (2005)
6. Gradus, R., Smulders, S.: The trade-off between environmental care and long-term growth-pollution in three prototype models. J. Econ. **58**, 25–51 (1993)
7. Hritonenko, N., Yatsenko, Y.: Modeling of environmental adaptation: amenity vs. productivity and modernization. Clim. Change. Econ. **4**:2, 1–24 (2013)
8. Intergovernmental Panel on Climate Change (IPCC): Climate change 2007, fourth assessment report. Cambridge University Press, Cambridge, New York (2007)
9. Jørgensen, S., Martín-Herrán, G., Zaccour, G.: Dynamic games in the economics and management of pollution. Environ. Model. Assess. **15**, 433–467 (2010)
10. 📖McGuffie, K., Henderson-Sellers, A.: A climate modelling primer, 3rd edn. Wiley, New York (2005)
11. Smulders, S., Gradus, R.: Pollution abatement and long-term growth. Eur. J. Polit. Econ. **12**, 505–532 (1996)
12. Stokey, N.: Are there limits to growth? Int. Econ. Rev. **39**, 1–31 (1998)
13. Tahvonen, O., Kuuluvainen, J.: Economic growth, pollution and renewable resources. J. Environ. Econ. Manage. **25**, 101–118 (1993)

[1] The book symbol 📖 means that the reference is a textbook recommended for further student reading.

Chapter 12
Models of Global Dynamics: From Club of Rome to Integrated Assessment

Human activities can cause negative global changes in the environment, such as global warming, excessive pollution of the environment, extinction of some species, degradation of entire geographic areas, and others. Section 12.1 discusses the modeling of global changes as a scientific problem of great complexity and importance. Section 12.2 describes the first global models—models of world dynamics developed by J. W. Forrester, D. H. Meadows, and M. D. Mesarovic in the 1970s. They represented first systematic attempts to analyze the global dynamics and produced outcomes that attracted significant public attention. These models evolved in the 1990s into a more quantitative approach known as integrated assessment models, which are explored in Sect. 12.3.

12.1 Global Trends and Their Modeling

In a narrow sense, the term *global change* is often used to refer to the *climate change* and related environmental issues, such as stratospheric ozone depletion or acid rains, explored in Chaps. 8 and 11. However, the principal meaning of global change is much more than the climate change. It reflects possible various changes in the environment, society, and economics, and their impacts on each other [9]. Human activities have caused many negative global-scale changes in the environment such as atmospheric and water pollution, extinction of several species, degradation of whole geographic areas, and others. Positive human impact on the environmental improvement and protection is rather limited. Reduced availability of clean water, fertile soils, and clean air may harm human populations as well as the well-being of other species.

Governmental and nongovernmental institutions, academia and private sectors are working intensively to identify and analyze major global trends that will shape the future of the world. The *global trends* provide a flexible framework to discuss

N. Hritonenko and Y. Yatsenko, *Mathematical Modeling in Economics, Ecology and the Environment*, Springer Optimization and Its Applications 88,
DOI 10.1007/978-1-4614-9311-2_12, © Springer Science+Business Media New York 2013

and debate the future of global change. Some of these trends will persist, others will become less important and change over time, and new global trends will appear. For example, globalization and technology have recently emerged as powerful components of the world development. A variety of theories and methods are used to analyze global trends. Extrapolation of existing trends and available statistics help to explore demographic and natural resource trends. Analytic mathematical models are frequently used to forecast economic and population growth, the development of science and technology, and conflicting trends in the environment.

12.1.1 Global Environmental Trends

Depending on goals and objectives, the global trends can be classified in several categories such as the following:

- Natural resources including the environment, food, water, and energy
- Climate change
- Global economy
- Demography, population health and diseases
- Science and technology
- Communication and transportation
- National and international governance

Examples of current global trends in the environment include the following:

- Average atmospheric concentrations of carbon dioxide CO_2 have climbed 20 % since its measurements began in 1959. CO_2 emissions from fossil fuels increased a record 4.5 % in 2004.
- The concentrations of several important greenhouse gases, such as CO_2, N_2O, and CH_4, have substantially increased in the environment. More nitrogen is now fixed into available forms through the production of fertilizers and burning of fossil fuels than is fixed naturally in all terrestrial systems.
- More than half of all accessible freshwater is used for human purposes.
- Almost 40 % of known oil reserves have already exhausted in the last 150 years. It took the nature hundred million years to generate this amount.
- Nearly 50 % of the land surface has been transformed by direct human actions, with significant consequences for biodiversity, nutrient cycling, and climate.
- Up to 50 % of fish stocks are fully exploited, 15–18 % are overexploited, and about 10 % have been depleted.

Specific environmental trends related to climate change are explored in detail in Chap. 11.

12.1.2 Global Demographic Trends

Populations in developed and developing countries are changing dramatically. These changes can radically affect future development of the world economy as well as local and global environments.

The average fertility rate in the developed world is 1.5 children per one woman and none of the developed countries has the replacement-level fertility. The fertility rate in the developing world has declined in the past 50 years from 6.2 to 3.3 children per one woman. However, it has increased in some countries, for instance, from 7.7 to 8 children in Niger. Even in countries with a decreasing fertility rate, the population continues to rise due to an increasing number of young people. This phenomenon is known as the *population momentum*. If the current fertility rates do not change, then the world population will increase and up to 12.8 billion people, including 11.6 billion in the developing world by 2050.

The life expectancy has increased all over the world: compare 25 and 60 years for babies born in India in 1881 and 2006. However, it is still different in rich and poor countries, e.g., 84 versus 46 years for females in Japan and Kenya. The current life expectancy in the developing world is 63 years. Almost 18 % of the developed-world population is 60 years or older compared to 6 % in 1900. The US Census Bureau predicts that the number will increase up to 28 % by 2050. The number of productive workers in developed countries is approaching the number of their dependants (retirees and youngsters) who are nonproductive consumers. This phenomenon in population age distribution is known as *squaring of demographic pyramid*. The globalization of the world economy increases international competition. To survive and prosper, developed countries have to innovate and provide a higher labor productivity using new efficient technologies (see Sect. 3.1).

Currently, 40 % of people in developing countries and 76 % in developed countries live in urban areas. By 2030, 60 % in the developing world is projected to live in cities. The urbanization trend and related changes in consumption patterns may also lead to new environmental problems.

12.1.3 Population and Environment

The global demographic trends create new challenges for the environment. High levels of consumption from freshwater to fossil fuels may damage the environment, put pressure on natural resources, and increase pollution. An impact on the environment varies from one region to another. In rich countries, transport emissions are growing most rapidly and consuming more energy. In developing countries, poor rural families engage in slash-and-burn agriculture and pollute local water and other resources.

Human population growth correlates with major environmental changes on the global long-time scale. There are basic human needs for food, water, shelter, community health, and employment that are related to the population size. The way in which society meets these needs determines environmental consequences at all scales. The pressure humans put on the environment depends on population, consumption, and technology. Both population and consumption will almost certainly rise. Hence, the development of new, environment friendly and more efficient technologies will have a greater importance than it has been assumed [2].

A new trend known as the environmental Kuznets curve states that, as nations industrialize, pollution levels initially rise, but then peak and decline (see Sect. 12.1.5). As a country gets richer, the middle class develops and demands a cleaner environment, which causes the replacement of older equipment and production methods with more expensive but environmentally cleaner ones.

Advances in science and technology may generate major breakthroughs in agriculture and health, protection of the environment and quality of life. On the other side, it is unclear whether the technology can compensate the environmental damages it causes. Some global trends in science and technology are discussed in Chaps. 3–5.

12.1.4 Modeling of Global Change

Mathematical modeling and computer simulation are essential in making forecasts, preventing possible negative impacts, studying interrelation and interaction among various ecological, economic, demographic, social, and other phenomena [1]. The global modeling involves three components:

- *modeling of human activity* (including economy, technology, social, and demographic processes);
- *modeling of biota* (biological communities);
- *modeling of inorganic nature* (the natural environment, excluding biota).

Modeling of human activity is the subject of mathematical economics (Chaps. 2–5) and social sciences. The influence of economic activities on the environment is considered by environmental economics (Chaps. 10 and 11). Mathematical models of biota have been successfully developed starting with the investigations of Malthus, Verhulst, Lotka, and Volterra during nineteenth and early twentieth centuries (see Chaps. 6 and 7). Modeling of inorganic nature is often identified with modeling of the environment as a whole. This area is usually separated into modeling of different components of the environment (atmosphere, water, soil, etc.). An important specific problem is modeling of pollution propagation, which is explored in Chaps. 8 and 9.

Historically, the first mathematical model of a global change was suggested by R. Malthus in "An Essay on the Principle of Population" in 1798, see Sect. 6.1.1. Following this model, the human population size would double every 25 years, and the food production as a land-limited resource could not possibly be increased fast enough to keep pace with the growing population. Fortunately, Malthus' predictions did not come true for various reasons, such as a quick increase in agricultural productivity and a smaller than projected birth rate.

An integrated modeling of the environmental, biologic, economic, and social components of a global system is a scientific problem of a great complexity. Many attempts to estimate the global development have failed. Nevertheless, large scientific projects on global modeling have been implemented since 1960s. The corresponding models are known as *global models*. Modern global models take into account delays and influence of different exogenous and endogenous factors. They use various mathematical tools, such as deterministic and stochastic models, differential, integral and difference equations, optimization and computer simulation, and so on. Different global modeling schools have developed common concepts along with their own theories and mathematical techniques for constructing and testing models. The system dynamics and complex adaptive methods are among the major techniques. There are thousands of papers on global modeling. In Sects. 12.2 and 12.3, we provide a brief description of two major generations of the global models.

12.1.5 Simplified Models of Human–Environmental Interaction

Ecological footprint is a semiquantitative aggregated indicator that measures the impact of a specific human population on the natural world. The ecological footprint is defined as the land area necessary to provide resources (grain, feed, wood, fish, and urban land) to and absorb emissions (carbon dioxide) of a given population (country, region, or the entire world). Its concept was developed by M. Wackernagel in the 1990s and has been calculated for many countries since 1961. The ecological footprint data are published biannually by the World Wide Fund for Nature.

IPAT identity: A highly simplified model of human impact on the environment is suggested by P. Ehrlich and J. Holdren:

$$I = P \times A \times T,$$

where the environmental impact I is the product of the population P, the affluence A, and the technology T (see Sect. 12.1.3).

Although this relation is not applicable to small space and time scales, it provides some practical insights on the role of human activities in large scales. Indeed, it is estimated that the population growth has accounted for 38 % of the emissions from the developed world and 22 % in the less developed world. The affluence A reflects the human wealth and material comfort, which generates higher levels of consumption. Thus, people in the developed world consume much more material and energy during their lives than do those in the less developed world. On average, developed nations consume twice as much grain, three times as much meat, nine times as much paper, and 11 times as much gasoline as the rest of the world. The environmental impact I correlates with the technology T, which increases the production efficiency and can reduce waste per output unit. The latter effect is usually associated with the environmental Kuznets curve.

The environmental Kuznets curve is an empirical relation between pollution and income suggested by S. Kuznets in 1950s. It states that the pollution per capita first rises when per capita income increases, then declines when the income exceeds a certain threshold (a *bell-shaped* curve). Experimental observations confirm that the Kuznets curve is valid for water pollutions and some air pollutions (sulfur dioxide, nitrogen oxides, and deforestation). Municipal waste, CO_2 emissions, and aggregate energy consumption do not follow this curve.

12.1.6 Aggregate Indicators in Global Models

In the economic–environmental modeling, it is necessary to measure the mutual influence of the environment and human society. The choice and formalization of operational parameters for this purpose are not easy. General formulations like "creation of favorable living conditions," "achieving high quality of human life," and "rational harmonious development" do not give enough information to construct formal mathematical models. More specific and practical approaches are based on achieving prescribed sanitary standards for water and air quality, radiation level, and others. However, the experience shows that such standards often do not match the required "life quality"; hence, they have to be subjected to a careful preliminary analysis.

Mathematical models need quantitative indicators of human welfare. The *gross domestic product (GDP) per capita* has been often used as such index, but it does not reflect human health. The average human life expectancy can be used, as a surrogate, to measure the quality of human life. It takes into account an aggregated influence of the environment on population health and combines qualitatively different environmental, economic, and social factors.

The United Nations Development Program (UNDP) has used the *Human Development Index (HDI)* as an aggregated measure of country's achievements in three basic dimensions of human development:

- a long and healthy life measured by the *life expectancy index*,
- knowledge presented by the *education index*, and
- living standard measured by the GDP per capita.

The HDI was developed by a Pakistani economist Mahbub-ul-Haq in 1990 and adopted by the UNDP in 1993. It has been used for a number of years for most countries and is listed annually in the Human Development Report. An HDI of 0.8 or more is considered to represent a high development. The report for 2005 showed that the HDI was improving for all countries, except for some post-Soviet states and

Sub-Saharan Africa. In 2005, Norway, Iceland, and Australia were among the three top-rated countries, while Niger and Sierra Leone were the bottom-rated countries with an HDI of 0.3 or less.

The "*quality-of-life index*" used by the global model "World 2" (Sect. 3.1) summarizes the impact of four factors: crowding, food, pollution, and material consumption. This choice was widely criticized, so the modified model "World 3" (Sect. 3.2) uses the more standard *human welfare index (HWI)* that approximates the HDI. A more recent *quality-of-life index* considers *healthiness* (determined by the life expectancy) as the first of their nine major indicators, followed by the *family life* and the *community life*, while the economic prosperity (*material well-being*) is the fourth indicator.

The next two sections discuss more complete and complex global models (world dynamics models and integrated assessment models) that involve various economic, environmental, and social factors.

12.2 Models of World Dynamics

The *models of world dynamics* were first developed by J. W. Forrester and D. H. Meadows in the 1970s as reports to *the Club of Rome*, an informal international organization of scientists concerned about the future development of human society.

The first model, known as the *World 2 computer model* (Sect. 12.2.1), was described by Forrester in "*World Dynamics*" [3] in 1971 (the first prototype *World 1* has never been published). The model was based on the *system dynamics method* developed by Forrester in 1961 for the analysis and design of industrial systems.

The modified *World 3 computer model* (Sect. 12.2.2) was developed by D. H. Meadows and his group and described in "*Limits to Growth*" in 1972. It kept the basic methodology unchanged but extended the quantitative data of the World 2 model and elaborated its structure. "Limits to Growth" was published in 30 languages and became a best seller in several countries. Two simulation updates of World 3 were published in "*Beyond the Limits*" in 1992 and in "*Limits to Growth: The 30-year Update*" [7] in 2004.

In the following 30 years, numerous global models were built in the tradition of system dynamics. The most significant global modeling projects of the 1970s and 1980s were "*Mankind at the turning point*" by M. D. Mesarovic and E. Pestel (see Sect. 12.2.3), "*The Future of the World Economy*" by W. Leontief, "*Reshaping the International Order*" by J. Tinbergen, and the *Bariloche model* by A. Herrero.

12.2.1 Forrester Model

Forrester applied the *method of system dynamics* to study the behavior of a complicated structure of interconnected variables. This method uses combinations of interacting feedback loops, nonlinear equations, and time delays to describe information flows through the system. The models may include highly aggregated parameters and variables, for which data is not available. The major focus is on the model structure, while less attention is paid to the estimation of parameters. A critical role is assigned to *feedbacks* (links) between interacting subsystems. The feedbacks can be *positive* or *negative*, for example:

- Population size ⇒ labor resource ⇒ product output ⇒ food consumption ⇒ birth rate ⇒ population growth (a *positive feedback*).
- Population size ⇒ environment contamination ⇒ death rate increase ⇒ population decrease (a *negative feedback*).

12.2.1.1 Structure of the Forrester Model "World 2"

The Forrester global model World 2 considers *five sectors* of the world system: population, industry, natural resources, agriculture, environment (pollution). These sectors are connected by multiple positive and negative feedbacks. The structure of the model is shown in Fig. 12.1. The model is a combination of separate modules from several modeling areas (economics, ecology, and others). The basic structure of this model is described by the following equations:

$$w = F_1(\kappa, K, R, N), \tag{12.1}$$

$$C = F_2(\kappa K, N, P), \tag{12.2}$$

$$dN/dt = [\eta_1(P, w, C, N) - \eta_2(P, w, C, N)]N, \tag{12.3}$$

$$\mathrm{d}K/\mathrm{d}t = a(w)N - \mu K, \quad \mu = \text{const}, \tag{12.4}$$

$$\mathrm{d}R/\mathrm{d}t = -r(w)N, \tag{12.5}$$

$$\mathrm{d}P/\mathrm{d}t = d(K, N) - \gamma(P)P, \tag{12.6}$$

$$\mathrm{d}\kappa/\mathrm{d}t = F_3(w, C)N, \tag{12.7}$$

where the major dynamic variables are

N—the human population size,
W—the quality-of-life index (see Sect. 12.1.5),
C—the output of consumption good (food production),
K—the capital amount,

R—the amount of natural resources,
P—the aggregated stock of the environmental pollution.

The relative part κ of capital used in agriculture, the birth and death rates η_1 and η_2, the accumulation of capital coefficient a, the resource consumption coefficient r, and the pollution coefficient d, the pollution deterioration rate γ reflect nonlinear links in the world system.

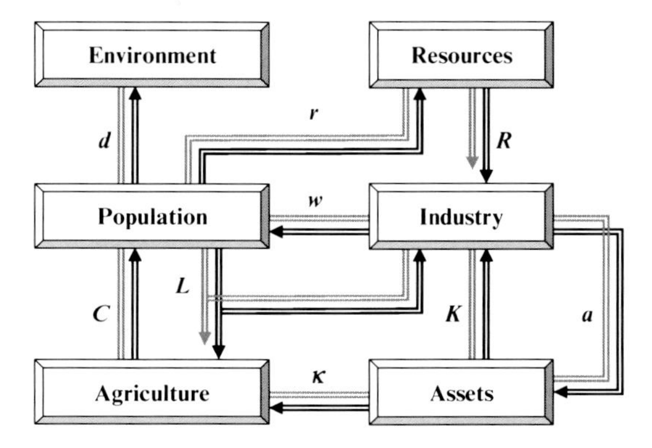

Fig. 12.1 The diagram of the Forrester global model "World 2"

It is easily to see that the Forrester model (12.1)–(12.7) is in fact a combination of the model components considered in Chaps. 2, 3, 6, 10 or their modifications. However, in contrast to the mentioned models, the coefficients $\kappa, \eta_1, \eta_2, a, r, d, \gamma$ in the model (12.1)–(12.7) are not constant and depend on other endogenous model variables.

12.2.1.2 Simulation Results

The model was used to simulate five global trends of the world development:

- an accelerated rate of industrialization,
- a fast growth of the human population,
- the extension of a food shortage zone,
- the exhaustion of nonrenewable resources,
- the environment contamination.

The simulation outcomes predicted major global trends, among them:

- the exhaustion of natural resources in the first half of the twenty-first century may lead to deceleration of industrial and agricultural growths, decline in the population size, and a global ecological catastrophe;
- in the case of possible discovery of unlimited resources, the catastrophe may occur due to excessive pollution of the environment;

- if the human society enforces effective conservation of nature, then the growth of the human population and industrial and agricultural production may continue until the land exhaustion occurs that may cause a new catastrophe.

The recommendations suggested a quick stabilization of the human population size, production and consumption; reduction of environmental pollution (four times smaller) and resource consumption (eight times smaller); development of only environment friendly activities (such as art, science, education, sport, etc.) that will not lead to high consumption of nonrenewable resources and environment degradation.

12.2.1.3 Limitations of the Forrester Model

First models of world dynamics attracted enormous scientific attention. Major critical comments on the Forrester model were:

- Too high aggregation level of model variables: the average growth rates of global population instead of in separate regions, the global environmental pollution instead of specific indices in different regions, and so on.
- The model appears to be very sensitive to variations of parameters; the initial conditions greatly influence simulation results.
- Technological change (Chap. 3) is not represented in the model.
- Equations of the model are reversible while the real world processes are irreversible.
- The model does not properly reflect human adaptation mechanisms in a case of undesirable global development; a social feedback is absent.

Modifications of the Forrester model were directed toward overcoming the mentioned drawbacks.

12.2.2 Meadows Models

The *Meadows model World 3* retains general features of the World 2 model such as the absence of regional division and the same sectors of the world system. At the same time, it considers more factors and interrelations in the world along with time delays and smoothing effects of some factors. For example, it takes into account the age-dependant birth and mortality rates, delays between the pollution emission and its negative impact of the environment, and so on.

12.2.2.1 Meadows Model "World 3"

The *World 3 model* describes interconnected dynamics of several key global variables such as the population size, industrial capital stock, persistent pollution stock, and the size of cultivated land. The dynamics of these variables is determined by such processes as births and deaths in the world population, investment and depreciation of the capital stock, pollution emission and assimilation, the land erosion, land development, and land removed for urban and industrial uses. The World 3 focuses on time delays in these processes and includes many feedback loops. For example, changes in population cause changes in the economy and, at the same time, the economic output affects birth and death rates.

The World 3 relationships are nonlinear. For instance, the influence of food per capita nonlinearly affects the human life expectancy, or the amount of land cultivated multiplied by the average land yield gives the total food production. The food production divided by the population size gives the food per capita. If the food per capita falls below a critical threshold, the death rate goes up. On the other side, if inadequately nourished people get more food, then their life expectancy increases. The increase of the daily food consumption from 1,000 to 2,000 Cal per person may increase life expectancy from 40 to 60 years. However, the increase of food consumption over 4,000 Cal provides a little gain in the life expectancy and a further increase starts to decrease it.

Delays and nonlinear feedback loops made the World 3 model dynamically advanced and complex. However, the model had many limitations:

- The model is a great simplification of the reality. The pollution description is very simplistic.
- The model does not distinguish different regions of the world. In reality, there are rich and poor nations, food shortage happens mostly in Africa, pollution contamination affects Europe, North America, and Asia, the land degradation occurs in the tropics.
- The model assumes a perfect market, successful technologies, and correct political decisions made without cost and delays.
- The model does not include a military sector that drains capital and resources from the productive economy. It has no wars that kill people, destroy capital, waste land, or generate pollution.
- The model has no corruption, floods, earthquakes, nuclear accidents, strikes, AIDS epidemics, or other failures.

12.2.2.2 "Limits to Growth"

The well-known "*Limits to Growth*" (1972) analyzed 12 scenarios from the World 3 model that show different global trends of world development from 1900 to 2100. The scenarios illustrated how population growth and natural resources interact with a variety of external constraints and limits. On a qualitative level, the simulation results remained the same as in the Forrester model. Namely, "Limits to Growth" reports that global ecological constraints (related to resource depletion and pollution emissions) would have significant influence on global developments in the twenty-first century. It warns that the humanity might have to divert much capital and manpower to battle these constraints; possibly so much that the average quality of life would decline. The simulated growth does not necessarily lead to a collapse of human society. A collapse follows growth only if it has led to an expansion in demands on the planet resources above sustainable levels. The collapse can be prevented by long-term world-scale planning and appropriate actions. The model calls for a proactive technological, cultural, and institutional changes in order to avoid an increase in the ecological footprint (see Sect. 12.1) of humanity beyond the carrying capacity of the Earth.

12.2.2.3 "Beyond the Limits"

D. H. Meadows and coauthors studied global developments between 1970 and 1990 and used this information to update the World 3 computer model of 1972. Their next book "*Beyond the Limits*" (1992) incorporated over 120 interdependent variables and presented 14 scenarios. The scenarios mainly supported the conclusions produced 20 years earlier.

A major new finding of the "Beyond the Limits" was that the limits of the Earth's capacity had already been "overshot". This conclusion was supported in the early 1990s by the growing evidence that humanity was moving further into unsustainable territory. For example, the rain forests were being cut at unsustainable rates, the global warming started, and the stratospheric ozone hole appeared. There were also concerns that the grain production could no longer keep up with the population growth.

12.2.2.4 "Limits to Growth: The 30-Year Update"

The 30-year update [7] of D.H.Meadows and his group work was published in 2004. Eleven alternative scenarios of the global trends until 2100 were evaluated on the basis of an updated World 3 model. The main conclusion is that the planet population, capital, pollution, and resource extraction grow exponentially. An exponentially growth rapidly exceeds any fixed limit. Technology and markets as the most powerful tools of the society may increase the economy's tendency to overshoot. If the primary goal was growth, then these tools would produce growth; however, if the primary goals were equity and sustainability, then they could serve those goals as well.

Once the population and economy have overshot physical limits of the Earth, there are only two ways back: an involuntary collapse caused by escalating shortages and crises, or a controlled reduction of the ecological footprint. Technological improvements combined with proper social policies can limit the growth and help to avoid the collapse.

12.2.3 Mesarovic–Pestel Model

The Mesarovic–Pestel world model is based on principles of the theory of hierarchical systems. In contrast to the previous models, the world is separated into ten regions according to their technical–economic and social–cultural characteristics. The regions interacted through export–import operations and population migration. Each region is divided into *five* interacting *hierarchical strata*:

- *ecological strata* that includes the natural environment;
- *technological strata* that includes technologies and their environment impact;
- *economic strata* that includes all economic activities;
- *social–political strata* that includes various organizations, governments, and informal organizations such as religious movements, parties, and so on;
- *individual strata* that describes human physical and psychological conditions.

The investigation technique uses a scenario-based analysis of the world development. The obtained major recommendation is a fast stabilization of population growth in critical regions. The modeled global trends include:

- population stabilization after 50 years would lead to the food crisis, whereas the stabilization during 25–30 years would help to avoid the crisis;
- possibility of several regional crises instead of the global one;
- decreasing danger of ecological crisis under a balanced development of the world system;
- preference for regional interactions.

12.2.4 Limitations of World Dynamics Models

The first models of world dynamics were incomplete and quite controversial. They underestimated the complexity of the global system and the capability of humans to adapt to changes. The model-based experiments concluded that the ongoing exhaustion of natural resources would result in a collapse of the world economy. However, the oil crises of the 1970s led to the intensification of exploration efforts and, subsequently, to the discovery and exploitation of additional reserves of resources. These crises also induced new investments in energy efficiency and renewable energy sources. The simulations made in 1971 included neither oil crises nor the responses they stimulated. In spite of their limitations, the world dynamics models represent the first systematic scientific attempt to analyze connections among global trends. Although no government has followed their recommendations and no serious study has been conducted to validate the modeling forecasts, their theoretical and methodological importance cannot be eliminated. Wide publicizing the outcomes of world dynamics models essentially contributed towards creating the general willingness of governments to consider environmental issues.

In the 1990s, because of criticisms about model validation and completeness, the world dynamics models gradually evolved into a more quantitative, integrated assessment models [5], which have become a mainstream in global modeling. Such models reflect interactions among various processes in the Earth global system paying a specific attention to model validation and identification of parameters.

12.3 Integrated Assessment Models: Structure and Results

The *integrated assessment* (IA) is a multidisciplinary approach to global modeling. It describes various relationships and interactions among global components (the environment, human society, economics, and others). The IA models elaborate on the Club of Rome traditions (Sect. 12.2) and are designed to analyze global changes and provide related policy recommendations.

The construction of a complete integrated model of the Earth's atmosphere, hydrosphere, and biosphere is impossible conceptually and technically. The IA methodology uses a *hierarchic structure* to overcome the complexity of global modeling. A set of reduced models (modeling modules) is used for each IA component. The modules are then linked to one another. The simplified nature of modules permits rapid prototyping of new concepts and exploration of their implications. At the lower level of aggregation, economy–energy models operate with regional political boundaries in multiyear time steps. These data are used in theme-specific models like *RAINS (Regional Acidification Information and Simulation)* or *IMAGE* (Sect. 12.3.3) at the next level of aggregation. Mega-models are set up on a higher level. The *TARGETS* (Sect. 12.3.3) is among the highest level of aggregation. The IA models provide a flexible framework of simulation tools for forecasting global trends and making policy decisions. An excellent overview of integral assessment models can be found in [5].

The IA models involve different levels of aggregation and integration. They can be global (DICE), consider several regions (AIM, Global 2100, MERGE), or reflect the impact of just one country (Global Trends 2015, Gemini, IMAGE). They can be deterministic (CETA, TARGETS, IMAGE) or probabilistic (PAGE, ICAM). Two key IA techniques are the *scenario analysis* (simulation) and *optimization*. Scenario analyses provide simple trend extrapolation without any human response. PAGE, IMAGE, TARGETS are of this type. Optimization is based on common economic assumptions that rationally acting agents have perfect knowledge about the system in question and are able to determine the optimal strategy for some future period (see Chaps. 2–4). This approach delivers valuable insights into efficient strategies but is limited by model assumptions. The CETA, DICE, MERGE, Global 2100 models (Sect. 12.3.2) are of the optimization type.

The IA models started with the focus on European acid rains in RAINS model developed by Alcamo and his team in 1990. Another set of IA models focuses on ecological economic developments on a regional scale. Among them are ISLAND developed by Engelen and his group in 1995, QUEST proposed by Biggs and his colleagues in 1996, Threshold 21 suggested by Millennium Institute in 1995, and others. The global climate change is considered by IMAGE, DICE, PAGE, ICAM, and others. Some of the most recognized IA models are discussed below.

12.3.1 Deterministic Models of Climate and Economy (DICE, RICE, WITCH)

Dynamic Integrated Model of Climate and the Economy (DICE) is an extended version of a traditional optimal growth model (Chaps. 2 and 3) with additional climate sector (Chap. 11), in which a single world producer–consumer chooses optimal levels of current consumption, investment in productive capital, and investment in the reduction of pollution emissions. The DICE model was developed by W. Nordhaus in 1994 and updated in 2007 [8]. In this model, the net output Y_N of the aggregate world product is determined as

$$Y_N(t) = Y_G(t)/(1 + GD(T)), \tag{12.8}$$

where the gross (without climate damage) product output Y_G is described by a standard Cobb–Douglas production function (Sect. 2.2) and the *damage function*

$$GD(T) = \alpha_0 T + \alpha_1 T^{\alpha}, \quad \alpha > 1, \tag{12.9}$$

that nonlinearly depends on the increase T of the average global temperature (compared to the 1900 level). The estimation and calibration of the damage function (12.9) is a crucial component of the model. The population growth and technological change are assumed to decline asymptotically to zero, which leads to stabilized population and productivity. Pollution emissions per unit output deteriorate exogenously at a fixed rate and can be further reduced by costly emission control measures. The objective is to maximize the discounted value of the logarithmic utility of per capita consumption.

Regional Integrated Model of Climate and the Economy (RICE) developed by Nordhaus and Yang is a regional version of the DICE model. In RICE, the world economy is disaggregated into *thirteen regions* with different initial capital stock, population, and technology. As in DICE, the production side of the economy is highly aggregated. Each region has its own central planner and produces one final product using capital, labor, and energy as inputs. Capital and labor are aggregated by a Cobb–Douglas production function, which is further aggregated with energy as CES technology, see (12.10) below. The final product is used for consumption and investments. The optimization horizon is until 2200. The capital market equalizes the real interest rate across regions. The RICE model can describe different strategies of different counties using three distinct approaches: uncontrolled market, cooperation, and nationalistic policies.

World Induced Technical Change Hybrid model (WITCH) is designed to assist in the study of the socioeconomic dimensions of climate change and to help policy makers to understand the economic consequences of climate policies. It modifies key features of the RICE model with the purpose of more detailed evaluation of the optimal responses of regional economies to climate change. All countries are grouped into *twelve regions*, which can strategically interact using a game-theoretic setup. Namely, the model can describe: (a) the *noncooperative behavior of regions*, when the regional planners maximize their own discounted utilities taking into account the interactions with other regions (*Nash equilibrium*); (b) the *cooperative behavior*, when a world central planner maximizes the sum of discounted regional utilities weighted by regional populations. The optimization horizon is until 2100.

The WITCH economic model expands the energy component of the aggregate production function to describe different energy-generating technologies and carbon mitigation options. The regions can choose their optimal investments into physical capital, R & D, and different energy technologies.

12.3.2 Deterministic Energy–Economy Models (Global 2100, CETA, MERGE, ECLIPSE)

Global 2100 of Manne and Richels (1991) is the energy–economy model that combines a macroeconomic model of overall economic activity with an energy-technology model. It considers *five world regions* with a single representative consumer–producer in each. The focus is on evaluating optimal paths of energy investments in the future. Fuel and technology choices are influenced by resource and technology availability as well as taxes and relative prices of different fuels.

Climate Emissions Trajectory Assessment (CETA) model proposed by Peck and Teisberg (1993) is an aggregated version of the Global 2100 model. The CETA considers one aggregated region, the world as a whole, and includes carbon cycles and climate change impact.

Model for Evaluating Regional and Global Effects (MERGE) [6] is a revised version of the Global 2100, being simulated up to 2200. It embodies a general equilibrium model with *five world regions*, in which regional consumers make savings and consumption decisions. The outputs of the energy sector and the rest of economy are modeled separately. The aggregated non-energy output is determined by a dynamic *nested CES production function*

$$Q = A\left[a\left(K^{\alpha}L^{1-\alpha}\right)^{\rho} + b\left(E_1^{\alpha_1}E_2^{\alpha_2}E_3^{\alpha_3}E_4^{\alpha_4}\right)^{\rho}\right]^{1/\rho}, \quad 0 < \rho < 1, \tag{12.10}$$

that combines the standard Cobb–Douglas production function $K^{\alpha}L^{1-\alpha}$ of capital K and labor L with the bundle of four energy-transport services: electricity E_1, heat E_2, other fuels E_3, and transport E_4. The parameter α, $0 < \alpha < 1$ is the share of capital and α_j, $0 < \alpha_j < 1$ is the share of the service j in the energy-transport bundle, $j = 1, 2, 3, 4$. Such nested CES production functions are frequent in modern applied economic research. They reflect distinct limited substitution possibilities for different services [3]. In particular, transport can be substituted with energy, for example when the agriculture output can be produced closer to consumers in artificial greenhouses or can be moved to a remote region with better climate conditions. The output Q is allocated among consumption, capital investment, and payment for energy. The objective is to maximize the aggregate discounted logarithmic utility of consumers over a future horizon. A simple climate model of MERGE represents the dynamics of major greenhouse gases, which yield global changes.

Energy and Climate Policy and Scenario Evaluation (ECLIPSE) model has many common features with MERGE but includes more details about the energy and economy of *eleven world regions*. It was designed to simulate the impact of changing energy prices (arising from policy or market changes) on predefined scenarios of economic growth and energy demand [10]. The basic economic model is similar to the one of MERGE. The simulation horizon is until 2100.

12.3.3 Scenario-Based Integrated Models (IMAGE, TARGETS)

12.3.3.1 Integrated Model to Assess the Greenhouse Effect (IMAGE)

The *IMAGE 1.0* links six autonomous modules: world energy/economy model, atmospheric chemistry model, carbon cycles model, climate model, sea level rise model, and a socioeconomic model of the Netherlands in 1990. The model is developed to evaluate long-term strategies to control global climate changes. On the basis of historical and predicted future emissions of greenhouse gases, it calculates the global temperature and the rise of sea levels. The climate change problem is described using the system dynamics method (Sect. 12.2) as a discrete dynamic system. The simulation period is from 1900 to 2100.

The *IMAGE 2* describes the dynamics of a geographically detailed integrated society–biosphere–climate system. It consists of three linked subsystems: the energy–industry system, the terrestrial environment system, and the atmosphere–ocean system. The modules are more process-oriented and contain fewer global parameterizations than in the first version. The energy–industry system computes the emissions of greenhouse gases in *thirteen world regions*, as a function of energy consumption and industrial production. The terrestrial environment models simulate the changes in a global land cover on a grid-scale through climatic and economic factors. The atmosphere–ocean models compute the accumulation of greenhouse gases in the atmosphere and the resulting regional and average temperatures and precipitation patterns. The model includes many important links among the subsystem models. The time horizon spans from 1900 to 2100 with time steps from 1 day to 5 years.

Tool to Assess Regional and Global Environmental and Health Targets for Sustainability (TARGETS) aims to analyze global changes and sustainable development from a synoptic perspective. It was constructed by Rotmans and de Vries in 1997. The TARGETS model consists of five interlinked horizontal modeling modules: population/health model, resource/economy model, biophysics model, land/soil model, and water model. Four vertical modules describe the dynamics of the system, different external impacts on the system, and various policy responses. The initial version of TARGETS employed a global scale, while subsequent versions split the world into *six regions*. The model explores the long-term dynamics of the human and environmental system which may shape the Earth over the next hundred years. The time horizon of the model spans 200 years, from 1900 to 2100, with the time step from 1 month to 1 year.

12.3.4 Probabilistic Integrated Models (PAGE, ICAM)

Policy Analysis of the Greenhouse Effect (PAGE) model was developed by C. Hope with European Union support in 1993. This is a probabilistic model with an emphasis on decision analysis. The model is designed to be simple in use and allow for extensive propagation of uncertainty. The PAGE generates estimates of the pollution concentration, global average temperature, abatement and damage cost for *four world regions* for the period of 1990–2200. A distinguished feature of the PAGE model is its treatment of uncertainty. All major parameterizations of pollution emissions, atmospheric climate, and economic impact are represented by triangular probability distributions whose parameters are set by users. These uncertainties are then propagated throughout the model.

Integrated Climate Assessment Model (ICAM) was developed at the Carnegie Mellon University in 1990s. Researchers from various disciplines focus on *three main research areas* where the role of uncertainty is emphasized: integrated assessment modeling, public and private decision-making and communication, and national and international policy-making. Different versions of the ICAM model provide increasingly sophisticated and detailed descriptions of climate change. Each modeling step quantifies uncertainties in model components and asks where additional research could contribute to solving a problem under study. This information is then used in the next iteration of the model. The ICAM model has been used to analyze a wide range of emissions of CO_2, N_2O, CH_4, and sulfate aerosols; generate regional population-growth projections with user-specified productivity/growth assumptions; simulate market and nonmarket effects of climate change; estimate the climate change impact on ecosystems; and contribute to decision-making.

12.3.5 Limitations of Integrated Assessment Models

The IA models have been intensively used to project global trends. Nevertheless, they have obvious methodological limitations, among them:

- too much focus on climate change;
- dependence on the background and expertise of modelers and their organizations;
- lack of transparency (integration of many components makes the model too complex);
- limited verification and validation of the global models by empirical data;
- incompleteness of knowledge (there are numerous missing links of knowledge which are hidden in the models);
- too high or non-adequate aggregation levels (in particular, regions do not match political units);
- too simplistic specification of policies (most climate change analyses are limited to a carbon tax policy).

A realistic concern of the IA-based climate modeling is an increasing mismatch between forecast and reality. Namely, the average global temperature during 2005–2013 has been flat and has not followed predictions of major climate models, despite roughly 100 billion tons of carbon added to the atmosphere between 2000 and 2010. It does not mean that the IA modeling is entirely wrong, but their climate blocks are not quite reliable. It may appear that the Earth responds to higher carbon concentrations in ways that have not been properly understood yet. Modern climate change models directly convert the increase of carbon concentration into the rise in global temperature, which becomes an input of the IA economic modeling components such as formula (12.9). If the global temperature increase does not happen, then forecasts of the described above IA models need to be corrected. It will have a profound significance for both climate science and environmental policy.

12.4 Global Modeling: A Look Ahead

A practical lesson from climate change modeling is that the global models need to be less politicized and focus more on other traditional global ecological problems of humanity: the shortage of food (it has always been very important); the exhaustion of natural resources (since nineteenth century), and pollution of the environment (arose in twentieth century).

There is no satisfactory mathematical theory of large, integrated, economic–environmental systems. Despite continuous improvement of mathematical tools, large-scale environmental processes are beyond a possibility of their adequate mathematical presentation. In many cases it is difficult to collect necessary information about the system structure and functions. In particular, the climate modeling has been developing intensively during the last decades. Sophisticated models have been elaborated to evaluate the influence of numerous human factors on the Earth's climate. However, none of the existing climate models can produce a reliable climate forecast of seasonal anomalies (a drought or a severe winter) for a 10–15 year horizon. Alternatively, there are no economic models that can predict frequent small recessions punctuated by rear depressions, seen in the world economic reality.

A further progress in the global modeling can be expected on the basis of synthesis of various economic–environmental models, including the integrated assessment methodology. A promising approach is to construct a hierarchical system of models that will reflect the hierarchical structure of economic–ecological interaction, from global models via models of subsystems to the models of elementary environmental and economic processes (macro-description via micro-description). Currently, the level of development of mathematical models substantially varies for different ecological problems.

Exercises

1. Compare the pollution accumulation equation (12.6) of the Forrester global model (12.1)–(12.7) with the pollution accumulation equation (11.1) of Chap. 11. Describe differences between these two equations. Determine necessary assumptions about the parameters of (12.6), under which (12.6) and (11.1) are equivalent.
2. Compare (12.5) of the natural resource R in the Forrester global model (12.1)–(12.7) with (10.1) of exhaustible resource dynamics of Chap. 10. Describe differences between these two equations. Discuss additional suggestions, under which (12.5) and (10.1) are equivalent.
3. Compare (12.3) of population dynamics in the Forrester global model (12.1)–(12.7) with the Malthus model (6.1) of Chap. 6. Describe differences between these two equations. Determine the necessary assumptions about the parameters of (12.3), under which (12.3) and (6.1) are equivalent.
4. Compare (12.4) of capital accumulation in the Forrester global model (12.1)–(12.7) with (2.33) of capital dynamics in the Solow–Swan model of Chap. 2. Describe differences between these two equations. Can you find assumptions about the parameters of (12.4), under which (12.4) and (2.33) are equivalent? If not, then why?
5. Use the Internet to obtain and discuss the most recent data about the dynamics of the Earth average temperature since 2005, discussed in Sect. 12.3.5. Has the

average temperature increased or decreased after 2013? What does it mean for climate modeling?
6. Use the Internet resources to get an update on the state-of-the-art global modeling. Which new models and modeling trends have appeared after publishing this textbook in 2013?

References[1]

1. 📖Brown, L.: Eco-economy. W.W. Norton, New York (2001)
2. Day, R.H.: The divergent dynamics of economic growth: studies in adaptive economizing, technological change, and economic development. Cambridge University Press, New York (2004)
3. Forrester, J.W.: World dynamics. Cambridge University, Cambridge, MA (1971)
4. Hritonenko, N., Yatsenko, Y.: Energy substitutability and modernization of energy-consuming technologies. Energy Economics **34**, 1548–1556 (2012)
5. 📖Janssen, M.: Modeling global change: the art of integrated assessment modeling. Edward Elgar, Cheltenham, UK (1998)
6. Manne, A., Mendelsohn, R., Richels, R.: MERGE: a model for evaluating regional and global effects of GHG reduction policies. Energy Policy **23**, 17–34 (1995)
7. Meadows, D., Randers, J., Meadows, D.: The limits to growth: the 30-year update. Chelsea Green, White River Junction, VT (2004)
8. Nordhaus, W.D.: Economic aspects of global warming in a post-Copenhagen environment. Proc. Natl. Acad. Sci. **107**, 11721–11726 (2010)
9. Steffen, W., Sanderson, A., Tyson, P., Jager, J., Matson, P., Moore III, B., Oldfield, F., Richardson, K., Schellnhuber, H., Turner II, B., Wasson, R.: Global change and the Earth system: a planet under pressure. Springer, Berlin (2004)
10. Turton, H.: ECLIPSE: an integrated energy-economy model for climate policy and scenario analysis. Energy **33**, 1754–1769 (2008)

[1] The book symbol 📖 means that the reference is a textbook recommended for further student reading.

Index

N. Hritonenko and Y. Yatsenko, *Mathematical Modeling in Economics, Ecology and the Environment*, Springer Optimization and Its Applications 88,
DOI 10.1007/978-1-4614-9311-2, © Springer Science+Business Media New York 2013

Q

R

S

Printed by Publishers' Graphics LLC
LMO140128.15.15.15